T0338529

HYPERGROUP THEORY

Other World Scientific Titles by the Author

Polygroup Theory and Related Systems
ISBN: 978-981-4425-30-8

Nearrings, Nearfields and Related Topics
ISBN: 978-981-3207-35-6

A Walk Through Weak Hyperstructures: Hv-Structures
ISBN: 978-981-3278-86-8

HYPERGROUP THEORY

Bijan Davvaz
Yazd University, Iran

Violeta Leoreanu-Fotea
Alexandru Ioan Cuza University of Iasi, Romania

NEW JERSEY · LONDON · SINGAPORE · BEIJING · SHANGHAI · HONG KONG · TAIPEI · CHENNAI · TOKYO

Published by

World Scientific Publishing Co. Pte. Ltd.

5 Toh Tuck Link, Singapore 596224

USA office: 27 Warren Street, Suite 401-402, Hackensack, NJ 07601

UK office: 57 Shelton Street, Covent Garden, London WC2H 9HE

British Library Cataloguing-in-Publication Data
A catalogue record for this book is available from the British Library.

HYPERGROUP THEORY

ISBN 978-981-124-938-9 (hardcover)
ISBN 978-981-124-939-6 (ebook for institutions)
ISBN 978-981-124-940-2 (ebook for individuals)

For any available supplementary material, please visit
https://www.worldscientific.com/worldscibooks/10.1142/12645#t=suppl

Desk Editor: Liu Yumeng

Printed in Singapore

Preface

Hypergroup theory is a field of algebra that appeared in the 1930s and was introduced by a French mathematician Marty, considered a new 20th century Galois. The theory has known various periods of flourishing: the 40s, then 70s and especially after the 90s the theory is studied on all continents, both theoretically and for the multitude of applications in various fields of knowledge: various chapters of mathematics, computer science, biology, physics, chemistry, sociology.

This book continues the series of monographs in the field, dedicated both to the theoretical part and to the applications.

The book presents an updated study of hypergroups, being structured on 12 chapters starting with the presentation of the basic notions in the domain: semihypergroups, hypergroups, classes of subhypergroups, types of homomorphisms, but also key notions: canonical hypergroups, join spaces and complete hypergroups. A detailed study is dedicated to the connections between hypergroups and binary relations, starting from connections established by Rosenberg and Corsini. Various types of binary relations are highlighted, in particular equivalence relations and the corresponding quotient structures, which enjoy certain properties: commutativity, cyclicity, solvability. A special attention is paid to the fundamental β relationship, which leads to a group quotient structure. In the finite case, the number of non-isomorphic Rosenberg hypergroups of small orders is mentioned. Also, the study of hypergroups associated with relations is extended to the case of hypergroups associated to n-ary relations. Then follows an applied excursion of hypergroups in important chapters in mathematics: lattices, Pawlak approximation, hypergraphs, topology, with various properties, characterizations, varied and interesting examples. The bibliography presented is an

v

updated one in the field, followed by an index of the notions presented in the book, useful in its study.

The book is addressed to specialists in the field, doctoral students, but also to mathematics enthusiasts.

Bijan Davvaz
Department of Mathematics, Yazd University,
Yazd, Iran

Violeta Leoreanu-Fotea
Faculty of Matematics,
Alexandru Ioan Cuza University of Iasi, Romania

Contents

Chapter 1

Semihypergroups

The concept of algebraic hyperstructures was introduced in 1934 by Marty [115] and has been studied in the following decades and nowadays by many mathematicians. In a classical algebraic structure, the composition of two elements is an element, while in an algebraic hyperstructure, the composition of two elements is a set. In this chapter we introduce the notion of a hypergroupoid and a semihypergroup. We present some examples to show that how these structures appear in natural phenomena.

1.1 Hyperoperations and hypergroupoids

We begin with the definition of a hypergroupoid.

Definition 1.1. Let H be a non-empty set. A mapping $\circ : H \times H \to \mathcal{P}^*(H)$, where $\mathcal{P}^*(H)$ denotes the family of all non-empty subsets of H, is called a *hyperoperation* on H. The couple $(H, \circ)$ is called a *hypergroupoid*.

In the above definition, if A and B are two non-empty subsets of H and $x \in H$, then we denote

$$A \circ B = \bigcup_{\substack{a \in A \\ b \in B}} a \circ b, \ x \circ A = \{x\} \circ A \ \text{ and } \ A \circ x = A \circ \{x\}.$$

In [57], several examples are provided from different biological points of view, and it is shown that the theory of hyperstructures exactly fits the inheritance issue. One of the interesting examples is *ABO* Blood Group Inheritance.

Example 1.1 *ABO* Blood Group Inheritance. In 1900, the Austrian physician Karl Landsteiner realized that human blood was of different

1

Table 1.1   Combinations of blood groups

$\otimes$	O	A	B	AB
O	O	O A	O B	A B
A	O A	O A	AB A B O	AB A B
B	O B	AB A B O	O B	AB A B
AB	A B	AB A B	AB A B	AB A B

types, and that only certain combinations were compatible. The International Society of Blood Transfusion (ISBT) recognizes 285 blood group antigens of which 245 are classified as one of 29 blood group systems. Each blood group system represents either a single gene or a cluster of two or three closely linked genes of related sequence with little or no recognized recombination occurring between them. Consequently, each blood group system is a genetically discrete entity.

Blood groups are inherited from both parents. The ABO blood type is controlled by a single gene (the ABO gene) with three alleles: I^A, I^B and i. The gene encodes glycosyltransferase that is an enzyme that modifies the carbohydrate content of the red blood cell antigens. The gene is located on the long arm of the ninth chromosome (9q34) [75; 174]. People with blood type A have antigen A on the surfaces of their blood cells, and may be of genotype $I^A I^A$ or $I^A i$. People with blood type B have antigen B on their red blood cell surfaces, and may be of genotype $I^B I^B$ or $I^B i$. People with the rare blood type AB have both antigens A and B on their cell surfaces, and are genotype $I^A I^B$. People with blood type O have neither antigen, and are genotype ii. A type A and a type B couple can also have a type O child if they are both heterozygous ($I^A i$ and $I^B i$, respectively). Considering $H = \{O, A, B, AB\}$ with Table 1.1, $(H, \otimes)$ is a hypergroupoid.

ABO blood type is often further differentiated by a + or −, which refers to another blood group antigen called the Rh factor. In this system, the Rh^+ phenotype (D allele) is dominant on the Rh^- phenotype

Table 1.2 The different crosses of Rhesus system

$\otimes$	Rh^+	Rh^-
Rh^+	Rh^+, Rh^-	Rh^+, Rh^-
Rh^-	Rh^+, Rh^-	Rh^-

(d allele). The antigen was originally identified in rhesus monkeys, hence the name [174]. For example:

$$P: \ Rh^+ \text{ (DD genotype)} \otimes Rh^- \text{ (dd } genotype)$$
$$\downarrow$$
$$F_1: \qquad Rh^+ \ (Dd \text{ genotype})$$

and

$$F_1 \otimes F_1: \qquad Rh^+ \ (Dd \text{ genotype}) \otimes Rh^+ \ (Dd \text{ genotype})$$
$$\downarrow$$
$$F_2: \quad Rh^+ \ (Dd \text{ genotype}), \ Rh^+ \ (Dd \text{ genotype}), \ Rh^- \ (dd \text{ genotype})$$

The different crosses of Rhesus system are indicated in Table 1.2. Considering $H = \{Rh^+, Rh^-\}$ with Table 1.2, $(H, \otimes)$ is a hypergroupoid.

Table 1.3 indicates different crosses of two blood antigenic phenotypes together (ABO and Rhesus systems). For example:

$$P: \qquad O^+ \ (iiDd \text{ genotype}) \otimes O^+ \ (iiDd \text{ genotype})$$
$$\downarrow$$
$$F_1: O^+ \ (iiDD \text{ genotype}), \ O^+ \ (iiDd \text{ genotype}), \ O^- (iidd \text{ genotype}).$$

Considering $H = \{O^+, O^-, A^+, A^-, B^+, B^-, AB^+, AB^-\}$ with Table 1.3, $(H, \otimes)$ is a hypergroupoid.

Now, we present another example of hyperstructures which arises from chemistry [55].

Example 1.2 Chain reaction. Chain reaction, in chemistry and physics, process yielding products that initiate further processes of the same kind, a self-sustaining sequence. Examples from chemistry are burning a fuel gas, the development of rancidity in fats, "knock" in internal-combustion engines, and the polymerization of ethylene to polyethylene. The best-known examples in physics are nuclear fissions brought about by neutrons. Chain reactions are in general very rapid but are also highly sensitive to reaction conditions, probably because the substances that sustain the reaction are easily affected by substances other than the reactants themselves. An atom or group of atoms possessing an odd (unpaired) electron is called radical. Radical species can be electrically neutral, in which case they are sometimes referred to as free radicals. Pairs of electrically neutral "free" radicals are formed via homolytic bond breakage. This can be achieved by heating in non-polar solvents or the vapor phase. At elevated temperature or under the influence ultraviolet light at room temperature, all molecular species will dissociate into radicals. Homolysis or homolytic bond fragmentation occurs when (in the language of Lewis theory) a two electron covalent bond breaks and one electron goes to each of the partner species.

For example, chlorine, Cl_2, forms chlorine radicals $(Cl^\bullet)$ and peroxides form oxygen radicals.

$$X\text{---}X \longrightarrow 2X^\bullet$$
$$Cl\text{---}Cl \longrightarrow 2Cl^\bullet$$
$$R\text{---}O\text{---}O\text{-}R \longrightarrow R\text{---}O^\bullet$$

Radical bond forming reactions (radical couplings) are rather rare processes. The reason is because radicals are normally present at low concentrations in a reaction medium, and it is statistically more likely they will abstract a hydrogen, or undergo another type of a substitution process, rather than reacting with each other by coupling. And as radicals are uncharged, there is little long range columbic attraction between two radical centers. Radical substitution reactions tend to proceed as chain reaction processes, often with many thousands of identical propagation steps. The propensity for chain reactivity gives radical chemistry a distinct feel compared with polar Lewis acid/base chemistry where chain reactions are less common. Methane can be chlorinated with chlorine to give chloromethane and hydrogen chloride. The reaction proceeds as a chain, radical, substitution mechanism. The process is a little more involved, and three steps are involved: initiation, propagation and termination:

(1) $Cl_2 \longrightarrow 2Cl^\bullet$
    (1) is called *Chain initiating step.*
(2) $Cl^\bullet + CH_4 \longrightarrow HCl + CH_3^\bullet$
(3) $CH_3^\bullet + Cl_2 \longrightarrow CH_3Cl + Cl^\bullet$
    then (2), (3), (2), (3), etc, until finally:
    (2) and (3) are called *Chain propagating steps.*
(4) $Cl^\bullet + Cl^\bullet \longrightarrow Cl_2$ or
(5) $CH_3^\bullet + CH_3^\bullet \longrightarrow CH_3CH_3$ or
(6) $CH_3^\bullet + Cl^\bullet \longrightarrow CH_3Cl.$
    (4), (5) and (6) are called *Chain-terminating steps.*

First in the chain of reactions is a chain-initiating step, in which energy is absorbed and a reactive particle generated; in the present reaction it is the cleavage of chlorine into atoms (Step 1). There are one or more chain-propagating steps, each of which consumes a reactive particle and generates another; there they are the reaction of chlorine atoms with methane (Step 2), and of methyl radicals with chlorine (Step 3).

A chlorine radical abstracts a hydrogen from methane to give hydrogen chloride and a methyl radical. The methyl radical then abstracts a chlorine atom (a chlorine radical) from Cl_2 to give methyl chloride and a chlorine radical which abstracts a hydrogen from methane and the cycle continues. Finally there are chain-terminating steps, in which reactive particles are consumed but not generated; in the chlorination of methane these would involve the union of two of the reactive particles, or the capture of one of them by the walls of the reaction vessel.

The halogens are all typical non-metals. Although their physical forms differ-fluorine and chlorine are gases, bromine is a liquid and iodine is a solid at room temperature, each consists of diatomic molecules; F_2, Cl_2, Br_2 and I_2. The halogens all react with hydrogen to form gaseous compounds, with the formulas HF, HCl, HBr, and HI all of which are very soluble in water. The halogens all react with metals to give halide.

$$: \ddot{F} \ - \ddot{F} :, \qquad : \ddot{C}l - \ddot{C}l :, \qquad : \ddot{B}r \ - \ddot{B}r :, \qquad : \ddot{I} \ - \ddot{I} :$$

The reader will find in [132] a deep discussion of chain reactions and halogens.

During chain reaction

$$A_2 + B_2 \overset{\text{Heat or Light}}{\longleftrightarrow} 2AB$$

there exist all molecules A_2, B_2, AB and of which fragment parts $A^{\bullet}, B^{\bullet}$ in experiment. Elements of this collection can combine with each other. All combinational probability for the set $H = \{A^{\bullet}, B^{\bullet}, A_2, B_2, AB\}$ to do without energy can be displayed as in Table 1.2.

Then, $(H, \oplus)$ is a hypergroupoid [55]. For instance, we can consider $A = H$ and $B \in \{F, CL, Br, I\}$.

We refer the readers to [5; 53; 57; 58] see more examples of hyperstructures in natural phenomena.

1.2   Semihypergroups

The concept of a semihypergroup is a generalization of the concept of a semigroup. Many authors studied different aspects of semihypergroups. In this section we study the main properties of semihypergroups. For more details we refer to [48].

Definition 1.2. A hypergroupoid $(H, \circ)$ is called a *semihypergroup* if for every $x, y, z \in H$, $x \circ (y \circ z) = (x \circ y) \circ z$, that is

$$\bigcup_{u \in y \circ z} x \circ u = \bigcup_{v \in x \circ y} v \circ z.$$

A non-empty subset A of H is called a *subsemihypergroup* if $x \circ y \subseteq A$ for all x, y in A.

A semihypergroup H is *finite* if it has only a finitely many elements. A semihypergroup H is *commutative* if it satisfies

$$x \circ y = y \circ x,$$

for all $x, y \in H$. Every semigroup is a semihypergroup.

Remark 1.1. The associativity for semihypergroups can be applied for subsets, i.e., if $(H, \circ)$ is a semihypergroup, then for all non-empty subsets A, B, C of H, we have $(A \circ B) \circ C = A \circ (B \circ C)$.

The element $a \in H$ is called *scalar* if

$$|a \circ x| = |x \circ a| = 1,$$

for all $x \in H$. An element e in a semihypergroup $(H, \circ)$ is called *scalar identity* if

$$x \circ e = e \circ x = \{x\},$$

for all $x \in H$. An element e in a semihypergroup $(H, \circ)$ is called *identity* if

$$x \in e \circ x \cap x \circ e,$$

for all $x \in H$. An element $a' \in H$ is called an *inverse* of $a \in H$ if there exists an identity $e \in H$ such that

$$e \in a \circ a' \cap a' \circ a.$$

An element 0 in a semihypergroup $(H, \circ)$ is called *zero element* if $x \circ 0 = 0 \circ x = \{0\}$, for all $x \in H$.

Example 1.3. Let $H = \{a, b, c, d\}$. Define the hyperoperation $\circ$ on H by the following table.

○	a	b	c	d
a	a	$\{a,b\}$	$\{a,c\}$	$\{a,d\}$
b	a	$\{a,b\}$	$\{a,c\}$	$\{a,d\}$
c	a	b	c	d
d	a	b	c	d

Then, $(H, \circ)$ is a semihypergroup.

Example 1.4. Let $H = \{a, b, c, d, e\}$. Define the hyperoperation $\circ$ on H by the following table.

○	a	b	c	d	e
a	a	$\{a,b,d\}$	a	$\{a,b,d\}$	$\{a,b,d\}$
b	a	b	a	$\{a,b,d\}$	$\{a,b,d\}$
c	a	$\{a,b,d\}$	$\{a,c\}$	$\{a,b,d\}$	$\{a,b,c,d,e\}$
d	a	$\{a,b,d\}$	a	$\{a,b,d\}$	$\{a,b,d\}$
e	a	$\{a,b,d\}$	$\{a,c\}$	$\{a,b,d\}$	$\{a,b,c,d,e\}$

Then, $(H, \circ)$ is a semihypergroup.

Example 1.5. Let H be the unit interval $[0,1]$. For every $x, y \in H$, we define

$$x \circ y = \left[0, \frac{xy}{2}\right].$$

Then, $(H, \circ)$ is a semihypergroup.

Example 1.6. Let $\mathbb{N}$ be the set of non-negative integers. We define the following hyperoperation on $\mathbb{N}$,

$$x \circ y = \{z \in \mathbb{N} \mid z \geq \max\{x, y\}\},$$

for all $x, y \in \mathbb{N}$. Then, $(\mathbb{N}, \circ)$ is a semihypergroup.

Example 1.7. The set of real numbers $\mathbb{R}$ with the following hyperoperation

$$a \circ b = \begin{cases} (a,b) & \text{if } a < b \\ (b,a) & \text{if } b < a \\ \{a\} & \text{if } a = b, \end{cases}$$

for all $a, b \in \mathbb{R}$ is a semihypergroup, where (a, b) is the open interval $\{x \mid a < x < b\}$.

Example 1.8. Let $(S, \cdot)$ be a semigroup and P a non-empty subset of S. We define the following hyperoperation on S,

$$x \circ_P y = x \cdot P \cdot y,$$

for all $x, y \in S$. Then, $(S, \circ_P)$ is a semihypergroup. The hyperoperation $\circ_P$ is called *P-hyperoperation*.

Definition 1.3. Let $(H_1, \circ)$ and $(H_2, \star)$ be two semihypergroups. A map $f : H_1 \to H_2$, is called

(1) a *homomorphism* or *inclusion homomorphism* if for all x, y of H_1, we have $f(x \circ y) \subseteq f(x) \star f(y)$;

(2) a *good (strong) homomorphism* if for all x, y of H_1, we have $f(x \circ y) = f(x) \star f(y)$;

(3) an *isomorphism* if it is a one to one and onto good homomorphism. If f is an isomorphism, then H_1 and H_2 are said to be *isomorphic*, which is denoted by $H_1 \cong H_2$.

Example 1.9. Let $H_1 = \{a, b, c\}$ and $H_2 = \{0, 1, 2\}$ be two semihypergroups with the following hyperoperations:

$\circ$	a	b	c
a	a	H_1	H_1
b	H_1	b	b
c	H_1	b	c

$\star$	0	1	2
0	0	H_2	H_2
1	H_2	1	1
2	H_2	1	$\{1, 2\}$

and let $f : H_1 \to H_2$ is defined by $f(a) = 0$, $f(b) = 1$ and $f(c) = 2$. Clearly, f is an inclusion homomorphism but it is not good homomorphism.

Lemma 1.1. *Let $(H_1, \circ)$ and $(H_2, \star)$ be two semihypergroups and $f : H_1 \to H_2$ be a good homomorphism. Then, Imf is a subsemihypergroup of H_2.*

Proof.   For every $a, b \in H_1$, we have $f(a) \star f(b) = f(a \circ b) \subseteq Imf$. $\square$

By using a certain type of equivalence relations, we can connect semihypergroups to semigroups. These equivalence relations are called strong regular relations. More exactly, by a given semihypergroup and by using a strong regular relation on it, we can construct a semigroup structure on the quotient set.

Notation 1.1. Let $(H, \circ)$ be a semihypergroup and ρ be an equivalence relation on H. If A and B are non-empty subsets of H, then

$A \overline{\rho} B$ means that for all $a \in A$, there exists $b \in B$ such that $a \rho b$ and
for all $b' \in B$, there exists $a' \in A$ such that $a' \rho b'$;
$A \overline{\overline{\rho}} B$ means that for all $a \in A$ and $b \in B$, we have $a \rho b$.

Definition 1.4. The equivalence relation ρ is called

(1) *regular on the right (on the left)* if for all x of H, from $a \rho b$, it follows that $(a \circ x) \overline{\rho} (b \circ x)$ $((x \circ a) \overline{\rho} (x \circ b))$ respectively);

(2) *strongly regular on the right* (*on the left*) if for all x of H, from $a\rho b$, it follows that $(a \circ x)\overline{\overline{\rho}}(b \circ x)$ $((x \circ a)\overline{\overline{\rho}}(x \circ b)$ respectively);

(3) ρ is called *regular* (*strongly regular*) if it is regular (strongly regular) on the right and on the left.

Theorem 1.1. *Let* $(H, \circ)$ *be a semihypergroup and* ρ *be an equivalence relation on* H.

(1) *If* ρ *is regular, then* H/ρ *is a semihypergroup, with respect to the following hyperoperation:* $\overline{x} \otimes \overline{y} = \{\overline{z} \mid z \in x \circ y\}$;

(2) *If the above hyperoperation is well-defined on* H/ρ, *then* ρ *is regular.*

Proof. (1) First, we check that the hyperoperation $\otimes$ is well-defined on H/ρ. Consider $\overline{x} = \overline{x_1}$ and $\overline{y} = \overline{y_1}$. We check that $\overline{x} \otimes \overline{y} = \overline{x_1} \otimes \overline{y_1}$. We have $x\rho x_1$ and $y\rho y_1$. Since ρ is regular, it follows that $(x \circ y)\overline{\overline{\rho}}(x_1 \circ y)$, $(x_1 \circ y)\overline{\overline{\rho}}(x_1 \circ y_1)$ whence $(x \circ y)\overline{\overline{\rho}}(x_1 \circ y_1)$. Hence, for all $z \in x \circ y$, there exists $z_1 \in x_1 \circ y_1$ such that $z\rho z_1$, which means that $\overline{z} = \overline{z_1}$. It follows that $\overline{x} \otimes \overline{y} \subseteq \overline{x_1} \otimes \overline{y_1}$ and similarly we obtain the converse inclusion. Now, we check the associativity of $\otimes$. Let $\overline{x}, \overline{y}, \overline{z}$ be arbitrary elements in H/ρ and $\overline{u} \in (\overline{x} \otimes \overline{y}) \otimes \overline{z}$. This means that there exists $\overline{v} \in \overline{x} \otimes \overline{y}$ such that $\overline{u} \in \overline{v} \otimes \overline{z}$. In other words, there exist $v_1 \in x \circ y$ and $u_1 \in v \circ z$, such that $v\rho v_1$ and $u\rho u_1$. Since ρ is regular, it follows that there exists $u_2 \in v_1 \circ z \subseteq x \circ (y \circ z)$ such that $u_1\rho u_2$. From here, we obtain that there exists $u_3 \in y \circ z$ such that $u_2 \in x \circ u_3$. We have $\overline{u} = \overline{u_1} = \overline{u_2} \in \overline{x} \otimes \overline{u_3} \subseteq \overline{x} \otimes (\overline{y} \otimes \overline{z})$. It follows that $(\overline{x} \otimes \overline{y}) \otimes \overline{z} \subseteq \overline{x} \otimes (\overline{y} \otimes \overline{z})$. Similarly, we obtain the converse inclusion.

(2) Let $a\rho b$ and x be an arbitrary element of H. If $u \in a \circ x$, then $\overline{u} \in \overline{a} \otimes \overline{x} = \overline{b} \otimes \overline{x} = \{\overline{v} \mid v \in b \circ x\}$. Hence, there exists $v \in b \circ x$ such that $u\rho v$, whence $(a \circ x)\overline{\overline{\rho}}(b \circ x)$. Similarly we obtain that ρ is regular on the left. $\qquad\square$

Notice that if ρ is regular on a semihypergroup $(H, \circ)$, then the canonical projection $\pi : H \to H/\rho$ is a good epimorphism. Indeed, for all x, y of H and $\overline{z} \in \pi(x \circ y)$, there exists $z' \in x \circ y$ such that $\overline{z} = \overline{z'}$. We have $\overline{z} = \overline{z'} \in \overline{x} \otimes \overline{y} = \pi(x) \otimes \pi(y)$. Conversely, if $\overline{z} \in \pi(x) \otimes \pi(y) = \overline{x} \otimes \overline{y}$, then there exists $z_1 \in x \circ y$ such that $\overline{z} = \overline{z_1} \in \pi(x \circ y)$.

Theorem 1.2. *If* $(H, \circ)$ *and* $(K, \star)$ *are semihypergroups and* $f : H \to K$ *is a good homomorphism, then the equivalence* ρ^f *associated with* f, *that is* $x\rho^f y \Leftrightarrow f(x) = f(y)$, *is regular and* $\varphi : f(H) \to H/\rho^f$, *defined by* $\varphi(f(x)) = \overline{x}$, *is an isomorphism.*

Proof. Let $h_1 \rho^f h_2$ and a be an arbitrary element of H. If $u \in h_1 \circ a$, then

$$f(u) \in f(h_1 \circ a) = f(h_1) \star f(a) = f(h_2) \star f(a) = f(h_2 \circ a).$$

Hence, there exists $v \in h_2 \circ a$ such that $f(u) = f(v)$, which means that $u \rho^f v$. Hence, ρ^f is regular on the right. Similarly, it can be shown that ρ^f is regular on the left. On the other hand, for all $f(x), f(y)$ of $f(H)$, we have

$$\varphi(f(x) \star f(y)) = \varphi(f(x \circ y)) = \{\overline{z} \mid z \in x \circ y\} = \overline{x} \otimes \overline{y} = \varphi(f(x)) \otimes \varphi(f(y)).$$

Moreover, if $\varphi(f(x)) = \varphi(f(y))$, then $x \rho^f y$, so φ is injective and clearly, it is also surjective. Finally, for all $\overline{x}, \overline{y}$ of H/ρ^f we have

$$\begin{aligned}\varphi^{-1}(\overline{x} \otimes \overline{y}) &= \varphi^{-1}(\{\overline{z} \mid z \in x \circ y\}) = \{f(z) \mid z \in x \circ y\}\\ &= f(x \circ y) = f(x) \star f(y) = \varphi^{-1}(\overline{x}) \star \varphi^{-1}(\overline{y}).\end{aligned}$$

Therefore, φ is an isomorphism. $\square$

By strongly regular relations we obtain a semigroup from a semihypergroup (see Figure 1.1).

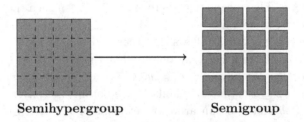

Semihypergroup   **Semigroup**

Fig. 1.1   A strongly regular relation gives us a semigroup

Theorem 1.3. *Let $(H, \circ)$ be a semihypergroup and ρ be an equivalence relation on H.*

(1) *If ρ is strongly regular, then H/ρ is a semigroup, with respect to the following operation: $\overline{x} \otimes \overline{y} = \overline{z}$, for all $z \in x \circ y$;*

(2) *If the above operation is well-defined on H/ρ, then ρ is strongly regular.*

Proof. (1) For all x, y of H, we have $(x \circ y)\overline{\overline{\rho}}(x \circ y)$. Hence, $\overline{x} \otimes \overline{y} = \{\overline{z} \mid z \in x \circ y\} = \{\overline{z}\}$, which means that $\overline{x} \otimes \overline{y}$ has exactly one element. Therefore, $(H/\rho, \otimes)$ is a semigroup.

(2) If $a\rho b$ and x is an arbitrary element of H, we check that $(a\circ x)\overline{\overline{\rho}}(b\circ x)$. Indeed, for all $u \in a\circ x$ and all $v \in b\circ x$ we have $\overline{u} = \overline{a}\otimes\overline{x} = \overline{b}\otimes\overline{x} = \overline{v}$, which means that $u\rho v$. Hence, ρ is strongly regular on the right and similarly, it can be shown that it is strongly regular on the left. □

Theorem 1.4. *If $(H, \circ)$ is a semihypergroup, $(S, \star)$ is a semigroup and $f : H \to S$ is a good homomorphism, then the equivalence ρ^f associated with f is strongly regular.*

Proof. Let $a\rho^f b$, $x \in H$ and $u \in a \circ x$. It follows that

$$f(u) = f(a) \star f(x) = f(b) \star f(x) = f(b \circ x).$$

Hence, for all $v \in b\circ x$, we have $f(u) = f(v)$, which means that $u\rho^f v$. Hence, ρ^f is strongly regular on the right and similarly, it is strongly regular on the left. □

Let α be a good homomorphism from a semihypergroup $(H, \circ)$ into a semihypergroup $(H', \star)$. The relation $\alpha^{-1} \circ \alpha$ is an equivalence relation ρ on H. This means that

$$a\rho b \iff \alpha(a) = \alpha(b).$$

The relation ρ is called the *kernel* of α. The natural mapping associated with ρ is $\varphi : H \to H/ker\alpha$, where $\varphi(a) = \rho(a)$. The mapping $\psi : H/\rho \to H'$, where $\psi(\rho(a)) = \alpha(a)$ is then the unique good homomorphism that makes the following diagram commutative.

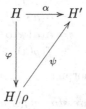

Theorem 1.5. *Let α_1 and α_2 be good homomorphisms of a semihypergroup $(H, \star)$ onto semihypergroups $(H_1, \star_1)$ and $(H_2, \star_2)$ respectively, such that $\alpha_1^{-1}\circ\alpha_1 \subseteq \alpha_2^{-1}\circ\alpha_2$. Then, a unique good homomorphism $\theta : H_1 \to H_2$ such that $\theta \circ \alpha_1 = \alpha_2$ exists. This means that the following diagram commutes.*

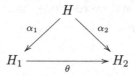

Proof. Let $a_1 \in H_1$ and a be an element of H such that $\alpha_1(a) = a_1$. Define $\theta(a_1) = \alpha_2(a)$. This is single valued, indeed if $\alpha_1(b) = a_1$ (b in H), we have $(a, b) \in \alpha_1^{-1} \circ \alpha_1 \subseteq \alpha_2^{-1} \circ \alpha_2$, so that $\alpha_2(a) = \alpha_2(b)$. It is clear that $\theta \circ \alpha_1 = \alpha_2$ and the assertion that θ is a good homomorphism follows from this

$$\theta(\alpha_1(a) \star_1 \alpha_1(b)) = \theta(\alpha_1(a \star b)) = \alpha_2(a \star b)$$
$$= \alpha_2(a) \star_2 \alpha_2(b) = \theta(\alpha_1(a)) \star_2 \theta(\alpha_1(b)).$$

The uniqueness of θ is evident, indeed if θ satisfies $\theta \circ \alpha_1 = \alpha_2$ we are compelled to define θ as above. $\qquad \square$

Theorem 1.6. *Let $(H, \star)$ and $(H', \star')$ be two semihypergroups and $\alpha : H \rightarrow H'$ be a good homomorphism. Then, $\rho = ker\alpha$ is a regular relation and there exists a good homomorphism $f : H/\rho \rightarrow H'$ such that $f \circ \varphi = \alpha$.*

Proof. Suppose that $a\rho b$. Then, for every $c \in H$ we have

$$\alpha(a \star c) = \alpha(a) \star' \alpha(c) = \alpha(b) \star' \alpha(c) = \alpha(b \star c).$$

Thus, for every $x \in a \star c$, there exists $y \in b \star c$ such that $\alpha(x) = \alpha(y)$ and so $x\rho y$. Therefore, ρ is a regular relation on H.

Now, let $\rho(a) \in H/\rho$ and define $f(\rho(a)) = \alpha(a)$. We note that if $b \in \rho(a)$, then $\alpha(a) = \alpha(b)$. So, f is well-defined. Moreover, f is a good homomorphism because if $\rho(a)$, $\rho(b) \in H/\rho$, then

$$f(\rho(a) \odot \rho(b)) = f(\{\rho(z) \mid z \in a \star b\})$$
$$= \{\alpha(z) \mid z \in a \star b\}$$
$$= \alpha(a \star b)$$
$$= \alpha(a) \star' \alpha(b)$$
$$= f(\rho(a)) \star' f(\rho(b)).$$

Therefore, f is a good homomorphism. $\qquad \square$

Theorem 1.7. *If ρ_1 and ρ_2 are strongly regular relations on a semihypergroup H such that $\rho_1 \subseteq \rho_2$, then there exists a good homomorphism from H/ρ_1 onto H/ρ_2.*

Proof. Let $\pi_1 : H \rightarrow H/\rho_1$ and $\pi_2 : H \rightarrow H/\rho_2$ be the canonical homomorphisms. Since $\rho_1 = \pi_1^{-1} \circ \pi_1$ and $\rho_2 = \pi_2^{-1} \circ \pi_2$, the hypotheses of Theorem 1.5 are satisfied and we conclude that there is a good homomorphism θ of H/ρ_1 onto H/ρ_2. $\qquad \square$

Table 1.3   The different crosses of two blood antigenic phenotypes together (ABO and Rhesus systems)

$\otimes$	O^+	O^-	A^+	A^-	B^+	B^-	AB^+	AB^-
O^+	O^+ O^-	O^+ O^-	A^+ A^- O^+ O^-	A^+ A^- O^+ O^-	B^+ B^- O^+ O^-	B^+ B^- O^+ O^-	A^+ A^- B^+ B^-	A^+ A^- B^+ B^-
O^-	O^+ O^-	O^-	A^+ A^- O^+ O^-	A^- O^-	B^+ B^- O^+ O^-	B^- O^-	A^+ A^- B^+ B^-	A^- B^-
A^+	A^+ A^- O^+ O^-	A^+ A^- O^+ O^-	A^+ A^- O^+ O^-	A^+ A^- O^+ O^-	AB^+ AB^- A^+ A^- B^+ B^- O^+ O^-	AB^+ AB^- A^+ A^- B^+ B^- O^+ O^-	AB^+ AB^- A^+ A^- B^+ B^-	AB^+ AB^- A^+ A^- B^+ B^-
A^-	A^+ A^- O^+ O^-	A^- O^-	A^+ A^- O^+ O^-	A^- O^-	AB^+ AB^- A^+ A^- B^+ B^- O^+ O^-	AB^- A^- B^- O^-	AB^+ AB^- A^+ A^- B^+ B^-	AB^- A^- B^-
B^+	B^+ B^- O^+ O^-	B^+ B^- O^+ O^-	AB^+ AB^- A^+ A^- B^+ B^- O^+ O^-	AB^+ AB^- A^+ A^- B^+ B^- O^+ O^-	B^+ B^- O^+ O^-	B^+ B^- O^+ O^-	AB^+ AB^- A^+ A^- B^+ B^-	AB^+ AB^- A^+ A^- B^+ B^-
B^-	B^+ B^- O^+ O^-	B^- O^-	AB^+ AB^- A^+ A^- B^+ B^- O^+ O^-	AB^- A^- B^- O^-	B^+ B^- O^+ O^-	B^- O^-	AB^+ AB^- A^+ A^- B^+ B^-	AB^- A^- B^-
AB^+	A^+ A^- B^+ B^-	A^+ A^- B^+ B^-	AB^+ AB^- A^+ A^- B^+ B^-	AB^+ AB^- A^+ A^- B^+ B^-	AB^+ AB^- A^+ A^- B^+ B^-	AB^+ AB^- A^+ A^- B^+ B^-	AB^+ AB^- A^+ A^- B^+ B^-	AB^+ AB^- A^+ A^- B^+ B^-
AB^-	A^+ A^- B^+ B^-	A^- B^-	AB^+ AB^- A^+ A^- B^+ B^-	AB^- A^- B^-	AB^+ AB^- A^+ A^- B^+ B^-	AB^- A^- B^-	AB^+ AB^- A^+ A^- B^+ B^-	AB^- A^- B^-

Table 1.4   Chain reactions

$\oplus$	$A^\bullet$	$B^\bullet$	A_2	B_2	AB
$A^\bullet$	$A^\bullet$ A_2	$A^\bullet$ $B^\bullet$ AB	$A^\bullet$ A_2	$A^\bullet$ B_2 $B^\bullet$ AB	$A^\bullet$ AB A_2 $B^\bullet$
$B^\bullet$	$A^\bullet$ $B^\bullet$ AB	$B^\bullet$ B_2	$A^\bullet$ $B^\bullet$ AB A_2	$B^\bullet$ B_2	$A^\bullet$ $B^\bullet$ AB B_2
A_2	$A^\bullet$ A_2	$A^\bullet$ $B^\bullet$ AB A_2	$A^\bullet$ A_2	$A^\bullet$ $B^\bullet$ A_2 B_2 AB	$A^\bullet$ $B^\bullet$ A_2 AB
B_2	$A^\bullet$ $B^\bullet$ B_2 AB	$B^\bullet$ B_2	$A^\bullet$ $B^\bullet$ A_2 B_2 AB	$B^\bullet$ B_2	$A^\bullet$ $B^\bullet$ B_2 AB
AB	$A^\bullet$ AB A_2 $B^\bullet$	$A^\bullet$ $B^\bullet$ AB B_2	$A^\bullet$ $B^\bullet$ A_2 AB	$A^\bullet$ $B^\bullet$ B_2 AB	$A^\bullet$ $B^\bullet$ A_2 B_2 AB

Chapter 2

Hypergroups

This chapter contains a brief description of the ideas, constructions and examples of hypergroups. The motivating example is the quotient of a group by any, not necessary normal, subgroup. Also, we present the definition of an H_v-group as a generalization of a hypergroup, which was introduced by Vougiouklis for the first time in 1990.

2.1 Hypergroups and some examples

We start with the following definition.

Definition 2.1. A hypergroupoid $(H, \circ)$ is called a *quasihypergroup* if for all a of H we have $a \circ H = H \circ a = H$. This condition is also called the *reproduction axiom*.

Definition 2.2. A hypergroupoid $(H, \circ)$ which is both a semihypergroup and a quasihypergroup is called a *hypergroup*.

A hypergroup for which the hyperproduct of any two elements has exactly one element is a group. Indeed, let $(H, \circ)$ be a hypergroup, such that for all x, y of H, we have $|x \circ y| = 1$. Then, $(H, \circ)$ is a semigroup, such that for all a, b in H, there exist x and y for which we have $a = b \circ x$ and $a = y \circ b$. It follows that $(H, \circ)$ is a group.

Now, we look at some examples of hypergroups.

Example 2.1. If H is a non-empty set and for all x, y of H, we define $x \circ y = H$, then $(H, \circ)$ is a hypergroup, called the *total hypergroup*.

Example 2.2. Let $(G, \cdot)$ be a group and P be a non-empty subset of G. For all x, y of G, we define $x \circ y = xPy$. Then, $(G, \circ)$ is a hypergroup, called a *P-hypergroup*.

15

Example 2.3. Let $(G, \cdot)$ be a group and for all $x, y \in G$, $\langle x, y \rangle$ denotes the subgroup generated by x and y. We define $x \circ y = \langle x, y \rangle$. Then, $(G, \circ)$ is a hypergroup.

Example 2.4. If $(G, \cdot)$ is a group, H is a normal subgroup of G and for all x, y of G, we define $x \circ y = xyH$, then $(G, \circ)$ is a hypergroup.

Example 2.5. Let $(G, \cdot)$ be a group and let H be a non-normal subgroup of it. If we denote $G/H = \{xH \mid x \in G\}$, then $(G/H, \circ)$ is a hypergroup, where for all xH, yH of G/H, we have $xH \circ yH = \{zH \mid z \in xHy\}$.

Example 2.6. If $(G, +)$ is an abelian group, ρ is an equivalence relation in G, which has classes $\overline{x} = \{x, -x\}$, then for all $\overline{x}, \overline{y}$ of G/ρ, we define $\overline{x} \circ \overline{y} = \{\overline{x+y}, \overline{x-y}\}$. Then, $(G/\rho, \circ)$ is a hypergroup.

Example 2.7. Let D be an integral domain and let F be its field of fractions. If we denote by U the group of the invertible elements of D, then we define the following hyperoperation on F/U: for all $\overline{x}, \overline{y}$ of F/U, we have $\overline{x} \circ \overline{y} = \{\overline{z} \mid \exists (u, v) \in U^2 \text{ such that } z = ux + vy\}$. We obtain that $(F/U, \circ)$ is a hypergroup.

Example 2.8. Let $(L, \wedge, \vee)$ be a lattice with a minimum element 0. If for all $a \in L$, $F(a)$ denotes the principal filter generated from a, then we obtain a hypergroup $(L, \circ)$, where for all a, b of L, we have $a \circ b = F(a \wedge b)$.

Example 2.9. Let $(L, \wedge, \vee)$ be a modular lattice. If for all x, y of L, we define $x \circ y = \{z \in L \mid z \vee x = x \vee y = y \vee z\}$, then $(L, \circ)$ is a hypergroup.

Example 2.10. Let $(L, \wedge, \vee)$ be a distributive lattice. If for all x, y of L, we define $x \circ y = \{z \in L \mid x \wedge y \leq z \leq x \vee y\}$, then $(L, \circ)$ is a hypergroup.

Example 2.11. Let H be a non-empty set and $\mu : H \longrightarrow [0, 1]$ be a function. If for all x, y of H we define $x \circ y = \{z \in L \mid \mu(x) \wedge \mu(y) \leq \mu(z) \leq \mu(x) \vee \mu(y)\}$, then $(H, \circ)$ is a hypergroup.

Example 2.12. Let H be a non-empty set and μ, λ be two functions from H to $[0, 1]$. For all x, y of H we define

$$x \circ y = \{u \in H \mid \mu(x) \wedge \lambda(x) \wedge \mu(y) \wedge \lambda(y) \leq \mu(u) \wedge \lambda(u)$$
$$\text{and } \mu(u) \vee \lambda(u) \leq \mu(x) \vee \lambda(x) \vee \mu(y) \vee \lambda(y)\}.$$

Then, the hyperstructure $(H, \circ)$ is a commutative hypergroup.

Example 2.13. Define the following hyperoperation on the real set $\mathbb{R}$: for all $x \in \mathbb{R}$, $x \circ x = x$ and for all different real elements x, y, $x \circ y$ is the open interval between x and y. Then, $(\mathbb{R}, \circ)$ is a hypergroup.

Similar to semihypergroups we can consider regular relations and strongly regular relations on hypergroups.

Theorem 2.1. *If $(H, \circ)$ is a hypergroup and R is an equivalence relation on H, then R is regular if and only if $(H/R, \otimes)$ is a hypergroup.*

Proof. If H is a hypergroup, then for all x of H we have $H \circ x = x \circ H = H$, whence we obtain $H/R \otimes \overline{x} = \overline{x} \otimes H/R = H/R$. According to Theorem 1.1, it follows that $(H/R, \otimes)$ is a hypergroup. $\qquad\square$

Theorem 2.2. *If $(H, \circ)$ is a hypergroup and R is an equivalence relation on H, then R is strongly regular if and only if $(H/R, \otimes)$ is a group.*

Proof. It is straightforward. $\qquad\square$

2.2 Conjugacy class and character hypergroups

In dealing with a symmetry group two symmetric operations belong to the same class if they present the same map with respect to (possibly) different coordinate systems where one coordinate system is converted into the other by a member of the group. In the language of group theory this means the elements a, b in a symmetric group G belong to the same class if there exists a $g \in G$ such that $a = gbg^{-1}$, i.e., a and b are conjugate.

Definition 2.3. Let a and b be elements of a group G. We say that a and b are *conjugate* in G (and call b a conjugate of a) if there exists $x \in G$ such that $b = x^{-1}ax$.

We write, for this, $a \sim_{Conj} b$ and refer to this relation as conjugacy.

Theorem 2.3. *Conjugacy is an equivalence relation on G.*

Proof. Since $a = e^{-1}ae$, for each $a \in G$, it follows that $a \sim_{Conj} a$, i.e., $\sim_{Conj}$ is reflexive.

If $a \sim_{Conj} b$, then there exists $x \in G$ such that $b = x^{-1}ax$. Hence, $a = (x^{-1})^{-1}bx^{-1}$. Since $x^{-1} \in G$, it follows that $b \sim_{Conj} a$, i.e., $\sim_{Conj}$ is symmetric.

Suppose that $a \sim_{Conj} b$ and $b \sim_{Conj} c$, where $a, b, c \in G$. Then, $b = x^{-1}ax$ and $c = y^{-1}by$, for some $x, y \in G$. Substituting for b in the expression for c we obtain $c = y^{-1}x^{-1}axy$. This shows that $c = (xy)^{-1}a(xy)$. Since $xy \in G$, it follows that $a \sim_{Conj} c$, i.e., $\sim_{Conj}$ is transitive. $\qquad\square$

Theorem 2.3 shows that we can divide a group G into equivalence classes under the relation of conjugacy. Each such equivalence class is called a *conjugate class*. The collection of all conjugacy classes of a group G is denoted by $Conj(G)$.

Theorem 2.4. *If G is a group, then $Conj(G)$ is a hypergroup where the product $A * B$ of conjugacy classes A and B consists of all conjugacy classes contained in the element-wise product AB.*

Proof.   It is straightforward. $\qquad\square$

This hypergroup was studied by Campaigne [14] and by Dietzman [68].
   Now, we illustrate some constructions.

Example 2.14. The symmetric group S_3 is the set of all bijective functions on the set $\{1, 2, 3\}$ which naturally arises as the set of permutations on X. We start by listing the elements of S_3. These elements are listed according to the cycle notation as follows:

$$e, \qquad\qquad r = (1\ 2),$$
$$a = (1\ 2\ 3), \qquad s = (1\ 3),$$
$$b = (1\ 3\ 2), \qquad t = (2\ 3).$$

Here, our group is just

$$S_3 = \{e,\ a,\ b,\ r,\ s,\ t\}.$$

The Cayley table for S_3 is shown as follows:

$\cdot$	e	a	b	r	s	t
e	e	a	b	r	s	t
a	a	b	e	s	t	r
b	b	e	a	t	r	s
r	r	t	s	e	b	a
s	s	r	t	a	e	b
t	t	s	r	b	a	e

The conjugacy classes of S_3 are $C_1 = \{e\}$, $C_2 = \{a,\ b\}$ and $C_3 = \{r,\ s,\ t\}$. Then, the hypergroup $Conj(S_3)$ is

*	$\mathcal{C}_1$	$\mathcal{C}_2$	$\mathcal{C}_3$
$\mathcal{C}_1$	$\mathcal{C}_1$	$\mathcal{C}_2$	$\mathcal{C}_3$
$\mathcal{C}_2$	$\mathcal{C}_2$	$\{\mathcal{C}_1, \mathcal{C}_2, \mathcal{C}_3\}$	$\mathcal{C}_2$
$\mathcal{C}_3$	$\mathcal{C}_3$	$\mathcal{C}_2$	$\{\mathcal{C}_1, \mathcal{C}_3\}$

Example 2.15. Consider the dihedral group D_4. This group is generated by a counter-clockwise rotation r of 90° and a horizontal reflection h. The group consists of the following 8 symmetries:

$$\{1 = r^0, \ r, \ r^2 = s, \ r^3 = t, \ h, \ hr = d, \ hr^2 = v, \ hr^3 = f\}.$$

Dihedral groups occur frequently in art and nature. Many of the decorative designs used on floor coverings, pottery, and buildings have one of Dihedral groups as a group of symmetry. In the case of D_4 there are five conjugacy classes: $\{1\}, \{s\}, \{r, \ t\}, \{d, \ f\}$ and $\{h, \ v\}$. Let us denote these classes by $\mathcal{C}_1, \ ..., \ \mathcal{C}_5$, respectively. Then the hypergroup $Conj(D_4)$ is

*	$\mathcal{C}_1$	$\mathcal{C}_2$	$\mathcal{C}_3$	$\mathcal{C}_4$	$\mathcal{C}_5$
$\mathcal{C}_1$	$\mathcal{C}_1$	$\mathcal{C}_2$	$\mathcal{C}_3$	$\mathcal{C}_4$	$\mathcal{C}_5$
$\mathcal{C}_2$	$\mathcal{C}_2$	$\mathcal{C}_1$	$\mathcal{C}_3$	$\mathcal{C}_4$	$\mathcal{C}_5$
$\mathcal{C}_3$	$\mathcal{C}_3$	$\mathcal{C}_3$	$\{\mathcal{C}_1, \mathcal{C}_2\}$	$\mathcal{C}_5$	$\mathcal{C}_4$
$\mathcal{C}_4$	$\mathcal{C}_4$	$\mathcal{C}_4$	$\mathcal{C}_5$	$\{\mathcal{C}_1, \mathcal{C}_2\}$	$\mathcal{C}_3$
$\mathcal{C}_5$	$\mathcal{C}_5$	$\mathcal{C}_5$	$\mathcal{C}_4$	$\mathcal{C}_3$	$\{\mathcal{C}_1, \mathcal{C}_2\}$

As a sample of how to calculate the table entries, consider $\mathcal{C}_3 \cdot \mathcal{C}_3$. To determine this product, compute the element-wise product of the conjugacy classes $\{r,t\}\{r,t\} = \{s,1\} = \mathcal{C}_1 \cup \mathcal{C}_2$. Thus $\mathcal{C}_3 \cdot \mathcal{C}_3$ consists of the two conjugacy classes $\mathcal{C}_1, \mathcal{C}_2$.

Closely related to the conjugacy classes of a finite group are its characters. Let $IrrG = \{\chi_1, \ \chi_2, \ ..., \ \chi_k\}$ be the collection of irreducible characters of a finite group G where χ_1 is the trivial character. The set $IrrG$ of G is a hypergroup, where the product $\chi_i * \chi_j$ is the set of irreducible components in the element-wise product $\chi_i\chi_j$. This hypergroup is called *character hypergroup*, and is investigated by Roth [159] who considered a duality between $IrrG$ and $ConjG$.

Before calculating $Irr(D_4)$ we need to know the five irreducible characters of Dihedral group D_4. These are given by the following character table. (Since characters are constant on conjugacy classes it is usual to list only the conjugacy classes across the top of the table.)

	C_1	C_2	C_3	C_4	C_5
$\chi_1 :$	1	1	1	1	1
$\chi_2 :$	1	1	-1	1	-1
$\chi_3 :$	1	1	-1	-1	1
$\chi_4 :$	1	1	1	-1	-1
$\chi_5 :$	2	-2	0	0	0

We illustrate the calculation of the hypergroup product of two characters by considering $\chi_5 * \chi_5$. The point-wise product of χ_5 with itself yields the following (non-irreducible) character:

	C_1	C_2	C_3	C_4	C_5
$\chi_5\chi_5 :$	4	4	0	0	0

This character can be written as a sum of irreducible characters in exactly one way: $\chi_5\chi_5 = \chi_1 + \chi_2 + \chi_3 + \chi_4$. This is indicated by the entry in the lower right hand corner of the hypergroup table for $\hat{D}_4$. In general the hypergroup product of two characters $\chi_i * \chi_j$ tell which are the irreducible characters in the product $\chi_i\chi_j$, but not the multiplicity. Using i in the place of the character χ_i the hypergroup $Irr D_4$ is

	1	2	3	4	5
1	1	2	3	4	5
2	2	1	4	3	5
3	3	4	1	2	5
4	4	3	2	1	5
5	5	5	5	5	$\{1,2,3,4\}$

2.3 H_v-groups as a generalization of hypergroups

H_v-groups are introduced by Vougiouklis at the Fourth AHA congress (1990) [181]. The concept of an H_v-group constitutes a generalization of a hypergroup.

Definition 2.4. Let H be a non-empty set and $\circ : H \times H \rightarrow \mathcal{P}^*(H)$ be a hyperoperation. The " $\circ$ " in H is called *weak associative* if

$$x \circ (y \circ z) \cap (x \circ y) \circ z \neq \emptyset, \text{ for all } x, y, z \in H.$$

The " ∘ " is called *weak commutative* if
$$x \circ y \cap y \circ x \neq \emptyset, \text{ for all } x, y \in H.$$
The hyperstructure $(H, \circ)$ is called an H_v-*semigroup* if " ∘ " is weak associative. An H_v-semigroup is called an H_v-*group* if
$$a \circ H = H \circ a = H, \text{ for all } a \in H.$$
In an obvious way, the H_v-subgroup of an H_v-group is defined.

All the weak properties for hyperstructures can be applied for subsets. For example, if $(H, \circ)$ is a weak commutative H_v-group, then for all non-empty subsets A, B and C of H, we have
$$(A \circ B) \cap (B \circ A) \neq \emptyset \text{ and } A \circ (B \circ C) \cap (A \circ B) \circ C \neq \emptyset.$$
To prove this, one has simply to take one element of each set.

Definition 2.5. Let $(H, \circ)$ and $(H, \star)$ be two H_v-groups defined on the same set H. We say that $\circ$ *less than or equal to* $\star$, and we write $\circ \leq \star$, if there is $f \in Aut(H, \star)$ such that $x \circ y \subseteq f(x \star y)$, for all $x, y \in H$.

A motivation to study the above structures is given by the following examples.

Example 2.16. Let $(G, \cdot)$ be a group and ρ be an equivalence relation on G. In G/ρ, the set of quotient, consider the hyperoperation $\odot$ defined by $\rho(x) \odot \rho(y) = \{\rho(z) \mid z \in \rho(x) \cdot \rho(y)\}$, where $\rho(x)$ denotes the equivalence class of the element x. Then, $(G, \odot)$ is an H_v-group which is not always a hypergroup.

Example 2.17. On the set $\mathbb{Z}_{mn}$ consider the hyperoperation $\oplus$ defined by setting $0 \oplus m = \{0, m\}$ and $x \oplus y = x + y$ for all $(x, y) \in \mathbb{Z}_{mn}^2 - \{(0, m)\}$. Then $(\mathbb{Z}_{mn}, \oplus)$ is an H_v-group. $\oplus$ is weak associative but not associative, since taking $k \notin m\mathbb{Z}$ we have
$$(0 \oplus m) \oplus k = \{0, m\} \oplus k = \{k, m + k\},$$
$$0 \oplus (m \oplus k) = 0 \oplus (m + k) = \{m + k\}.$$
Moreover, it is weak commutative but not commutative.

Example 2.18. Consider the group $(\mathbb{Z}^n, +)$ and take $m_1, ..., m_n \in \mathbb{N}$. We define a hyperoperation $\oplus$ in $\mathbb{Z}^n$ as follows:
$$(m_1, 0, ..., 0) \oplus (0, 0, ..., 0) = \{(m_1, 0, ..., 0), (0, 0, ..., 0)\},$$
$$(0, m_1, ..., 0) \oplus (0, 0, ..., 0) = \{(0, m_1, ..., 0), (0, 0, ..., 0)\},$$
$$(0, 0, ..., m_n) \oplus (0, 0, ..., 0) = \{(0, 0, ..., m_n), (0, 0, ..., 0)\},$$
and $\oplus = +$ in the remaining cases. Then, $(\mathbb{Z}^n, \oplus)$ is an H_v-group.

For more details about H_v-groups we refere the readers to [63; 183].

Chapter 3

Subhypergroups

In general, we are not interested in arbitrary subsets of a hypergroup H for which the hypergroups structure is not preserved. Any subsets we do consider they are subhypergroups, that preserve algebraic properties derived from those of H. The study of several types of subhypergroups was begun by Marty, Dresher, Ore, Krasner, and was continued by Sureau, Koskas, Corsini, Freni, Davvaz, Leoreanu, Račková, and many others.

3.1 Closed, invertible, ultraclosed and conjugable subhypergroups

First we begin with the definition of a subhypergroup of a hypergroup.

Definition 3.1. A non-empty subset K of a hypergroup $(H, \circ)$ is called a *subhypergroup* if it is a hypergroup.

Hence, a non-empty subset K of a hypergroup $(H, \circ)$ is a subhypergroup if for all a of K we have $a \circ K = K \circ a = K$.

There are several kinds of subhypergroups. In what follows, we introduce closed, invertible, ultraclosed and conjugable subhypergroups and some connections among them.

Among the mathematicians who studied this topic, we mention Marty, Dresher, Ore, Krasner who analyzed closed and invertible subhypergroups. Koskas considered another type of subhypergroups which are complete parts. Later Sureau studied ultraclosed, invertible and conjugable subhypergroups. Corsini obtained important results about ultraclosed and complete parts. Also, Leoreanu studied and obtained other interesting results on subhypergroups.

Let us present now the definition of these types of subhypergroups.

Definition 3.2. Let $(H, \circ)$ be a hypergroup and $(K, \circ)$ be a subhypergroup of it. We say that K is:

- *closed on the left* (*on the right*) if for all k_1, k_2 of K and x of H, from $k_1 \in x \circ k_2$ ($k_1 \in k_2 \circ x$, respectively), it follows that $x \in K$;
- *invertible on the left* (*on the right*) if for all x, y of H, from $x \in K \circ y$ ($x \in y \circ K$), it follows that $y \in K \circ x$ ($y \in x \circ K$, respectively);
- *ultraclosed on the left* (*on the right*) if for all x of H, we have $K \circ x \cap (H \backslash K) \circ x = \emptyset$ ($x \circ K \cap x \circ (H \backslash K) = \emptyset$);
- *conjugable on the right* if it is closed on the right and for all $x \in H$, there exists $x' \in H$ such that $x' \circ x \subseteq K$. Similarly, we can define the notion of *conjugable on the left*.

We say that K is *closed* (*invertible, ultraclosed, conjugable*) if it is closed (invertible, ultraclosed, conjugable, respectively) on the left and on the right.

In any hypergroup a hyperoperation has two inverses. Let $(H, \circ)$ be a hypergroup. The sets
$$a/b = \{x \mid a \in x \circ b\},$$
$$b\backslash a = \{x \mid a \in b \circ x\}.$$
are called the *right and left extensions*, respectively. Let A and B be two non-empty subsets of H. Each of A/B and $B\backslash A$ is defined by $A/B = \bigcup \{a/b \mid a \in A, b \in B\}$ and $B\backslash A = \bigcup \{b\backslash a \mid a \in A, b \in B\}$. For two sets A and B, the symbol $A \approx B$ means that A and B are incident, i.e., $A \cap B \neq \emptyset$. By using the above notations we can say that a subhypergroup $(K, \circ)$ of a hypergroup $(H, \circ)$ is closed if $a/b \subseteq K$ and $b\backslash a \subseteq K$, for all $a, b \in K$; and it is invertible if $a/b \approx K$ implies $b/a \approx K$, and $b\backslash a \approx K$ implies $a\backslash b \approx K$, for all $a, b \in H$.

A subhypergroup K is called *reflexive* if $a K = K\backslash a$, for all $a \in H$.

Example 3.1. Let $(A, \circ)$ be a hypergroup, $H = A \cup T$, where T is a set with at least three elements and $A \cap T = \emptyset$. We define the hyperoperation $\otimes$ on H, as follows:

if $(x, y) \in A^2$, then $x \otimes y = x \circ y$;
if $(x, t) \in A \times T$, then $x \otimes t = t \otimes x = t$;
if $(t_1, t_2) \in T \times T$, then $t_1 \otimes t_2 = t_2 \otimes t_1 = A \cup (T \setminus \{t_1, t_2\})$.

Then, $(H, \otimes)$ is a hypergroup and $(A, \otimes)$ is an ultraclosed, non-conjugable subhypergroup of H.

Example 3.2. Let $(A, \circ)$ be a total hypergroup, with at least two elements and let $T = \{t_i\}_{i \in \mathbb{N}}$ such that $A \cap T = \emptyset$ and $t_i \neq t_j$ for $i \neq j$. We define the hyperoperation $\otimes$ on $H = A \cup T$ as follows:

> if $(x, y) \in A^2$, then $x \otimes y = A$;
> if $(x, t) \in A \times T$, then $x \otimes t = t \otimes x = (A \setminus \{x\}) \cup T$;
> if $(t_i, t_j) \in T \times T$, then $t_i \otimes t_j = t_j \otimes t_i = A \cup \{t_{i+j}\}$.

Then, $(H, \otimes)$ is a hypergroup and $(A, \otimes)$ is a non-closed subhypergroup of H.

Example 3.3. Let us consider the group $(\mathbb{Z}, +)$ and the subgroups $S_i = 2^i \mathbb{Z}$, where i is a non-negative integer. For any $x \in \mathbb{Z} \setminus \{0\}$, there exists a unique integer $n(x)$, such that $x \in S_{n(x)} \setminus S_{n(x)+1}$. Define the following commutative hyperoperation on $\mathbb{Z} \setminus \{0\}$:

> if $n(x) < n(y)$, then $x \circ y = x + S_{n(y)}$;
> if $n(x) = n(y)$, then $x \circ y = S_{n(x)} \setminus \{0\}$;
> if $n(x) > n(y)$, then $x \circ y = y + S_{n(x)}$.

Notice that if $n(x) < n(y)$, then $n(x + y) = n(x)$. Then, $(\mathbb{Z} \setminus \{0\}, \circ)$ is a hypergroup and for all $i \in \mathbb{N}$, $(S_i \setminus \{0\}, \circ)$ is an invertible subhypergroup of $\mathbb{Z} \setminus \{0\}$.

Lemma 3.1. *A subhypergroup K is invertible on the right if and only if $\{x \circ K\}_{x \in H}$ is a partition of H.*

Proof. If K is invertible on the right and $z \in x \circ K \cap y \circ K$, then $x, y \in z \circ K$, whence $x \circ K \subseteq z \circ K$ and $y \circ K \subseteq z \circ K$. It follows that $x \circ K = z \circ K = y \circ K$. Conversely, if $\{x \circ K\}_{x \in H}$ is a partition of H and $x \in y \circ K$, then $x \circ K \subseteq y \circ K$, whence $x \circ K = y \circ K$ and so we have $x \in y \circ K = x \circ K$. Hence, for all x of H we have $x \in x \circ K$. From here, we obtain that $y \in y \circ K = x \circ K$. $\qquad\square$

Similar to Lemma 3.1, we can give a necessary and sufficient condition for invertible subhypergroups on the left.

The following theorems present some connections among the above types of subhypergroups.

If A and B are subsets of H such that we have $H = A \cup B$ and $A \cap B = \emptyset$, then we denote $H = A \oplus B$.

Theorem 3.1. *If a subhypergroup K of a hypergroup $(H, \circ)$ is ultraclosed, then it is closed and invertible.*

Proof. First we check that K is closed. For $x \in K$, we have $K \cap x \circ (H \setminus K) = \emptyset$ and from $H = x \circ K \cup x \circ (H \setminus K)$, we obtain $x \circ (H \setminus K) = H \setminus K$, which means that $K \circ (H \setminus K) = H \setminus K$. Similarly, we obtain $(H \setminus K) \circ K = H \setminus K$, hence K is closed. Now, we show that $\{x \circ K\}_{x \in H}$ is a partition of H. Let $y \in x \circ K \cap z \circ K$. It follows that $y \circ K \subseteq x \circ K$ and $y \circ (H \setminus K) \subseteq x \circ K \circ (H \setminus K) = x \circ (H \setminus K)$. From $H = x \circ K \oplus x \circ (H \setminus K) = y \circ K \oplus y \circ (H \setminus K)$, we obtain $x \circ K = y \circ K$. Similarly, we have $z \circ K = y \circ K$. Hence, $\{x \circ K\}_{x \in H}$ is a partition of H, and according to the above lemma, it follows that K is invertible on the right. Similarly, we can show that K is invertible on the left. $\square$

Theorem 3.2. *If a subhypergroup K of a hypergroup $(H, \circ)$ is invertible, then it is closed.*

Proof. Let $k_1, k_2 \in K$. If $k_1 \in x \circ k_2 \subseteq x \circ K$, then $x \in k_1 \circ K \subseteq K$. Similarly, from $k_1 \in k_2 \circ x$, we obtain $x \in K$. $\square$

We denote the set $\{e \in H \mid \exists x \in H, \text{ such that } x \in x \circ e \cup e \circ x\}$ by I_p and we call it the *set of partial identities* of H.

Theorem 3.3. *A subhypergroup K of a hypergroup $(H, \circ)$ is ultraclosed if and only if K is closed and $I_p \subseteq K$.*

Proof. Suppose that K is closed and $I_p \subseteq K$. First, we show that K is invertible on the left. Suppose there are x, y of H such that $x \in K \circ y$ and $y \notin K \circ x$. Hence, $y \in (H \setminus K) \circ x$, whence $x \in K \circ (H \setminus K) \circ x \subseteq (H \setminus K) \circ x$, since K is closed. We obtain that $I_p \cap (H \setminus K) \neq \emptyset$, which is a contradiction. Hence, K is invertible on the left. Now, we check that K is ultraclosed on the left. Suppose that there are a and x in H such that $a \in K \circ x \cap (H \setminus K) \circ x$. It follows that $x \in K \circ a$, since K is invertible on the left. We obtain $a \in (H \setminus K) \circ x \subseteq (H \setminus K) \circ K \circ a \subseteq (H \setminus K) \circ a$, since K is closed. This means that $I_p \cap (H \setminus K) \neq \emptyset$, which is a contradiction. Therefore, K is ultraclosed on the left and similarly it is ultraclosed on the right.

Conversely, suppose that K is ultraclosed. According to Theorem 3.1, K is closed. Now, suppose that $I_p \cap (H \setminus K) \neq \emptyset$, which means that there is $e \in H \setminus K$ and there is $x \in H$, such that $x \in e \circ x$, for instance. We obtain $x \in (H \setminus K) \circ x$, whence $K \circ x \subseteq (H \setminus K) \circ x$, which contradicts that K is ultraclosed. Hence, $I_p \subseteq K$. $\square$

Theorem 3.4. *If a subhypergroup K of a hypergroup $(H, \circ)$ is conjugable, then it is ultraclosed.*

Proof. Let $x \in H$. Denote $B = x \circ K \cap x \circ (H \setminus K)$. Since K is conjugable it follows that K is closed and there exists $x' \in H$, such that $x' \circ x \subseteq K$. We obtain

$$\begin{aligned}
x' \circ B &= x' \circ (x \circ K \cap x \circ (H \setminus K)) \\
&\subseteq K \cap x' \circ x \circ (H \setminus K) \\
&\subseteq K \cap K \circ (H \setminus K) \\
&\subseteq K \cap (H \setminus K) = \emptyset.
\end{aligned}$$

Hence, $B = \emptyset$, which means that K is ultraclosed on the right. Similarly, we check that K is ultraclosed on the left. $\qquad\square$

Definition 3.3. A subhypergroup N of a hypergroup $(H, \circ)$ is *normal* if $a \circ N = N \circ a$.

Theorem 3.5. *In any hypergroup, if N is a subhypergroup, then N is normal if and only if $N \setminus a = a/N$.*

Proof. Consider the two pairs (a) and (b) of equivalent statements (1) and (2):

(a) (1) $x \approx N \setminus y$, (2) $y \approx n \circ x$,
(b) (1) $x \approx y/N$, (2) $y \approx x \circ N$.

Suppose that N is normal. Then parts (2) of (a) and (b) are equivalent. Thus so are parts (1). Therefore, $N \setminus a = a/N$. On the other hand, if $N \setminus a = a/N$, then parts (1) are equivalent, so parts (2). Consequently, N is normal. $\qquad\square$

Both the defining condition for normality and the condition of the theorem hold for sets; that is, if N is normal then $A \circ N = N \circ A$ and $N \setminus A = A/N$.

3.2 Hypergroups induced by quasiordered groups

In [153], Račková gave a description of invertible, closed, normal, reflexive subhypergroups in hypergroups induced by quasiordered groups.

A *quasiorder* is a binary relation that is reflexive and transitive. A *quasiordered group* is a triple $(H, \cdot, \leq)$, where $(H, \cdot)$ is a group and $\leq$ is a quasiordering on H having the substitution property on $(H, \cdot)$, that is for any $a, b, c, d \in H$ such that $a \leq b$ and $c \leq d$, we have $a \cdot c \leq b \cdot d$.

A hyperoperation is naturally induced on each quasiordered group $(H, \cdot, \leq)$. For $x \in H$, let us denote $[x]_\leq$ the principal end generated by

x, that is $[x)_\leq = \{y \in H \mid x \leq y\}$. Analogously the principal beginning $(x]_\leq$ is defined. For $x, y \in H$, let us denote $x \circ y = [x \cdot y)_\leq$. Then $(H, \circ)$ is a hypergroupoid associated with $(H, \cdot, \leq)$.

Theorem 3.6. *Let* $(H, \cdot, \leq)$ *be a quasiordered group and* $(H, \circ)$ *the hypergroupoid associated with it. Then* $(H, \circ)$ *is a hypergroup.*

Proof.   It is straightforward. $\square$

Let $(H, \cdot, \leq)$ be a quasiordered group, and $(H, \circ)$, where $a \circ b = [a \cdot b)_\leq$, be the induced hypergroup. Let $(K, \circ)$ be its subhypergroup, that is, $[x \cdot y)_\leq \subseteq K$ holds for $x, y \in K$, namely $x \cdot y \in S$. For the right and left extensions we have

$$a/b = \{x \in H \mid a \in x \circ b\} = \{x \in H \mid x \cdot b \leq a\} = \{x \in H \mid x \leq a \cdot b^{-1}\},$$
$$b \backslash a = \{x \in H \mid a \in b \circ x\} = \{x \in H \mid b \cdot x \leq a\} = \{x \in H \mid x \leq b^{-1} \cdot a\}.$$

Theorem 3.7. *A subhypergroup* $(K, \circ)$ *is closed if and only if* $(K, \cdot)$ *is a subgroup of the group* $(H, \cdot)$ *and* $(a]_\leq \cup [a)_\leq \subseteq K$, *for all* $a \in K$.

Proof.   Suppose that $(K, \circ)$ is closed. Let a, b be arbitrary elements of K. Then, we have $a/a = (e]_\leq$, where e is the neutral element. This implies that $e \in K$. Since $a, b \in K$, it follows that $a/b \subseteq K$ or $(a \cdot b^{-1}]_\leq K$. This implies that $a \cdot b^{-1} \in K$. Since $a \in K$, it follows that $a/e = (a]_\leq$; and so $(a]_\leq \subseteq K$. On the other hand, since $(K, \circ)$ is a subhypergroup, we obtain $a \circ e = [a)_\leq K$.

Conversely, suppose that $(K, \cdot)$ is a subgroup of $(H, \cdot)$. Let a, b be arbitrary elements of K. Since K is a subgroup, it follows that $a \cdot b^{-1} \in K$. this yields that $(a \cdot b^{-1}]_\leq = a/b \subseteq K$. Similarly, we obtain $b \backslash a \subseteq K$. $\square$

Theorem 3.8. *A subhypergroup* $(K, \circ)$ *of* $(H, \circ)$ *is closed if and only if it is invertible.*

Proof.   By Theorem 3.2, each invertible subhypergroup is closed.

Now, suppose that $(K, \circ)$ is closed. If $a/b \approx K$, then there is $x \in K$ such that $x \leq a \cdot b^{-1}$. By Theorem 3.7, we conclude that $a \cdot b^{-1} \in K$. Moreover, $(a \cdot b^{-1})^{-1} = b \cdot a^{-1} \in K$. This yields that $b/a \subseteq K$. In particular, we have $b/a \approx K$. Similarly, we can show that $b \backslash K$. $\square$

Theorem 3.9. *A subhypergroup* $(K, \circ)$ *is reflexive if and only if the following property holds: If for* $x, y \in H$ *the element* $x \cdot y$ *is covered by an element of* K, *then* $y \cdot x$ *is also covered by an element of* K.

Proof. It is easy to see that

$$a \backslash K = \bigcup_{b \in K} a \backslash b = \bigcup_{b \in K} (a^{-1} \cdot b]_{\leq},$$

$$K/a = \bigcup_{b \in K} b/a = \bigcup_{b \in K} (b \cdot a^{-1}]_{\leq}.$$

Thus, we have

$$x \in a \backslash K \ \Leftrightarrow \ \exists b \in K \text{ such that } a \cdot x \leq b,$$
$$x \in K/a \ \Leftrightarrow \ \exists c \in K \text{ such that } x \cdot a \leq c.$$

This completes the proof. □

Corollary 3.1. *A closed subhypergroup $(K, \circ)$ is reflexive if and only if the following property holds: For each $x, y \in H$, if $x \cdot y \in K$, then also $y \cdot x \in K$.*

Theorem 3.10. *A subhypergroup $(K, \circ)$ is normal if and only if the following property holds: For $x, y \in H$, if $x \cdot y$ covers an element of K, then $y \cdot x$ also covers an element of K.*

Proof. Clearly, we have

$$a \circ K = \bigcup_{b \in K} [a \cdot b)_{\leq} \text{ and } K \circ a = \bigcup_{b \in K} [b \cdot a)_{\leq}.$$

Therefore, we can write

$$x \in a \circ K \ \Leftrightarrow \ \exists b \in K \text{ such that } b \leq a^{-1} \cdot x,$$
$$x \in K \circ a \ \Leftrightarrow \ \exists c \in K \text{ such that } c \leq x \cdot a^{-1}.$$

This completes the proof. □

Corollary 3.2. *A closed subhypergroup $(K, \circ)$ is normal if and only if the following property holds: for each $x, y \in H$, if $x \cdot y \in K$, then also $y \cdot x \in K$.*

Corollary 3.3. *Let $(K, \circ)$ be a closed subhypergroup. Then $(K, \circ)$ is normal if and only if it is reflexive.*

3.3 Subhypergroups in a partially ordered algebra

In [50], Davvaz introduced a natural interpretation of subhypergroups in a partially ordered algebra. He studied their connection with corresponding crisp concepts through their newly defined Q-cuts. The proven theorems also highly generalized the existing ones. The main reference for this section is [50].

Let $P = (P, *, 1, \leq)$ be a partially ordered algebra. Therefore $(P, *)$ is a monoid, where 1 is the unity for $*$, and $*$ is isotone in both variables; and

$(P, \leq)$ is a complete lattice, i.e., $\leq$ is a partial order on P such that for any $S \subseteq P$, infimum and supremum of S exist and these will be denoted by $\bigwedge_{s \in S} \{s\}$ and $\bigvee_{s \in S} \{s\}$, respectively. Further on P always denotes such a structure. Let X be a non-empty set. A map $\mu : X \to P$ is called a P-*subset* of X. A subset Q of an ordered set $(P, \leq)$ is called a *right segment* of $(P, \leq)$ if and only if

$$\forall q \in Q, \ \forall p \in P : \ (q \leq p \Rightarrow p \in Q).$$

Clearly any closed interval $[p, 1]$ in P is a right segment of $(P, \leq)$.

Definition 3.4. Let H be a hypergroup and μ be a P-subset of H. Then, μ is called a P-*subhypergroup* of H, if

1) $\mu(x) * \mu(y) \leq \bigwedge_{z \in x \circ y} \{\mu(z)\}$ for all $x, y \in H$;

2) for all $x, a \in H$ there exists $y \in H$ such that $x \in a \circ y$ and

$$\mu(a) * \mu(x) \leq \mu(y);$$

3) for all $x, a \in H$ there exists $z \in H$ such that $x \in z \circ a$ and

$$\mu(x) * \mu(a) \leq \mu(z).$$

Example 3.4. Consider $H = \{e, \ a, \ b\}$ and define $\circ$ on H with the help of the following table:

$\circ$	e	a	b
e	e	H	H
a	H	a	a
b	H	a	$\{a, b\}$

Then $(H, \circ)$ is a hypergroup. Suppose that $(P, *, 1, \leq)$ is a partially ordered algebra such that P is a lattice and $* =$ meet. Now, we define $\mu : H \to P$ by $\mu(a) = \mu(b) \leq \mu(e)$. Then μ is a P-subhypergroup of H.

Definition 3.5. Let Q be a right segment of P. Then, by the Q-*cut* μ_Q of some P-subset μ of X we mean the following subset of X:

$$\mu_Q = \{x \in X | \ \mu(x) \in Q\}.$$

Theorem 3.11. *Let H be a hypergroup and μ be a P-subset of H. If each non-empty Q-cut μ_Q of μ is a subhypergroup of H, then μ is a P-subhypergroup of H.*

Proof. Take any element $x, y \in H$ and consider the right segment of P as follows:

$$Q = [\mu(x) * \mu(y), 1].$$

Since

$$\mu(x) * \mu(y) \leq \mu(x) * 1 = \mu(x) \leq 1,$$
$$\mu(x) * \mu(y) \leq 1 * \mu(y) = \mu(y) \leq 1,$$

it follows that $\mu(x), \mu(y) \in Q$. Hence, $x, y \in \mu_Q$. Since μ_Q is a subhypergroup of H, it follows that for every $z \in x \circ y$ we have $z \in \mu_Q$ and so $\mu(z) \in Q$. Therefore, $\mu(x) * \mu(y) \leq \mu(z)$ which implies that

$$\mu(x) * \mu(y) \leq \bigwedge_{z \in x \circ y} \{\mu(z)\},$$

and in this way the condition (1) of Definition 12.2 is verified. To verify the second condition, if $x, a \in H$, we consider the right segment of P as follows:

$$Q = [\mu(a) * \mu(x), 1].$$

Then, $\mu(a), \mu(x) \in Q$, and so $x, a \in \mu_Q$. Hence, there exists $y \in \mu_Q$ such that $x \in a \circ y$. Since $y \in \mu_Q$, it follows that $\mu(a) * \mu(x) \leq \mu(y)$. In the similar way the third condition of Definition 12.2 is valid. $\square$

Theorem 3.12. *Let H be a hypergroup, μ be a P-subhypergroup of H, and Q be a right segment of P. If Q is closed under $*$, then μ_Q is a subhypergroup of H.*

Proof. Consider a right segment Q satisfying the given condition. Then for any elements $x, y \in \mu_Q$ we have $\mu(x), \mu(y) \in Q$, and so $\mu(x) * \mu(y) \in Q$. For every $z \in x \circ y$, we have $\mu(x) * \mu(y) \leq \mu(z)$. Since Q is a right segment, it follows that $\mu(z) \in Q$ or $z \in \mu_Q$ which means that $x \circ y \subseteq \mu_Q$.

Now, let $x, a \in \mu_Q$, we have $\mu(x), \mu(a) \in Q$ and so $\mu(a) * \mu(x) \in Q$. Since $x, a \in H$, it follows that there exists $y \in H$ such that $x \in a \circ y$ and $\mu(a) * \mu(x) \leq \mu(y)$. Since Q is a right segment, it follows that $\mu(y) \in Q$ or $y \in \mu_Q$. Therefore, we have $a \circ \mu_Q = \mu_Q$. Similarly we can show that $\mu_Q \circ a = \mu_Q$. This yields that μ_Q is a subhypergroup of H. $\square$

Let $(H_1, \circ)$ and $(H_2, \bullet)$ be two hypergroups. Then we can define a hyperproduct on $H_1 \times H_2$ as follows:

$$(x_1, x_2) \otimes (y_1, y_2) = \{(a, b) | a \in x_1 \circ x_2, \ b \in y_1 \bullet y_2\}.$$

Clearly $H_1 \times H_2$ equipped with $\otimes$ is a hypergroup.

Definition 3.6. Let H_1, H_2 be two hypergroups and μ, λ be P-subsets of H_1, H_2 respectively. The *P-product* of μ, λ is defined as follows:

$$(\mu \times \lambda)(x, y) = \mu(x) * \lambda(y).$$

Theorem 3.13. *Let H_1, H_2 be two hypergroups, and μ, λ be P-subhypergroups of H_1, H_2 respectively. If P is an abelian monoid, then the P-product $\mu \times \lambda$ is a P-subhypergroup of $H_1 \times H_2$.*

Proof. Suppose that $(x_1, x_2), (y_1, y_2) \in H_1 \times H_2$. For every $(z_1, z_2) \in (x_1, x_2) \otimes (y_1, y_2)$ we have

$$\begin{aligned}
\mu \times \lambda(x_1, x_2) * \mu \times \lambda(y_1, y_2) &= \mu(x_1) * \lambda(x_2) * \mu(y_1) * \lambda(y_2) \\
&= (\mu(x_1) * \mu(y_1)) * (\lambda(x_2) * \lambda(y_2)) \\
&\leq \mu(z_1) * \lambda(z_2) \\
&= \mu \times \lambda(z_1, z_2).
\end{aligned}$$

Therefore, the condition (1) of Definition 12.2 is satisfied. Now, for every (x_1, x_2) and $(a_1, a_2) \in H_1 \times H_2$ there exists $(y_1, y_2) \in H_1 \times H_2$ such that $x_1 \in a_1 \circ y_1$, $x_2 \in a_2 \bullet y_2$ and $\mu(a_1) * \mu(x_1) \leq \mu(y_1)$, $\lambda(a_2) * \lambda(x_2) \leq \lambda(y_2)$. Therefore, we have $(x_1, x_2) \in (a_1, a_2) \otimes (y_1, y_2)$ and

$$\begin{aligned}
\mu \times \lambda(a_1, a_2) * \mu \times \lambda(x_1, x_2) &= \mu(a_1) * \lambda(a_2) * \mu(x_1) * \lambda(x_2) \\
&= (\mu(a_1) * \mu(x_1)) * (\lambda(a_2) * \lambda(x_2)) \\
&\leq \mu(y_1) * \lambda(y_2) \\
&= \mu \times \lambda(y_1, y_2).
\end{aligned}$$

The proof of condition (3) of Definition 12.2 is similar to the proof of second condition. $\square$

Corollary 3.4. *Let H_1, H_2 be two hypergroups, and μ, λ be P-subhypergroups of H_1 H_2 respectively. If Q_1, Q_2 are right segments of P and closed under $*$, then*

$$(\mu \times \lambda)_{Q_1 \times Q_2} = \mu_{Q_1} \times \lambda_{Q_2}.$$

Definition 3.7. A P-subset $r : X \times X \to P$ is called a *P-relation* on P-subset μ, if it satisfies the following property:

$$r(x, y) \leq \mu(x) * \mu(y) \quad \text{for all } x, y \in X.$$

A P-relation r on a P-subset μ is said to be

1) reflexive, if $r(x, y) = \mu(x) * \mu(y)$ for all $x \in X$;
2) symmetric, if $r(x, y) = r(y, x)$ for all $x, y \in X$;

3) transitive, if for any $x, z \in X$

$$r(x, y) * r(y, z) \leq r(x, z) \text{ for all } y \in X.$$

A reflexive, symmetric and transitive P-relation r on a P-subset μ is called P-similarity.

Definition 3.8. Let H be a hypergroup and μ a P-subhypergroup of H. A P-relation r on μ is called a P-compatible relation on μ if

$$r(x_1, x_2) * r(y_1, y_2) \leq \bigwedge_{\substack{z_1 \in x_1 \circ y_1 \\ z_2 \in x_2 \circ y_2}} \{r(z_1, z_2)\} \text{ for all } x_1, x_2, y_1, y_2 \in H.$$

A P-compatible P-similarity relation is called a P-strongly regular relation on the P-subhypergroup μ.

Lemma 3.2. Let r be a P-relation on a P-subset μ. If each Q-cut r_Q is an equivalence relation on μ_Q for any right segment Q of P, then r is a P-similarity on μ.

Proof. It is straightforward. $\square$

Now, we consider the converse of Lemma 3.2.

Lemma 3.3. Let r be a P-similarity on a P-subset μ, and let Q be a right segment of P. If Q is closed under $*$, then r_Q is an equivalence relation on μ_Q.

Proof. It is straightforward. $\square$

Theorem 3.14. Let H be a hypergroup, μ be a P-subset of H, and r be a P-relation on μ. If for all non-empty right segment Q of P, μ_Q is a subhypergroup and r_Q is a strongly regular relation on μ_Q, then r is a P-strongly regular relation on μ.

Proof. By Lemma 3.2, it is enough to show that r is a P-compatible relation. Suppose that $x_1, x_2, y_1, y_2 \in H$, we put

$$Q = [r(x_1, x_2) * r(y_1, y_2), 1],$$

then $r(x_1, x_2), r(y_1, y_2) \in Q$ and so $(x_1, x_2) \in r_Q$, $(y_1, y_2) \in r_Q$. Since

$$r(x_1, x_2) \leq \mu(x_1) * \mu(x_2) \leq \mu(x_1) * 1 = \mu(x_1),$$
$$r(x_1, x_2) \leq \mu(x_1) * \mu(x_2) \leq 1 * \mu(x_2) = \mu(x_2),$$

it follows that $x_1, x_2 \in \mu_Q$, similarly we obtain $y_1, y_2 \in \mu_Q$. Since r_Q is a strongly regular relation on the subhypergroup μ_Q, it follows that $x_1 \circ y_1 \overline{\overline{R}} x_2 \circ y_2$, and consequently for all $z_1 \in x_1 \circ y_1$ and $z_2 \in x_2 \circ y_2$ we have $z_1 r_Q z_2$ or $(z_1, z_2) \in r_Q$. Therefore, $r(x_1, x_2) * r(y_1, y_2) \le r(z_1, z_2)$, and so

$$r(x_1, x_2) * r(y_1, y_2) \le \bigwedge_{\substack{z_1 \in x_1 \circ y_1 \\ z_2 \in x_2 \circ y_2}} \{r(z_1, z_2)\}.$$

This completes the proof. $\qquad\qquad\qquad\qquad\qquad\qquad\qquad\qquad\square$

Theorem 3.15. *Let H be a hypergroup, μ be a P-subhypergroup of H, and r be a P-strongly regular relation on μ. If a right segment Q of P is closed under $*$, then r_Q is a strongly regular relation on μ_Q.*

Proof. Suppose that Q is some right segment satisfying the given condition. By Theorem 3.12, μ_Q is a subhypergroup of H and using Lemma 3.3, r_Q is an equivalence relation on μ_Q. Suppose $(x_1, x_2), (y_1, y_2) \in r_Q$ then $r(x_1, x_2) \in Q$ and $r(y_1, y_2) \in Q$ where x_1, x_2, y_1, y_2 are elements of H. Since Q is closed under $*$, it follows that

$$r(x_1, x_2) * r(y_1, y_2) \in Q.$$

Since r is a P-strongly regular relation, it follows that

$$\bigwedge_{\substack{z_1 \in x_1 \circ y_1 \\ z_2 \in x_2 \circ y_2}} \{r(z_1, z_2)\} \in Q$$

and so $r(z_1, z_2) \in Q$ for all $z_1 \in x_1 \circ y_1$ and $z_2 \in x_2 \circ y_2$, which implies $(z_1, z_2) \in r_Q$. Hence, $x_1 \circ y_1 \overline{\overline{r_Q}} x_2 \circ y_2$. On the other hand, $(x_1, x_2), (y_1, y_2) \in r_Q$ imply that $x_1, x_2, y_1, y_2 \in \mu_Q$. Therefore, r_Q is a strongly regular relation on μ_Q. $\qquad\qquad\qquad\square$

Chapter 4

Homomorphisms and Isomorphisms

Homomorphisms of hypergroups were studied by Dresher, Ore, Krasner, Kuntzmann, Koskas, Jantosciak, Corsini, Freni, Leoreanu, Davvaz and many others. In this chapter, we present several kinds of homomorphisms.

4.1 Homomorphisms

In this section we study homomorphisms between hypergroups. A homomorphism is a structure-preserving map between two hypergroups. Several kinds of homomorphisms between hypergroups exist. The main references for this chapter are [22; 89].

Definition 4.1. Let $(H_1, \circ)$ and $(H_2, *)$ be two hypergroups. A function $f : H_1 \to H_2$, is called

(1) a *homomorphism* or *inclusion homomorphism* if for all x, y of H_1, we have $f(x \circ y) \subseteq f(x) * f(y)$;

(2) a *good homomorphism* if for all x, y of H_1, we have $f(x \circ y) = f(x) * f(y)$;

(3) an *isomorphism* if it is a one to one and onto good homomorphism. If f is an isomorphism, then H_1 and H_2 are said to be *isomorphic*, which is denoted by $H_1 \cong H_2$.

Theorem 4.1. *Let $(H_1, \circ)$ and $(H_2, *)$ be two hypergroups and $f : H_1 \to H_2$ be a good homomorphism. Then, $Im f$ is a subhypergroup of H_2.*

Proof. For every $a, b \in H_1$, we have $f(a) * f(b) = f(a \circ b) \subseteq Im f$. Moreover, there exist $x, y \in H_1$ such that $a \in x \circ b$ and $a \in b \circ y$ and consequently $f(a) \in f(x) * f(b)$ and $f(a) \in f(b) * f(y)$. $\qquad \square$

For simplicity we use notation $x_f = f^{-1}(f(x))$ and for a subset A of H_1, $A_f = f^{-1}(f(A)) = \cup\{x_f \mid x \in A\}$.

Notice that the defining condition for an inclusion homomorphism is equivalent to

$$x \circ y \subseteq f^{-1}(f(x) * f(y)).$$

It is also clear for an inclusion homomorphism that

$$(x \circ y)_f \subseteq f^{-1}(f(x) * f(y)).$$

The defining condition for an inclusion homomorphism is also valid for sets. That is, if A, B are non-empty subsets of H_1, then it follows that

$$f(A \circ B) \subseteq f(A) * f(B).$$

Applying the above relation for $A = x_f$ and $B = y_f$, we obtain

$$x_f \circ y_f \subseteq f^{-1}(f(x) * f(y))$$

and

$$(x_f \circ y_f)_f \subseteq f^{-1}(f(x) * f(y)).$$

Homomorphisms having various types of properties are defined and studied in the literature. Each of these properties is expressed using $f^{-1}(f(x) * f(y))$. We consider four types of homomorphisms in the following definition.

Definition 4.2. Let $(H_1, \circ)$ and $(H_2, *)$ be two hypergroups and $f : H_1 \to H_2$ be a mapping. Then, given $x, y \in H_1$, f is called a *homomorphism of*

> *type 1*, if $f^{-1}(f(x) * f(y)) = (x_f \circ y_f)_f$;
> *type 2*, if $f^{-1}(f(x) * f(y)) = (x \circ y)_f$;
> *type 3*, if $f^{-1}(f(x) * f(y)) = x_f \circ y_f$;
> *type 4*, if $f^{-1}(f(x) * f(y)) = (x \circ y)_f = x_f \circ y_f$.

Note that $x \circ y \subseteq (x \circ y)_f$, $x \circ y \subseteq x_f \circ y_f$ and that $(x \circ y)_f \subseteq (x_f \circ y_f)_f$, $x_f \circ y_f \subseteq (x_f \circ y_f)_f$. Hence, a homomorphism of any type 1–4 is indeed an inclusion homomorphism. Observe that a one to one homomorphism of H_1 onto H_2 of any type 1–4 is an isomorphism.

Theorem 4.2. *Let $(H_1, \circ)$ and $(H_2, *)$ be two hypergroups, A, B are non-empty subsets of H_1 and $f : H_1 \to H_2$ be a mapping. Then, f is a homomorphism of*

> (1) *type 1 implies $f^{-1}(f(A) * f(B)) = (A_f \circ B_f)_f$;*
> (2) *type 2 implies $f^{-1}(f(A) * f(B)) = (A \circ B)_f$;*

(3) *type 3 implies* $f^{-1}(f(A) * f(B)) = A_f \circ B_f$;

(4) *type 4 implies* $f^{-1}(f(A) * f(B)) = (A \circ B)_f = A_f \circ B_f$.

Proof. It can be easily checked. ∎

Theorem 4.3. *Let* $(H_1, \circ)$ *and* $(H_2, *)$ *be two hypergroups and* $f : H_1 \to H_2$ *be a mapping. Then, f is a homomorphism of*

(1) *type 4 if and only if f is a homomorphism of type 2 and type 3;*

(2) *type 1 if f is a homomorphism of type 2 or type 3.*

Proof. (1) It is trivial.

(2) Suppose that $x, y \in H_1$ and f is a homomorphism of type 2. Then,

$$(x \circ y)_f \subseteq (x_f \circ y_f)_f \subseteq f^{-1}(f(x) * f(y)) = (x \circ y)_f.$$

Similarly, if f is a homomorphism of type 3, then

$$x_f \circ y_f \subseteq (x_f \circ y_f)_f \subseteq f^{-1}(f(x) * f(y)) = x_f \circ y_f.$$

Hence, in either case, f is a homomorphism of type 1. Thus, (2) holds. ∎

Homomorphism Types

The defining condition for a homomorphism of type 1 or type 2 can easily be simplified if the homomorphism is onto.

Theorem 4.4. *Let* $(H_1, \circ)$ *and* $(H_2, *)$ *be two hypergroups and* $f : H_1 \to H_2$ *be an onto mapping. Then, given $x, y \in H_1$, f is a homomorphism of*

(1) *type 1 if and only if $f(x_f \circ y_f) = f(x) * f(y)$;*

(2) *type 2 if and only if $f(x \circ y) = f(x) * f(y)$.*

Proof. It is straightforward. ∎

Corollary 4.1. *Let* $(H_1, \circ)$ *and* $(H_2, *)$ *be two hypergroups, A, B be nonempty subsets of H_1 and $f : H_1 \to H_2$ be an onto mapping. Then, f is a homomorphism of*

(1) *type 1 implies $f(A_f \circ B_f) = f(A) * f(B)$;*

(2) *type 2 implies $f(A \circ B) = f(A) * f(B)$.*

Example 4.1. Consider $H_1 = \{0, 1, 2\}$ and $H_2 = \{a, b\}$ together with the following hyperoperations:

∘	0	1	2
0	0	$\{0,1\}$	$\{0,2\}$
1	$\{0,1\}$	1	$\{1,2\}$
2	$\{0,2\}$	$\{1,2\}$	2

*	a	b
a	a	$\{a,b\}$
b	$\{a,b\}$	b

and suppose that $f : H_1 \to H_2$ is defined by $f(0) = f(1) = a$ and $f(2) = b$. Then, f is a good homomorphism of type 4.

In a hypergroup H, we deal with equivalence relations for which the family of equivalence classes forms a hypergroup under the hyperopertation induced by that one of H. For an equivalence relation ρ on H, we may use x_ρ, $\overline{x}$ or $\rho(x)$ to denote the equivalence class of $x \in H$. Moreover, generally, if A is a non-empty subset of H, then $A_\rho = \cup\{x_\rho \mid x \in A\}$.

Let H/ρ (read H modulo ρ) denote the family $\{x_\rho \mid x \in H\}$ of classes of ρ. The hyperoperation on H induces a hyperoperation $\otimes$ on H/ρ defined by

$$x_\rho \otimes y_\rho = \{z_\rho \mid z \in x_\rho \circ y_\rho\},$$

where $x, y \in H$. The structure $(H/\rho, \otimes)$ is known as a *factor* or *quotient structure*. Note that in the definition of $\otimes$, the condition $z \in x_\rho \circ y_\rho$ maybe replaced by $z \in (x_\rho \circ y_\rho)_\rho$ or $z_\rho \subseteq (x_\rho \circ y_\rho)_\rho$. Obviously, $\cup(x_\rho \otimes y_\rho) = (x_\rho \circ y_\rho)_\rho$.

Theorem 4.5. *Let $(H, \circ)$ be a hypergroup. Then, $(H/\rho, \otimes)$ is a hypergroup if and only if for all $x, y, z \in H$,*

$$\left((x_\rho \circ y_\rho)_\rho \circ z_\rho\right)_\rho = \left(x_\rho \circ (y_\rho \circ z_\rho)_\rho\right)_\rho.$$

Proof. In H/ρ, we have

$$\begin{aligned}
(x_\rho \otimes y_\rho) \otimes z_\rho &= \{u_\rho \mid u \in x_\rho \circ y_\rho\} \otimes z_\rho \\
&= \{t_\rho \mid t \in u_\rho \circ z_\rho, \, u \in x_\rho \circ y_\rho\} \\
&= \{t_\rho \mid t \in (x_\rho \circ y_\rho)_\rho \circ z_\rho\}.
\end{aligned}$$

Similarly, $x_\rho \otimes (y_\rho \otimes z_\rho) = \{t_\rho \mid t \in x_\rho \circ (y_\rho \circ z_\rho)_\rho\}$. Therefore, $\otimes$ is associative. Reproducibility in $(H/\rho, \otimes)$ is a consequence of reproducibility in H. Suppose that x_ρ, $y_\rho \in H/\rho$. Let $u, v \in H$ such that $y \in x \circ u$, $v \circ x$. Then, obviously, $y_\rho \in x_\rho \otimes u_\rho$, $v_\rho \otimes x_\rho$. Hence, the proposition holds. □

Definition 4.3. Let ρ be an equivalence relation on a hypergroup $(H, \circ)$. Then, given $x, y \in H$, ρ is said to be of *type 1*, if H/ρ is a hypergroup;

type 2, if $x_\rho \circ y_\rho \subseteq (x \circ y)_\rho$;

type 3, if $(x \circ y)_\rho \subseteq x_\rho \circ y_\rho$;

type 4, if $x_\rho \circ y_\rho = (x \circ y)_\rho$.

Observe that being of type 2 is equivalent to $(x_\rho \circ y_\rho)_\rho = (x \circ y)_\rho$, and of type 3 to $(x_\rho \circ y_\rho)_\rho = x_\rho \circ y_\rho$. Note that for an equivalence of type 2, $x_\rho \otimes y_\rho = \{z_\rho \mid z \in x \circ y\}$.

Theorem 4.6. *Let ρ be an equivalence relation on a hypergroup $(H, \circ)$. Then, ρ is of*

(1) *type 4 if and only if ρ is of type 2 and type 3;*

(2) *type 1 if ρ is of type 2 or type 3.*

Proof. (1) It is straightforward.

(2) Let $x, y, z \in H$. Suppose that ρ is of type 2. Then, we obtain

$$\left((x_\rho \circ y_\rho)_\rho \circ z_\rho\right)_\rho = ((x \circ y) \circ z)_\rho \quad \text{and} \quad \left(x_\rho \circ (y_\rho \circ z_\rho)_\rho\right)_\rho = (x \circ (y \circ z))_\rho.$$

Suppose that ρ is of type 3. Then, we have

$$\left((x_\rho \circ y_\rho)_\rho \circ z_\rho\right)_\rho = (x_\rho \circ y_\rho) \circ z_\rho \quad \text{and} \quad \left(x_\rho \circ (y_\rho \circ z_\rho)_\rho\right)_\rho = x_\rho \circ (y_\rho \circ z_\rho).$$

Hence, in either case, Theorem 4.5 applies and yields that H/ρ is a hypergroup. Therefore, ρ is of type 1 and (2) holds. $\qquad \square$

Homomorphisms between hypergroups and equivalence relations on hypergroups are closely related. The following results are fundamental.

Theorem 4.7. *Let $(H_1, \circ)$ and $(H_2, *)$ be two hypergroups and $f : H_1 \to H_2$ be an onto mapping. We denote also by f the equivalence relation by f on H_1 of which classes are the family $\{x_f \mid x \in H_1\}$. Then, for $n = 1, 2, 3$ or 4, f is an equivalence relation of type n on H_1 for which H_1/f is canonically isomorphic to H_2 if and only if f is a homomorphism of type n.*

Proof. Let $n = 1$. Suppose that f is a homomorphism of type 1. In order to show f is an equivalence relation of type 1 on H_1, Theorem 4.5 is employed. Let $x, y, z \in H_1$. By Corollary 4.1 (1),

$$\begin{aligned}
\left((x_f \circ y_f)_f \circ z_f\right)_f &= f^{-1}\left(f((x_f \circ y_f)_f \circ z_f)\right) \\
&= f^{-1}(f(x_f \circ y_f) * f(z)) \\
&= f^{-1}(f(x) * f(y) * f(z)).
\end{aligned}$$

Similarly, we obtain $\left(x_f \circ (y_f \circ z_f)_f\right)_f = f^{-1}(f(x) * f(y) * f(z))$. Thus, f is an equivalence relation of type 1. The canonical mapping θ of H_1/f onto

H_2 is given for $x \in H_1$ by $\theta(x_f) = f(x)$. It is clearly well defined and one to one. Moreover, for $x, y \in H_1$,

$$\theta(x_f \otimes y_f) = \theta(\{z_f \mid z \in x_f \circ y_f\}) = \{f(z) \mid z \in x_f \circ y_f\} = f(x_f \circ y_f)$$

and

$$\theta(x_f) * \theta(y_f) = f(x) * f(y).$$

Therefore, Theorem 4.4 (1) yields that H_1/f is canonically isomorphic to H_2 if and only if f is a homomorphism of type 1. The theorem is then established for $n = 1$.

Now, let $n > 1$. If f is a homomorphism of type n, then Theorem 4.3 and the theorem for $n = 1$ imply that H_1/f is canonically isomorphic to H_2. On the other hand, if H_1/f is canonically isomorphic to H_2, then the above relations yields for $x, y \in H_1$ that $f^{-1}(f(x) * f(y)) = (x_f \circ y_f)_f$. Therefore, for $n = 2, 3$ or 4, f is an equivalence relation of type n if and only if f is a homomorphism of type n. $\qquad\square$

Corollary 4.2. *Let $(H, \circ)$ be a hypergroup. Let θ be an equivalence relation on H. We denote also by θ the canonical mapping of H onto H/θ. Then, for $n = 1, 2, 3$ or 4, θ is an equivalence relation of type n on H if and only if H/θ is a hypergroup and θ is a homomorphism of type n of H onto H/θ.*

Proof. By Theorem 4.7, if θ is an equivalence relation of type $1, 2, 3$ or 4, then H/θ is a hypergroup. Note that for $x \in H$, $\theta(x) = x_\theta$, an element of H/θ, and that $\theta^{-1}(\theta(x)) = x_\theta$, a subset of H. This compatibility allows the theorem to be applied and yields the corollary. $\qquad\square$

Three equivalence relations on a hypergroup are of such importance that they will be referred to as the fundamental equivalences. They arise as one tries to discriminate between pairs of elements by means of the hyperoperation.

Definition 4.4. Let $(H, \circ)$ be a hypergroup. Let $x, y \in H$. Then, x and y are said to be *operationally equivalent* or *$\circ$-equivalent* if $x \circ a = y \circ a$ and $a \circ x = a \circ y$ for every $a \in H$. The elements x and y are said to be *inseparable* or *i-equivalent* if $x \in a \circ b$ when and only when $y \in a \circ b$ for every $a, b \in H$. Also, x and y are said to be *essentially indistinguishable* or *e-equivalent* if they are both operationally equivalent and inseparable.

Obviously, the relations o-, i- and e-equivalence, denoted respectively by o, i, e (or by $\sim_o, \sim_i, \sim_e$) are equivalence relations on a hypergroup H.

For $x \in H$, the o-, i- and e-equivalence classes of x are hence denoted by x_o, x_i and x_e, respectively.

Theorem 4.8. *Let* $(H, \circ)$ *be a hypergroup and* $x, y \in H$.

(1) $x_o \circ y_o = x \circ y$; *o-equivalence is of type 2.*
(2) $(x \circ y)_i = x \circ y$; *i-equivalence is of type 3.*
(3) $x_e = x_o \cap x_i$; $x_e \circ y_e = (x \circ y)_e = x \circ y$; *e-equivalence is of type 4.*

Proof. It is an immediate consequence of definition. $\qquad\square$

Corollary 4.3. *Given that* H *is a hypergroup,* H/o, H/i *and* H/e *are hypergroups. The canonical mappings of* H *onto* H/o, H/i *and* H/e *are homomorphisms of type* $2, 3$ *and* 4, *respectively.*

Definition 4.5. A hypergroup H in which $x_o = x$ for each $x \in H$, $x_i = x$ for each $x \in H$ or $x_e = x$ for each $x \in H$ is said respectively to be *o-reduced hypergroup*, *i-reduced hypergroup* or *e-reduced*. An e-reduced hypergroup is simply said to be *reduced*.

Example 4.2. Let $H = \{a, b, c, d\}$. Let the hyperoperation $\circ$ on H be given by the following table:

$\circ$	a	b	c	d
a	$\{a, b\}$	$\{a, b\}$	$\{c, d\}$	$\{c, d\}$
b	$\{a, b\}$	$\{a, b\}$	$\{c, d\}$	$\{c, d\}$
c	$\{c, d\}$	$\{c, d\}$	a	b
d	$\{c, d\}$	$\{c, d\}$	b	a

One easily checks that $(H, \circ)$ is a hypergroup. The o-equivalence classes are seen to be $\{a, b\}$, c and d, whereas the i-equivalence classes are a, b and $\{c, d\}$. Therefore, the e-equivalence classes are all singletons, that is to say, H is e-reduced.

Example 4.3. Let $H = \mathbb{Z} \times \mathbb{Z}^*$, where $\mathbb{Z}^*$ is the set of non-zero integers. Let ρ be the equivalence relation that puts equivalent fractions into the same class, that is, for $(x, y) \in H$,

$$(x, y)_\rho = \{(u, v) \mid xv = yu\}.$$

The hyperoperation $\circ$ on H is given for $(w, x), (y, z) \in H$ by

$$(w, x) \circ (y, z) = (wz + xy, xz)_\rho.$$

Then, $(H, \circ)$ is a hypergroup in which

$$(x, y)_o = (x, y)_i = (x, y)_e = (x, y)_\rho,$$

for $(x, y) \in H$. Furthermore, $(H/\rho, \otimes) \cong (\mathbb{Q}, +)$.

Theorem 4.9. *Let $(H_1, \circ)$, $(H_2, *)$ and $(H_3, \bullet)$ be hypergroups. For $n =$ $1, 2, 3$ or 4, let f be a homomorphism of type n of H_1 onto H_2 and g be a homomorphism of type n of H_2 onto H_3. Then, gf is a homomorphism of type n of H_1 onto H_3.*

Proof. Suppose that $x, y \in H_1$. We first observe that for $z \in H_1$,

$$z_{gf} = f^{-1}g^{-1}gf(z) = f^{-1}(f(z)_g).$$

Let $n = 1$. By the above relation, we obtain

$$gf(x_{gf} \circ y_{gf}) = gf\left(f^{-1}(f(x)_g) \circ f^{-1}(f(y)_g)\right).$$

Since f is onto, Corollary 2.4.8 (1) applies and yields

$$gf\left(f^{-1}(f(x)_g) \circ f^{-1}(f(y)_g)\right) = g(f(x)_g * f(y)_g).$$

Then, Theorem 4.4 (1) gives

$$g\left(f(x)_g * f(y)_g\right) = gf(x) \bullet gf(y).$$

Hence, gf is a homomorphism of type 1 by the above relations and Theorem 4.4 (1).

Let $n = 2$. Similar to the previous case, but simpler.

Let $n = 3$. Since g is of type 3,

$$f^{-1}g^{-1}(gf(x) \bullet gf(y)) = f^{-1}(f(x)_g * f(y)_g).$$

Since f is onto, Theorem 4.2 (3) applies and gives

$$f^{-1}(f(x)_g * f(y)_g) = f^{-1}(f(x)_g) \circ f^{-1}(f(y)_g).$$

Now, we obtain

$$f^{-1}(f(x)_g) \circ f^{-1}(f(y)_g) = x_{gf} \circ y_{gf}.$$

So, by the above relations, we conclude that gf is a homomorphism of type 3.

Now, suppose that $n = 4$. Then, gf is a homomorphism of type 4 by Theorem 4.3 (1). $\qquad\square$

In the following definition, we introduce more types of homomorphisms of hypergroups that appeared in the literature under various names.

Recall that a, b, we denote $a/b = \{x \mid a \in x \circ b\}$ and $b \backslash a = \{y \mid a \in b \circ y\}$.

Definition 4.6. Let $(H_1, \circ)$ and $(H_2, *)$ be two hypergroups and $f : H_1 \to H_2$ be a mapping. We say that f is a homomorphism of *type 5*, if for all $x, y \in H_1$, f is a good homomorphism and furthermore

$$(1)\ f(x/y) = f(x/f^{-1}(f(y))),$$
$$(2)\ f(y\backslash x) = f(f^{-1}(f(y)\backslash x));$$

type 6, if for all $x, y \in H_1$, f is a good homomorphism and furthermore

$$(3)\ f(x/f^{-1}(f(y))) = f(x)/f(y),$$
$$(4)\ f(f^{-1}(f(y))\backslash x) = f(y)\backslash f(x);$$

type 7, if for all $x, y \in H_1$, f is a good homomorphism and furthermore

$$(5)\ f(x/y) = f(x)/f(y),$$
$$(6)\ f(y\backslash x) = f(y)\backslash f(x).$$

Theorem 4.10. *If $f : H_1 \to H_2$ is a homomorphism of type 7, then f is a homomorphism of type 4.*

Proof. In general, if f is an inclusion homomorphism, then for every $x, y \in H_1$,

$$f^{-1}(f(x)) \circ f^{-1}(f(y)) \subseteq f^{-1}(f(x) * f(y)).$$

Now, let f be a homomorphism of type 7. Suppose that $z \in f^{-1}(f(x) * f(y))$. Then, $f(z) \in f(x) * f(y)$, which implies that $f(y) \in f(x)\backslash f(z)$ and so $f(y) \in f(x\backslash z)$. Thus, there exists $y' \in x\backslash z$ such that $f(y) = f(y')$, consequently $z \in x \circ y' \subseteq f^{-1}(f(x)) \circ f^{-1}(f(y))$. Therefore, $f^{-1}(f(x) * f(y)) \subseteq f^{-1}(f(x)) \circ f^{-1}(f(y))$. $\qquad\square$

Theorem 4.11. *If $f : H_1 \to H_2$ is an onto homomorphism of type 4, then f is a homomorphism of type 6.*

Proof. We know that an onto homomorphism of type 4 is a good homomorphism. Suppose that $u \in f(z)/f(x)$. Then, there exists y such that $f(y) = u$ and so $f(z) \in f(y) * f(x)$. Consequently, $z \in f^{-1}(f(y) * f(x)) = f^{-1}(f(y)) \circ f^{-1}(f(x))$, it follows that there exist a, b such that $z \in a \circ b$, where $a \in f^{-1}(f(y))$ and $b \in f^{-1}(f(x))$. Hence, $f(y) = f(a)$, $f(x) = f(b)$ and so $u = f(a) \in f(z/b) \subseteq f(z/f^{-1}(f(x)))$. Therefore, $f(z)/f(x) \subseteq f(z/f^{-1}(f(x)))$. Note that the inverse inclusion is always true. Similarly, we can prove $f(f^{-1}(f(y))\backslash x) = f(y)\backslash f(x)$. $\qquad\square$

Notice that, in general, a homomorphism of type 4 is not good.

Example 4.4. Let $\mathcal{H}(\mathbb{Z})$ be the hypergroup $(\mathbb{N}, \circ)$, where for all x, y,

$$x \circ y = \{x + y, \ |x - y|\}.$$

Let $f : \mathcal{H}(\mathbb{Z}) \to \mathcal{H}(\mathbb{Z})$ be the function defined by

$$f(x) = \begin{cases} 0 \text{ if } x \in 2\mathbb{N} \\ 1 \text{ if } x \in 2\mathbb{N} + 1. \end{cases}$$

Then, f is a homomorphism of type 4 but is not a good homomorphism.

4.2 Homomorphisms of hypergroupoids associated with $\mathcal{L}$-maps

Several types of connections between hypergroupoids and lattices have been considered in the last 40 years. We mention here some papers where these connections are studied: Nakano [136]; Varlet [176]; Mittas; Konstantinidou; Serafimidis [130; 98], Comer [17]; Kehagias [95; 96], Calugareanu; Radu; Leoreanu [13; 110], Tofan; Volf [175] and lately, Rosenberg; Leoreanu-Fotea [111; 105].

In this section, we characterize some types of homomorphisms of hypergroupoids associated with $\mathcal{L}$-maps, in particular, when $\mathcal{L}$ is a chain or a bounded lattice. These results are contained in [112].

Let $\mathcal{L} = \langle L; \vee, \wedge \rangle$ be a lattice, whose order is denoted by $\leq$, and H a non-empty set. An $\mathcal{L}$-*map* is a map $\mu : H \to L$. The most common case is $\mathcal{L} = \langle [0, 1]; \max, \min \rangle$, where $[0, 1]$ is the real closed interval, and μ is then called a *fuzzy set*.

For $a, b \in L$, set

$$\lfloor a \rfloor_\mu = \mu^{-1}(a) = \{h \in H : \mu(h) = a\}, \tag{4.1}$$

$$(a, b)_\mu = \operatorname{im} \mu \cap [a \wedge b, a \vee b] \tag{4.2}$$

where, as usual, for $a, b \in L$, with $a < b$, the symbol $[a, b]$ denotes the closed interval $\{\ell \in L : a \leq \ell \leq b\}$ in $\mathcal{L}$ and $\operatorname{im} \mu$ is the image of μ. If μ is clear from the context, we drop the subscript in the above equalities and write h' for $\mu(h)$. Next, define a map $(x, y) \mapsto x \circ_\mu y$ from H^2 into the powerset $\mathcal{P}(H)$ by setting for all $x, y \in H$

$$x \circ y = x \circ_\mu y = \bigcup_{i \in (x', y')} \lfloor i \rfloor, \tag{4.3}$$

where on the right-hand side we have also omitted the subscripts μ. For all $x, y \in H$ clearly $x', y' \in (x', y')$ and hence

$$\{x, y\} \subseteq \lfloor x' \rfloor \cup \lfloor y' \rfloor \subseteq x \circ y. \tag{4.4}$$

Thus the cardinality of $x \circ y$ is at least 1 and hence $x \circ y$ is always a nonvoid subset of H. Then $\mathbb{H}_\mu = \langle H; \circ_\mu \rangle$, or shortly $\mathbb{H} = \langle H; \circ \rangle$, is a *hypergroupoid*, i.e., $\circ$ maps H^2 into the set $\mathcal{P} = \mathcal{P}(H) \setminus \{\emptyset\}$ of nonvoid subsets of H. We extend $\circ$ to $\mathcal{P}(H)$ by setting for arbitrary $X, Y \subseteq H$

$$X \circ Y = \bigcup_{\substack{x \in X \\ y \in Y}} x \circ y.$$

For $h \in H$ and $X \subseteq H$ we abbreviate $\{h\} \circ X$ and $X \circ \{h\}$ by $h \circ X$ and $X \circ h$ respectively. It is immediate from (4.2) and (4.3) that $\mathbb{H}_\mu$ is commutative; i.e., $x \circ y = y \circ x$ for all $x, y \in H$. From (4.4) it follows that $\mathbb{H}_\mu$ satisfies the reproductive law: $h \circ H = H = H \circ h$ for every $h \in H$.

Let $\mathbb{H} = \langle H; \circ \rangle$ and $\mathbb{H}' = \langle H'; \bullet \rangle$ be two hypergroupoids.

In the next theorem we consider two $\mathcal{L}$-maps μ and ν on H and H' respectively and characterize the homomorphismsand the good homomorphisms between the corresponding hypergroupoids $\mathbb{H}_\mu$ and $\mathbb{H}'_\nu$. As usual, the *kernel* of a map $g : A \to B$ is the equivalence relation $\ker g = \{(x, y) \in A^2 \mid g(x) = g(y)\}$ which partitions A into blocks (also called equivalence classes) of $\ker g$.

Theorem 4.12. *Let μ and ν be two $\mathcal{L}$-maps on H and H' and $f : H \to H'$. Then f is a homomorphism (good homomorphism) from $\mathbb{H}_\mu$ into $\mathbb{H}'_\nu$ if and only if*

(1) *f maps each block of $\ker \mu$ into (onto) a block of $\ker \nu$, and*
(2) *the map $f^* : z \mapsto z^*$ with $z^* = \nu(f(\mu^{-1}(z)))$ from $\operatorname{im} \mu$ into $\operatorname{im} \nu$ satisfies for all $a, b \in \operatorname{im} \mu$*

$$(a, b)^*_\mu \subseteq (a^*, b^*)_\nu, \tag{4.5}$$

$$((a, b)^*_\mu = (a^*, b^*)_\nu) \tag{4.6}$$

(here for $A \subseteq \operatorname{im} \mu$ the symbol A^* denotes the image of A under f^*).

Note that the map f^* in Theorem 4.12 (2) is well defined provided f satisfies (1). Indeed, let $z \in \operatorname{im} \mu$ and $B = \lfloor z \rfloor_\mu$. Now, by (1), f maps the block B of $\ker \mu$ into a block B' of $\ker \nu$, hence $\nu(B') = \{t\}$ and $f^*(z) = t$.

Proof. Let $\mathbb{H}_\mu = \langle H; \circ \rangle$ and $\mathbb{H}'_\nu = \langle H'; \bullet \rangle$. For $h \in H$, set $h' = \mu(h)$.

($\Rightarrow$) Let f be a homomorphism from $\mathbb{H}_\mu$ into $\mathbb{H}'_\nu$. Let $x \in H$ be arbitrary. From (4.4) we get $x \circ x = \lfloor \mu(x) \rfloor_\mu$. Similarly, $f(x) \bullet f(x) = \lfloor \nu(f(x)) \rfloor_\nu$ and so

$$f(\lfloor \mu(x) \rfloor_\mu) = f(x \circ x) \subseteq f(x) \bullet f(x) = \lfloor \nu(f(x)) \rfloor_\nu \qquad (4.7)$$

proving (1).

In order to prove (2), let $a, b \in L$ be such that $\lfloor a \rfloor_\mu \neq \emptyset \neq \lfloor b \rfloor_\mu$ and $x \in \lfloor a \rfloor_\mu$, $y \in \lfloor b \rfloor_\mu$. For every $h \in H$ set $g(h) = \nu(f(h))$. Notice that due to (1) clearly $g(x) = a^*$ and $g(y) = b^*$. From (4.3) and the definition of a homomorphism

$$\bigcup_{z \in (a,b)_\mu} f\lfloor z \rfloor = f\left(\bigcup_{z \in (a,b)_\mu} \lfloor z \rfloor \right) = f(x \circ y) \subseteq f(x) \bullet f(y)$$

$$= \bigcup_{t \in (g(x),g(y))_\nu} \lfloor t \rfloor_\nu = \bigcup_{t \in (a^*,b^*)_\nu} \lfloor t \rfloor_\nu. \qquad (4.8)$$

By (1), for every $z \in \operatorname{im} \mu$, we get $f\lfloor z \rfloor \subseteq \lfloor z^* \rfloor_\nu$. Now, (4.8) shows that $z^* \in (a^*, b^*)_\nu$ proving (4.5).

Let f be a good homomorphism. Then we have equality in both (4.7) and (4.8) and thus f maps each block of $\ker \mu$ onto a block of $\ker \nu$ and (4.6) holds.

($\Leftarrow$) Let f map each block of $\ker \mu$ into a block of $\ker \nu$.

(a) Let f satisfy (4.5). Let $x, y \in H$ be arbitrary. Set $a = \mu(x)$ and $b = \mu(y)$. Using (4.3), (4.5), and (1) we have

$$f(x \circ y) = f\left(\bigcup_{z \in (a,b)_\mu} \lfloor z \rfloor \right)$$

$$= \bigcup_{z \in (a,b)_\mu} f\lfloor z \rfloor \subseteq \bigcup_{z \in (a,b)_\mu} \lfloor z^* \rfloor_\nu \subseteq \bigcup_{t \in (a^*,b^*)_\mu} \lfloor t \rfloor_\nu = f(x) \bullet f(y) \qquad (4.9)$$

proving that f is a homomorphism.

(b) Let f map each block of $\ker \mu$ onto a block of $\ker \nu$ and let f satisfy (4.6). Then both inclusions in (4.9) are equalities and f is a good homomorphism. $\qquad \square$

Let $\mathbb{H} = (H, \circ)$ and $\mathbb{H}' = (H, \bullet)$ be hypergroupoids. Recall that a map $f : H \to H'$ is a 2-*homomorphism* if for all $x, y \in H$

$$f(x \circ y) = \operatorname{im} f \cap (f(x) \circ f(y)).$$

We can characterize 2-homomorphisms between $\mathcal{L}$-maps.

Corollary 4.4. *Let μ and ν be $\mathcal{L}$-fuzzy sets on H and H' respectively and $f : H \to H'$. Then f is a 2-homomorphism from $\mathbb{H}_\mu$ to $\mathbb{H}'_\nu$ if and only if f is a homomorphism.*

Proof. Clearly, a 2-homomorphism is a homomorphism.

For the converse, let f be a homomorphism from $\mathbb{H}_\mu$ into $\mathbb{H}'_\nu$. Let $x, y \in H$ be arbitrary and $a = \mu(x)$, $b = \mu(y)$. Then $f(x \circ_\mu y)$ consists of $f(B)$ for a block B of $\ker\mu$ such that $\mu(B) = \{z\}$ where $z \in (a, b)_\mu$. Clearly $f(B) \subseteq B'$, where $B' \in \ker\nu$ and $\nu(B') = \{z^*\}$. This in fact shows that $f(x \circ_\mu y) = \operatorname{im} f \cap (f(x) \circ_\nu f(y))$. $\qquad\square$

As usual, the situation is simpler for chains $\mathcal{L}$, i.e., if the order $\leq$ of $\mathcal{L}$ is a chain (also called total or linear order). For orders $(M, \leq)$ and $(N, \leq)$ a map $g : M \to N$ is *isotone* (*dually isotone*) if $g(a) \leq g(b)$ whenever $a < b$ ($a > b$). We need the following simple and probably known lemma. As usual, in an order $(Q, \leq)$ an element c is *between* a and b if $a \leq c \leq b$ or $a \geq c \geq b$. We denote by B_{ab} the set of all elements between a and b.

Lemma 4.1. *Let $(M, \leq)$ and $(N, \leq)$ be two chains and $\varphi : m \mapsto m^*$ a map from M into N preserving the betweeness relation. Then φ is either isotone or dually isotone.*

Proof. Suppose to the contrary that there are $x, y, z, t \in M$ with $x < y$, $z < t$, $x^* < y^*$ and $z^* > t^*$. In the chain $(M, \leq)$ the least of x, y, z, t is x or z while the greatest among x, y, z, t is either y or t. Accordingly, we have four cases.

(1) Suppose that $x \leq z < t \leq y$. Then $x^* \leq t^* < z^* \leq y^*$ in contradiction to $z^* \in B_{x^* t^*}$.

(2) Suppose that $x \leq y$, $z \leq t$. Then $y, z \in B_{xt}$ implies that $y^*, z^* \in B_{x^* t^*}$. Here $t^* < z^*$ shows $t^* \leq x^*$ but then $y^* \notin B_{x^* t^*}$ due to $x^* < y^*$, a contradiction.

(3) Suppose that $z \leq x$, $t \leq y$. Then $x^*, t^* \in B_{z^* y^*}$. Here $x^* < y^*$ shows $z^* < y^*$ in contradiction to $t^* < z^*$.

(4) Finally, let $z \leq x < y \leq t$. Then $x^*, y^* \in B_{z^* t^*}$, whence $t^* \leq x^* < y^* \leq z^*$, in contradiction to $y^* \in B_{x^* t^*}$.

This proves the lemma. $\qquad\square$

Corollary 4.5. *Let μ and ν be $\mathcal{L}$-maps on H and H' respectively such that $(\operatorname{im}\mu, \leq)$ and $(\operatorname{im}\nu, \leq)$ are chains. Then $f : H \to H'$ is a homomorphism (good homomorphism) from $\mathbb{H}_\mu$ into $\mathbb{H}'_\nu$ if and only if*

(1) *f maps each block of $\ker\mu$ into (onto) a block of $\ker\nu$, and*
(2) *the map $f^* : z \mapsto z^*$ with $z^* = \nu(f(\mu^{-1}(z)))$ from $\operatorname{im}\mu$ into $\operatorname{im}\nu$ is an isotone map or a dually isotone map (surjective isotone map*

or surjective dually isotone map) from the chain $(\operatorname{im}\mu, \leq)$ *into the chain* $(\operatorname{im}\nu, \leq)$.

Proof.   This follows from Theorem 4.12 and Lemma 4.1. $\square$

In the following lemma and corollary we assume the axiom of choice (AC) in the form which guarantees that every chain bounded from above is contained in a maximal chain (where a chain C is maximal if there is no chain D such that $C \subsetneq D \subsetneq M$).

Lemma 4.2. (AC) *Let* $(M, \vee, \wedge)$ *be a bounded lattice, let* $(N, \leq)$ *be an order and* $\varphi : m \mapsto m^*$ *a map from* M *into* N *preserving the betweeness relation. Then* φ *is either isotone or dually isotone.*

Proof.   From Lemma 4.1 it follows that φ is isotone or dually isotone on any chain of $(M, \vee, \wedge)$. Let $\mathcal{C}$ be a maximal chain of $(M, \vee, \wedge)$, which clearly contains also 0 and 1. Since φ preserves the betweeness relation, we distinguish two situations:

(1) If $0^* \leq 1^*$, then $a^* \leq b^*$ for all $a, b \in \mathcal{C}$, with $a \leq b$. Moreover, we have $a^* \leq b^*$ for all $a, b \in M$, with $a \leq b$, since any such two elements belong to at least one maximal chain of $(M, \vee, \wedge)$. In other words, φ is isotone.

(2) As above, if $0^* \geq 1^*$, then $a^* \geq b^*$ for all $a, b \in M$, $a \leq b$. Hence, φ is dually isotone. $\square$

Corollary 4.6. (AC) *Let* μ *and* ν *be* $\mathcal{L}$-*maps on* H *and* H' *respectively such that* $(\operatorname{im}\mu, \leq)$ *is a bounded lattice and* $(\operatorname{im}\nu, \leq)$ *is an order. Let* $f : H \to H'$ *be such that the map* $f^* : z \mapsto z^*$ *with* $z^* = \nu(f(\mu^{-1}(z))$ *from* $\operatorname{im}\mu$ *into* $\operatorname{im}\nu$ *preserves the betweeness relation. Then* f *is a homomorphism (good homomorphism) from* $\mathbb{H}_\mu$ *into* $\mathbb{H}'_\nu$, *if and only if*

(1) *f maps each block of* $\ker\mu$ *into (onto) a block of* $\ker\nu$, *and*

(2) *the map* f^* *is an isotone map or a dually isotone map (surjective isotone map or surjective dually isotone map) from the bounded lattice* $(\operatorname{im}\mu, \leq)$ *into the order* $(\operatorname{im}\nu, \leq)$.

Proof.   This follows from Theorem 4.12 and Lemma 4.2. $\square$

4.3   Isomorphisms of hypergroups associated with $\mathcal{L}$-maps

Let $\mathbb{H} = (H; \circ)$ and $\mathbb{H}' = (H'; \bullet)$ be two hypergroups. A bijective good homomorphism from $\mathbb{H}$ onto $\mathbb{H}'$ is an *isomorphism* from H onto $\mathbb{H}'$.

Corollary 4.7. *Let μ and ν be two $\mathcal{L}$-maps on H and H' respectively and f a bijection from H onto H'. Then f is an isomorphism from $\mathbb{H}_\mu$ onto $\mathbb{H}'_\nu$ if and only if*

(1) *f maps every block of $\ker\mu$ onto a block of $\ker\nu$, and*
(2) *the map $f^* : z \mapsto z^*$ with $z^* = \nu(f(\mu^{-1}(z)))$ is a bijection from $\operatorname{im}\mu$ onto $\operatorname{im}\nu$ carrying every $(a,b)_\mu$ onto $(a^*,b^*)_\nu$.*

Proof. Condition (1) is just condition (1) of Theorem 4.12 for a bijection. The map f^* is a bijection from $\operatorname{im}\mu$ onto $\operatorname{im}\nu$ due to (1). By (4.6), the map φ carries every $(a,b)_\mu$ onto $(a^*,b^*)_\nu$. $\qquad\square$

Corollary 4.8. *Let μ and ν be two $\mathcal{L}$-maps on H and H' respectively such that $(\operatorname{im}\mu, \leq)$ and $(\operatorname{im}\nu, \leq)$ are chains, and let f be a bijection from H onto H'. Then f is an isomorphism from $\mathbb{H}_\mu$ onto $\mathbb{H}'_\nu$ if and only if*

(1) *f maps every block of $\ker\mu$ onto a block of $\ker\nu$, and*
(2) *the map $f^* : z \mapsto z^*$ with $z^* = \nu(f(\mu^{-1}(z)))$ is an order isomorphism from $(\operatorname{im}\mu, \leq)$ onto $(\operatorname{im}\nu, \leq)$.*

We may ask when $\mathbb{H}_\mu = \mathbb{H}'_\nu$. Consider now $f = \operatorname{id}_H$. We obtain the next result:

Corollary 4.9. *Let μ and ν be two $\mathcal{L}$-maps on $H = H'$. Then $\mathbb{H}_\mu = \mathbb{H}'_\nu$ if and only if*

(1) *$\ker\mu = \ker\nu$, and*
(2) *the map $z \mapsto z^*$ with $z^* = \nu(\mu^{-1}(z))$ is a bijection from $\operatorname{im}\mu$ onto $\operatorname{im}\nu$ carrying every $(a,b)_\mu$ onto $(a^*,b^*)_\nu$.*

Corollary 4.10. *Let μ and ν be two $\mathcal{L}$-maps on H and H' respectively such that $(\operatorname{im}\mu, \leq)$ is a bounded lattice and $(\operatorname{im}\nu, \leq)$ is an order and let f be a bijection from H onto H', such that the map $f^* : z \mapsto z^*$ with $z^* = \nu(f(\mu^{-1}(z)))$ preserves the betweeness relation. Then f is an isomorphism from $\mathbb{H}_\mu$ onto $\mathbb{H}'_\nu$ if and only if*

(1) *f maps every block of $\ker\mu$ onto a block of $\ker\nu$, and*
(2) *the map f^* is an order isomorphism from $(\operatorname{im}\mu, \leq)$ onto $(\operatorname{im}\nu, \leq)$.*

Notice that the condition that the map $f^* : z \mapsto z^*$ with $z^* = \nu(f(\mu^{-1}(z)))$ preserves the betweeness relation is essential in the above corollary. We present some examples of isomorphisms $f : \mathbb{H}_\mu \to \mathbb{H}'_\nu$, for which f^* are not order isomorphisms. Clearly, for each one of these examples, f^* does not preserve the betweeness relation.

Example 4.5. Let $n \geq 2$ and $H = H' = L = \{\{0,1\}^n, \preceq\}$, $\mu = \nu = id_H$, where for $x = (x_1, ..., x_n)$, $y = (y_1, ..., y_n) \in \{0,1\}^n$ we get $x \preceq y$ provided $x_i \leq y_i$ for $i = 1, ..., n$ (and $\leq$ is the natural order on $\{0,1\}$).

Let $I \subset \{1, ..., n\}$ be fixed. Define the self-map $f_I : x \mapsto x^*$ of L by setting $x^* = (y_1, ..., y_n)$ for every $x = (x_1, ..., x_n) \in \{0,1\}^n$, where $y_i = \bar{x}_i \; (= 1 - x_i)$ if $i \in I$ and $y_j = x_j$ otherwise.

We claim that for all $a, b \in L$, $a \preceq b$ we have

$$[a, b]^* = [a^*, b^*]. \tag{4.10}$$

For notational simplicity let $I = \{1, ..., k\}$ where $1 \leq k \leq n$. First

$$[a, b] = \{c = (c_1, ..., c_n) \in \{0,1\}^n : a_i \wedge b_i \leq c_i \leq a_i \vee b_i, \; i = 1, ..., n\}$$

where the meet and join are in $(\{0,1\}, \leq)$. Using de Morgan's laws we get $\overline{a_i \wedge b_i} = \bar{a}_i \vee \bar{b}_i$ and $\overline{a_i \vee b_i} = \bar{a}_i \wedge \bar{b}_i$. We obtain

$$[a, b]^* = \{c \in \{0,1\}^n : \bar{a}_i \wedge \bar{b}_i \leq c_i \leq \bar{a}_i \vee \bar{b}_i, \; (i = 1, ..., k),$$

$$a_i \wedge b_i \leq c_i \leq a_i \vee b_i, \; (i = k+1, ..., n)\}. \tag{4.11}$$

Clearly, $a^* = (\bar{a}_1, ... \bar{a}_k, a_{k+1}, ... a_n)$, $a^* = (\bar{b}_1, ... \bar{b}_k, b_{k+1}, ... b_n)$ and

$$[a^*, b^*] = \{c \in \{0,1\}^n : \bar{a}_i \wedge \bar{b}_i \leq c_i \leq \bar{a}_i \vee \bar{b}_i, \; (i = 1, ..., k),$$

$$a_i \wedge b_i \leq c_i \leq a_i \vee b_i, \; (i = k+1, ..., n)\}. \tag{4.12}$$

Together (4.11) and (4.12) give (4.10). For $n = 2$ and $I = \{1\}$ this gives the map $f_{\{1\}}$ given by the following table:

x	(0,0)	(0,1)	(1,0)	(1,1)
x^*	(1,0)	(1,1)	(0,0)	(0,1)

Example 4.6. In the situation of Example 4.5, let π be a permutation of $\{1, ..., n\}$. Define the self-map $f_\pi : x \mapsto x^*$ of L by setting $x^* = (x_{\pi(1)}, ..., x_{\pi(n)})$, for every $x = (x_1, ..., x_n) \in \{0,1\}^n$. Proceeding as in Example 4.5 one can show the validity of (4.10).

Combining Examples 4.5 and 4.6, one gets $2^n n!$ maps satisfying (4.10).

Chapter 5

Fundamental Relations

The main tools connecting the class of hyperstructures with the classical algebraic structures are the fundamental relations. The fundamental relation has an important role in the study of semihypergroups and especially of hypergroups. We study several fundamental relations which give us group, abelian group, cyclic group and solvable group structures.

5.1 The β relation on semihypergroups

In this section, we introduce the β-relation. This relation was introduced by Koskas [101] and studied mainly by Corsini [22], Davvaz [49], Davvaz and Leoreanu-Fotea [61], Freni [73], Vougiouklis [183], and many others.

Definition 5.1. Let $(H, \circ)$ be a semihypergroup and $n > 1$ be a natural number. We define the relation β_n on H as follows:

$$x \beta_n y \text{ if there exists } a_1, a_2, \ldots, a_n \text{ in } H, \text{ such that } \{x, y\} \subseteq \prod_{i=1}^{n} a_i,$$

and let $\beta = \bigcup_{n \geq 1} \beta_n$, where $\beta_1 = \{(x, x) \mid x \in H\}$ is the diagonal relation on H.

Clearly, the relation β is reflexive and symmetric.

Example 5.1. If $(H, \circ)$ is a semihypergroup of cardinal 2, then the relation β is transitive.

In general, denote by β^* the transitive closure of β.

Remark 5.1. A relation ρ^* is the *transitive closure* of a relation ρ if and only if

(1) ρ^* is transitive,

(2) $\rho \subseteq \rho^*$,

(3) for any relation θ, if $\rho \subseteq \theta$ and θ is transitive, then $\rho^* \subseteq \theta$, i.e., ρ^* is the smallest relation that satisfies (1) and (2).

Remark 5.2. For any relation ρ, the transitive closure of ρ always exists.

Theorem 5.1. β^* *is the smallest strongly regular relation on H.*

Proof.   We show that:

(1) β^* is a strongly regular relation on H;

(2) If ρ is a strongly regular relation on H, then $\beta^* \subseteq \rho$.

(1) Let $a\beta^*b$ and x be an arbitrary element of H. It follows that there exist $x_0 = a, x_1, \ldots, x_n = b$ such that for all $i \in \{0, 1, \ldots, n-1\}$ we have $x_i\beta x_{i+1}$. Let $u_1 \in a \circ x$ and $u_2 \in b \circ x$. We check that $u_1\beta^*u_2$. From $x_i\beta x_{i+1}$ it follows that there exists a hyperproduct P_i, such that $\{x_i, x_{i+1}\} \subseteq P_i$ and so $x_i \circ x \subseteq P_i \circ x$ and $x_{i+1} \circ x \subseteq P_i \circ x$, which means that $x_i \circ x \overline{\overline{\beta}} x_{i+1} \circ x$. Hence for all $i \in \{0, 1, \ldots, n-1\}$ and for all $s_i \in x_i \circ x$ we have $s_i\beta s_{i+1}$. If we consider $s_0 = u_1$ and $s_n = u_2$, then we obtain $u_1\beta^*u_2$. Then β^* is strongly regular on the right and similarly, it is strongly regular on the left.

(2) We have $\beta_1 = \{(x, x) \mid x \in H\} \subseteq \rho$, since ρ is reflexive. Suppose that $\beta_{n-1} \subseteq \rho$ and show that $\beta_n \subseteq \rho$. If $a\beta_n b$, then there exist $x_1, \ldots, x_n$ in H, such that $\{a, b\} \subseteq \prod\limits_{i=1}^{n} x_i$. Hence, there exists u, v in $\prod\limits_{i=1}^{n-1} x_i$, such that $a \in u \circ x_n$ and $b \in v \circ x_n$. We have $u\beta_{n-1}v$ and according to the hypothesis, we obtain $u\rho v$. Since ρ is strongly regular, it follows that $a\rho b$. Hence $\beta_n \subseteq \rho$. By induction, it follows that $\beta \subseteq \rho$, whence $\beta^* \subseteq \rho$. $\square$

Hence the relation β^* is the smallest equivalence relation on H, such that the quotient H/β^* is a semigroup.

Definition 5.2. β^* is called the *fundamental relation* on H and H/β^* is called the *fundamental semigroup*.

Therefore, H/β^* is a semigroup with respect to an operation defined by

$$\beta^*(x) \odot \beta^*(y) = \beta^*(z), \text{ where } x, y \in H \text{ and } z \in x \circ y.$$

The β^* plays a critical role in semihypergroup theory; it permits to define a covariant functor between the category of semihypergroups and the category of semigroups. For this reason the relation β^* has been an intensely studied relation. A remarkable result concerning this relation in hypergroups was established by Freni [73]. He proved that the relation β is transitive in any hypergroup, i.e., in hypergroups $\beta = \beta^*$.

There exists semihypergroups in which the relation β is not transitive.

Example 5.2. Let $H = \{a,\ b,\ c,\ d\}$ be a semihypergroup with the following hyperoperation.

$\circ$	a	b	c	d
a	$\{b,c\}$	$\{b,d\}$	$\{b,d\}$	$\{b,d\}$
b	$\{b,d\}$	$\{b,d\}$	$\{b,d\}$	$\{b,d\}$
c	$\{b,d\}$	$\{b,d\}$	$\{b,d\}$	$\{b,d\}$
d	$\{b,d\}$	$\{b,d\}$	$\{b,d\}$	$\{b,d\}$

Then, it is easy to see that $c\beta^*d$ but not $c\beta d$.

5.2 Complete parts

Complete parts were introduced and studied for the first time by Koskas [101]. Later, this topic was analyzed by Corsini [22] and Sureau [173] mostly in the general theory of hypergroups. De Salvo studied complete parts from a combinatorial point of view. A generalization of them, called n-complete parts, was introduced by R. Migliorato [123]. Other mathematicians gave a contribution to the study of complete parts and of the heart of a hypergroup. Among them, V. Leoreanu analyzed the structure of the heart of a hypergroup in her Ph.D. Thesis.

We present now the definitions.

Definition 5.3. Let $(H, \circ)$ is a semihypergroup and A be a non-empty subset of H. We say that A is a *complete part* of H if for any non-zero natural number n and for all $a_1, \ldots, a_n$ of H, the following implication holds:

$$A \cap \prod_{i=1}^{n} a_i \neq \emptyset \ \Rightarrow \ \prod_{i=1}^{n} a_i \subseteq A.$$

Theorem 5.2. *If $(H, \circ)$ is a semihypergroup and ρ is a strongly regular relation on H, then for all z of H, the equivalence class of z is a complete part of H.*

Proof. Let $a_1, \ldots, a_n$ be elements of H, such that

$$\overline{z} \cap \prod_{i=1}^{n} a_i \neq \emptyset.$$

Then there exists $y \in \prod_{i=1}^{n} a_i$, such that $y \rho z$. The homomorphism $\pi_H : H \to$
H/ρ is good and H/ρ is a semigroup. It follows that $\pi_H(y) = \pi_H(z) =$
$\pi_H \left(\prod_{i=1}^{n} a_i \right) = \prod_{i=1}^{n} \pi_H(a_i)$. This means that $\prod_{i=1}^{n} a_i \subseteq \bar{z}$. $\qquad \square$

Definition 5.4. Let A be a non-empty part of H. The intersection of the parts of H which are complete and contain A is called the *complete closure* of A in H; it will be denoted by $\mathcal{C}(A)$.

Theorem 5.3. *Let $(H, \circ)$ be a semihypergroup. The following conditions are equivalent:*

(1) *For all $x, y \in H$ and $a \in x \circ y$, $\mathcal{C}(a) = x \circ y$;*
(2) *For all $x, y \in H$, $\mathcal{C}(x \circ y) = x \circ y$.*

Proof. $(1 \Rightarrow 2)$: We have

$$\mathcal{C}(x \circ y) = \bigcup_{a \in x \circ y} \mathcal{C}(a) = x \circ y.$$

$(2 \Rightarrow 1)$: From $a \in x \circ y$, we obtain $\mathcal{C}(a) \subseteq \mathcal{C}(x \circ y) = x \circ y$. This means that $\mathcal{C}(a) \cap x \circ y \neq \emptyset$, whence $x \circ y \subseteq \mathcal{C}(a)$. Therefore, $\mathcal{C}(a) = x \circ y$. $\qquad \square$

Definition 5.5. A semihypergroup is *complete* if it satisfies one of the above equivalent conditions. A hypergroup is *complete* if it is a complete semihypergroup.

Lemma 5.1. *The following properties hold.*

(1) $A \subseteq \mathcal{C}(A)$;
(2) $\mathcal{C}(\mathcal{C}(A)) = \mathcal{C}(A)$;
(3) $A \subseteq A'$ *implies* $\mathcal{C}(A) \subseteq \mathcal{C}(A')$;
(4) *If $B \subseteq A$ and A is a complete part of H, then $\mathcal{C}(B) \subseteq A$.*

Proof. It is straightforward. $\qquad \square$

Denote $K_1(A) = A$ and for all $n \geq 1$ denote

$$K_{n+1}(A) = \left\{ x \in H | \exists p \in \mathbb{N}, \exists (h_1, \ldots, h_p) \in H^p : x \in \prod_{i=1}^{p} h_i, K_n(A) \cap \prod_{i=1}^{p} h_i \neq \emptyset \right\}.$$

Then, $(K_n(A))_{n \geq 1}$ is an increasing chain of subsets of H.
    Let $K(A) = \bigcup_{n \geq 1} K_n(A)$.

Theorem 5.4. *We have $\mathcal{C}(A) = K(A)$.*

Proof. Notice that $K(A)$ is a complete part of H. Indeed, if we suppose $K(A) \cap \prod_{i=1}^{p} x_i \neq \emptyset$, then there exists $n \geq 1$, such that $K_n(A) \cap \prod_{i=1}^{p} x_i \neq \emptyset$, which means that $\prod_{i=1}^{p} x_i \subseteq K_{n+1}(A)$.

Now, if $A \subseteq B$ and B is a complete part of H, then we show that $K(A) \subseteq B$. We have $K_1(A) \subseteq B$ and suppose $K_n(A) \subseteq B$. We check that $K_{n+1}(A) \subseteq B$. Let $z \in K_{n+1}(A)$, which means that there exists a hyperproduct $\prod_{i=1}^{p} x_i$, such that $z \in \prod_{i=1}^{p} x_i$ and $K_n(A) \cap \prod_{i=1}^{p} x_i \neq \emptyset$. Hence $B \cap \prod_{i=1}^{p} x_i \neq \emptyset$, whence $\prod_{i=1}^{p} x_i \subseteq B$. We obtain $z \in B$. Therefore, $\mathcal{C}(A) = K(A)$. $\square$

If $x \in H$, we denote $K_n(\{x\}) = K_n(x)$. This implies that

$$K_n(A) = \bigcup_{a \in A} K_n(a).$$

Theorem 5.5. *If x is an arbitrary element of a semihypergroup $(H, \circ)$, then*

    (1) *For all $n \geq 2$ we have $K_n(K_2(x)) = K_{n+1}(x)$;*
    (2) *The next equivalence holds: $x \in K_n(y) \Leftrightarrow y \in K_n(x)$.*

Proof. (1) We check the equality by induction. We have

$$K_2(K_2(x)) = \left\{ z \in H | \exists q \in \mathbb{N}, \exists (a_1, \ldots, a_q) \in H^q : z \in \prod_{i=1}^{q} a_i, K_2(x) \cap \prod_{i=1}^{q} a_i \neq \emptyset \right\}$$
$$= K_3(x).$$

Suppose that $K_{n-1}(K_2(x)) = K_n(x)$. Then,

$$K_n(K_2(x))$$
$$= \left\{ z \in H | \exists q \in \mathbb{N}, \exists (a_1, \ldots, a_q) \in H^q : z \in \prod_{i=1}^{q} a_i, K_{n-1}(K_2(x)) \cap \prod_{i=1}^{q} a_i \neq \emptyset \right\}$$
$$= K_{n+1}(x).$$

    (2) We check the equivalence by induction. For $n = 2$, we have

$$x \in K_2(y)$$
$$= \left\{ z \in H | \exists q \in \mathbb{N}, \exists (a_1, \ldots, a_q) \in H^q : z \in \prod_{i=1}^{q} a_i, K_1(y) \cap \prod_{i=1}^{q} a_i \neq \emptyset \right\}.$$

Hence $\{y, x\} \subseteq \prod_{i=1}^{q} a_i$, whence $y \in K_2(x)$.

Suppose that the following equivalence holds:

$$x \in K_{n-1}(y) \iff y \in K_{n-1}(x)$$

and we check $x \in K_n(y) \iff y \in K_n(x)$. If $x \in K_n(y)$ then there exists $\prod_{i=1}^{p} a_i$ with $x \in \prod_{i=1}^{p} a_i$ and there exists $v \in \prod_{i=1}^{p} a_i \cap K_{n-1}(y)$. It follows that $v \in K_2(x)$ and $y \in K_{n-1}(v)$. Hence, $y \in K_{n-1}(K_2(x)) = K_n(x)$. Similarly, we obtain the converse implication. $\square$

Corollary 5.1. $x \in \mathcal{C}(y)$ *if and only if* $y \in \mathcal{C}(x)$.

Corollary 5.2. *The binary relation defined as follows:*

$$x K y \iff \exists n \geq 1, \ x \in K_n(y)$$

is an equivalence relation.

Theorem 5.6. *The equivalence relations* K *and* β^* *coincide.*

Proof. If $x \beta y$, then x and y belong to the same hyperproduct and so, $x \in K_2(y) \subseteq K(y)$. Hence $\beta \subseteq K$, whence $\beta^* \subseteq K$. Now, if we have $x K y$ and $x \neq y$, then there exists $n \geq 1$, such that $x K_{n+1} y$, which means that there exists a hyperproduct P_1, such that $x \in P_1$ and $P_1 \cap K_n(y) \neq \emptyset$. Let $x_1 \in P_1 \cap K_n(y)$. Hence $x \beta x_1$. From $x_1 \in K_n(y)$ it follows that there exists a hyperproduct P_2, such that $x_1 \in P_2$ and $P_2 \cap K_{n-1}(y) \neq \emptyset$. Let $x_2 \in P_2 \cap K_{n-1}(y)$. Hence $x_1 \beta x_2$ and $x_2 \in K_{n-1}(y)$. After a finite number of steps, we obtain that there exist x_{n-1}, x_n such that $x_{n-1} \beta x_n$ and $x_n \in K_{n-(n-1)}(y) = \{y\}$. Therefore $x \beta^* y$. $\square$

Theorem 5.7. *Let* $(H, \circ)$ *is a hypergroup. If A is a non-empty subset of H, then*

$$\mathcal{C}(A) = \bigcup_{a \in A} \mathcal{C}(a).$$

Proof. Clearly, for all $a \in A$, we have $\mathcal{C}(a) \subseteq \mathcal{C}(A)$. On the other hand, $\mathcal{C}(A) = \bigcup_{n \geq 1} K_n(A)$. We shall prove by induction. For $n = 1$, we have $K_1(A) = \bigcup_{a \in A} K_1(a)$. Suppose that $K_n(A) \subseteq \bigcup_{a \in A} K_n(a)$. If $z \in K_{n+1}(A)$, then there exists a hyperproduct P such that $z \in P$ and $K_n(A) \cap P \neq \emptyset$, whence there exists $a \in A$ such that $K_n(a) \cap P \neq \emptyset$. Hence, $z \in K_{n+1}(a)$. We obtain $K_{n+1}(A) \subseteq \bigcup_{a \in A} K_{n+1}(a)$. Therefore, we conclude that $\mathcal{C}(A) = \bigcup_{a \in A} \mathcal{C}(A)$. $\square$

Definition 5.6. Let $\varphi : H \to H/\beta^*$ be the canonical projection. Then, the *heart* (or *core*) of H is defined as follows:

$$\omega_H = \{x \in H \mid \varphi(x) = 1\},$$

where 1 is the identity element of H/β^*.

Theorem 5.8. *Let $(H, \circ)$ be a hypergroup. If A is a non-empty subset of H, then $\omega_H \circ A = A \circ \omega_H = \varphi^{-1}\varphi(A)$.*

Proof. Clearly, we have

$$\varphi^{-1}\varphi(A) = \{x \in H \mid \exists a \in A \text{ such that } \varphi(x) = \varphi(a)\}.$$

Assume that $y \in \varphi^{-1}\varphi(A)$. Then, there exists $a \in A$ such that $\varphi(y) = \varphi(a)$. Since H is a hypergroup, it follows that there exists $z \in H$ such that $y \in a \circ z$. This yields that $\varphi(y) = \varphi(a) \odot \varphi(z)$, which implies that $z \in \omega_H$. Hence, we conclude that $\varphi^{-1}\varphi(A) \subseteq A \circ \omega_H$.

Conversely, let $y \in A \circ \omega_H$. Then, we obtain $\varphi(y) \in \varphi(A)$. This implies that $A \circ \omega_H \subseteq \varphi^{-1}\varphi(A)$.

Thus, we obtain $A \circ \omega_H = \varphi^{-1}\varphi(A)$. Similarly, we can prove that $\omega_H \circ A = \varphi^{-1}\varphi(A)$. $\square$

Theorem 5.9. *Let $(H, \circ)$ be a hypergroup. If A is a non-empty subset of H, then*

$$\omega_H \circ A = A \circ \omega_H = \mathcal{C}(A).$$

Proof. It is not difficult to see that

$$\varphi^{-1}\varphi(A) = \{x \in H \mid \exists a \in A \text{ such that } x \in \mathcal{C}(a)\} = \bigcup_{a \in A} \mathcal{C}(a).$$

Now, by Theorems 5.7 and 5.8, the result follows. $\square$

Theorem 5.10. *Let $(H, \circ)$ be a hypergroup. If A is a complete part and B is a non-empty subset of H, then $A \circ B$ and $B \circ A$ are complete parts.*

Proof. By Theorem 5.9, we can write

$$\mathcal{C}(A \circ B) = A \circ B \circ \omega_H = A \circ \omega_H \circ B = A \circ B.$$

This completes the proof. $\square$

Theorem 5.11. *Let $(H, \circ)$ be a hypergroup and A be a non-empty subset of H. Ten, A is a complete part of H if and only if $A \circ \omega_H = A$.*

Proof. The result follows from Theorem 5.9. $\square$

Theorem 5.12. ω_H *is a complete part of H.*

Proof. We have $\omega_H \circ \omega_H = \omega_H$, which means that ω_H is a complete part of H. $\qquad\square$

Theorem 5.13. *Any complete part subhypergroup K of $(H, \circ)$ is invertible.*

Proof. Let $y \in K \circ x$. We obtain $\varphi(y) = \varphi(k) \odot \varphi(x)$ for some $k \in K$. Since $\varphi(K)$ is a subgroup of $(H/\beta^*, \odot)$, it follows that $\varphi(x) = \varphi(k)^{-1} \odot \varphi(y) = \varphi(K) \odot \varphi(y) = \varphi(K \circ y)$. Hence $x \in \varphi^{-1}(\varphi(K \circ y)) = C(K \circ y) = K \circ y$. Therefore, K is left invertible. Similarly, it can be shown that K is right invertible. $\qquad\square$

Theorem 5.14. *The heart of a hypergroup H is the intersection of all subhypergroups of H which are complete parts.*

Proof. Denote the class of all complete parts subhypergroups of H by $CPS(H)$. Since $\omega_H \in CPS(H)$, it follows that $\bigcap_{K \in CPS(H)} K \subseteq \omega_H$. Now, we show that for all $K \in CPS(H)$ we have $\omega_H \subseteq K$. Let $x \in \omega_H$. Since $K \circ \omega_H = K$, it follows that there are a, b in K, such that $b \in a \circ x \subseteq K \circ x$. Since K is invertible, it follows that $x \in K \circ b = K$. Hence $\omega_H \subseteq \bigcap_{K \in CPS(H)} K$. $\qquad\square$

Theorem 5.15. *Each complete part subhypergroup K of $(H, \circ)$ is ultra-closed.*

Proof. First of all, K is closed since it is invertible. On the other hand, for all $x \in K$, there exists $e \in H$ such that $x \in x \circ e$. From here, we obtain that $e \in \omega_H$. Moreover, there exists $x' \in H$ such that $e \in x' \circ x$. Since ω_H is a complete part, it follows that $x' \circ x \subseteq \omega_H$. Denote $B = x \circ K \cap x \circ (H \setminus K)$. We obtain

$$x' \circ B \subseteq x' \circ (x \circ K \cap x \circ (H \setminus K)) \subseteq \omega_H \circ K \cap \omega_H \circ (H \setminus K) \subseteq K \cap K \circ (H \setminus K) = \emptyset$$

since K is closed. Hence $B = \emptyset$, which means that K is ultraclosed on the right. Similarly, it is ultraclosed on the left. $\qquad\square$

To each binary relation R on a non-empty set H, a partial hypergroupoid $(H, \circ_R)$ is associated as follows:

$$x \circ_R x = \{y \in H \mid (x, y) \in R\},$$
$$x \circ_R z = x \circ_R x \cup z \circ_R z,$$

for all $x, z \in H$. Hence, we obtain

$$x \circ_R x \circ_R x = \bigcup_{a \in x \circ_R x} a \circ_R a \cup x \circ_R x.$$

Theorem 5.16. *Let R be a relation on H, such that $R \subseteq R^2$. Then R is transitive if and only if for all $x \in H$, we have $x \circ_R x \circ_R x = x \circ_R x$.*

Proof. It is straightforward. □

Theorem 5.17. *Let $(H, \circ)$ be a semihypergroup. The relation β is transitive if and only if for all x of H, we have $C(x) = K_2(x)$.*

Proof. First, denote by $Pr(H)$ the set of all hyperproducts of H. According to Theorem 5.16, the relation β is transitive if and only if for all $x \in H$, $x \circ_\beta x \circ_\beta x = x \circ_\beta x$. We have

$$x \circ_\beta x \circ_\beta x = \{t \in H \mid t \in K_2(a), \ a \in K_2(x)\} \cup K_2(x) = K_3(x) \cup K_2(x).$$

Hence, β is transitive if and only if for all $x \in H$, we have $K_3(x) \cup K_2(x) = K_2(x)$, which means that for all $x \in H$, we have $K_3(x) \subseteq K_2(x)$. We show that for every positive integer n, $K_{n+1} \subseteq K_n(x)$. Indeed, if we suppose that $K_s(x) \subseteq K_{s-1}(x)$ where $s \in \mathbb{Z}^+$, then

$$\begin{aligned}
K_{s+1}(x) &= \bigcup\{P_0 \in Pr(H) \mid P_0 \cap K_s(x) \neq \emptyset\} \\
&\subseteq \bigcup\{P_0 \in Pr(H) \mid P_0 \cap K_{s-1}(x) \neq \emptyset\} \\
&= K_s(x).
\end{aligned}$$

Since $C(x) = \bigcup_{i \in \mathbb{Z}^+} K_i(x)$, it follows that β is transitive if and only if for all x of H, we have $C(x) = K_2(x)$. In other words, β is transitive if and only if for all x of H, $K_2(x)$ is a complete part of H. □

Theorem 5.18. *If $(H, \circ)$ is a hypergroup, then for all x of ω_H, we have $K_2(x) = \omega_H$.*

Proof. Clearly, for all x of ω_H, we have $K_2(x) \subseteq \omega_H$. Now, it is sufficient to show that $K_2(x)$ is a complete part of H. Suppose $v \in K_2(x) \cap P$, where $P \in Pr(H)$. We check that $P \subseteq K_2(x)$. There exists $P_0 \in Pr(H)$ such that $x \in P_0$, $v \in P_0 \cap P$. On the other hand, there exists $e \in \omega_H$, such that $P \subseteq P \circ e$. Moreover, there are a, b in ω_H, such that $x \in v \circ a$, $e \in b \circ x$. We have

$$P \subseteq P \circ e \subseteq P \circ b \circ x \subseteq P \circ b \circ v \circ a \subseteq P \circ b \circ P_0 \circ a = P_2 \in Pr(H)$$
$$x \in v \circ a \subseteq P \circ a \subseteq P \circ e \circ a \subseteq P \circ b \circ x \circ a \subseteq P \circ b \circ P_0 \circ a = P_2.$$

Hence, we conclude that $P \subseteq K_2(x)$. □

Theorem 5.19. *If* $(H, \circ)$ *is a hypergroup, then* $\beta = \beta^*$.

Proof. We check that for all $x \in H$, we have $K_3(x) \subseteq K_2(x)$. Let $y \in K_3(x)$. Then there exists $z \in H$ and there exist P, Q in $Pr(H)$ such that $\{y, z\} \subseteq P$ and $\{z, x\} \subseteq Q$. Hence $\varphi(y) = \varphi(z) = \varphi(x)$. There exist $u, e \in \omega_H$ such that $y \in z \circ u$, $x \in x \circ e$. By the above theorem, we obtain $u \in K_2(e)$ and so there exists $T \in Pr(H)$ such that $\{u, t\} \subseteq T$. We have $y \in Q \circ u \subseteq Q \circ T$ and $x \in x \circ e \subseteq Q \circ T$. Hence $y \in K_2(x)$, since x, y belong to the same hyperproduct. Therefore, β is transitive, so it is an equivalence relation. $\qquad\qquad\square$

5.3 Abelian groups derived from hypergroups

Freni in [74] introduced the relation α as a generalization of the relation β. The letter γ already has been used for the corresponding fundamental relation on hyperrings by Vougiouklis (1991, 1994). Thus, there is a confusion on the symbolism. Therefore in this section we use the symbol α instead of γ for semihypergroups. In this section we would like the fundamental semigroup to be commutative. Notice that we use the Greek letter α for the relation because of the "A" in: Abelian. The main reference for this section is [74]. Therefore, we give the following definition.

Definition 5.7. Let H be a semihypergroup. Then, we set

$$\alpha_1 = \{(x, x) \mid x \in H\}$$

and for every integer $n > 1$, α_n is the relation defined as follows:

$$x \alpha_n y \Leftrightarrow \exists (z_1, \ldots, z_n) \in H^n, \exists \sigma \in \mathbb{S}_n : x \in \prod_{i=1}^{n} z_i, \ y \in \prod_{i=1}^{n} z_{\sigma(i)}.$$

Obviously, for $n \geq 1$, the relations α_n are symmetric, and the relation $\alpha = \bigcup_{n \geq 1} \alpha_n$ is reflexive and symmetric.

Let α^* be the transitive closure of α. If H is a hypergroup, then $\alpha = \alpha^*$ [74].

Theorem 5.20. *The relation* α^* *is a strongly regular relation.*

Proof. Clearly, α^* is an equivalence relation. In order to prove that it is strongly regular, we have to show first that

$$x \alpha y \Rightarrow (x \circ a) \,\overline{\overline{\alpha}}\, (y \circ a) \text{ and } (a \circ x) \,\overline{\overline{\alpha}}\, (a \circ y),$$

for every $a \in H$. If $x \alpha y$, then there is $n \in \mathbb{N}$ such that $x \alpha_n y$. Hence, there exist $(z_1, \ldots, z_n) \in H^n$ and $\sigma \in \mathbb{S}_n$ such that $x \in \prod_{i=1}^{n} z_i$ and $y \in \prod_{i=1}^{n} z_{\sigma(i)}$. For every $a \in H$, set $a = z_{n+1}$ and let τ be a permutation of $\mathbb{S}_{n+1}$ such that

$$\tau(i) = \sigma(i), \text{ for all } i \in \{1, 2, \ldots, n\};$$
$$\tau(n+1) = n+1.$$

For all $v \in x \circ a$ and for all $w \in y \circ a$, we have $v \in x \circ a \subseteq \prod_{i=1}^{n} z_i \circ a = \prod_{i=1}^{n+1} z_i$ and $w \in y \circ a \subseteq \prod_{i=1}^{n} z_{\sigma(i)} \circ a = \prod_{i=1}^{n} z_{\sigma(i)} \circ z_{n+1} = \prod_{i=1}^{n+1} z_{\tau(i)}$. So, $v \alpha_{n+1} w$ and hence $v \alpha w$. Thus, $(x \circ a) \, \overline{\overline{\alpha}} \, (y \circ a)$. In the same way, we can show that $(a \circ x) \, \overline{\overline{\alpha}} \, (a \circ y)$.

Moreover, if $x \alpha^* y$, then there exist $m \in \mathbb{N}$ and

$$(w_0 = x, w_1, \ldots, w_{m-1}, w_m = y) \in H^{m+1}$$

such that $x = w_0 \alpha w_1 \alpha \ldots \alpha w_{m-1} \alpha w_m = y$. Now, we obtain

$$x \circ a = w_0 \circ a \, \overline{\overline{\alpha}} \, w_1 \circ a \, \overline{\overline{\alpha}} \, w_2 \circ a \, \overline{\overline{\alpha}} \, \ldots \, \overline{\overline{\alpha}} \, w_{m-1} \circ a \, \overline{\overline{\alpha}} \, w_m \circ a = y \circ a.$$

Finally, for all $v \in x \circ a = w_0 \circ a$ and for all $w \in w_m \circ a = y \circ a$, taking $z_1 \in w_1 \circ a$, $z_2 \in w_2 \circ a, \ldots, z_{m-1} \in w_{m-1} \circ a$, we have $v \alpha z_1 \alpha z_2 \alpha \ldots \alpha z_{m-1} \alpha w$, and so $v \alpha^* w$. Therefore, $x \circ a \, \overline{\overline{\alpha^*}} \, y \circ a$. Similarly, we obtain $a \circ x \, \overline{\overline{\alpha^*}} \, a \circ y$. Hence, α^* is strongly regular. $\qquad \square$

Corollary 5.3. *The quotient H/α^* is a commutative semigroup. Furthermore, if H is a hypergroup, then H/α^* is an abelian group.*

Proof. Since α^* is a strongly regular relation, the quotient H/α^* is a semigroup under the following operation:

$$\alpha^*(x_1) \odot \alpha^*(x_2) = \alpha^*(z), \text{ for all } z \in x_1 \circ x_2.$$

Moreover, if H is a hypergroup, then H/α^* is a group. Finally, if σ is the cycle of $\mathbb{S}_2$ such that $\sigma(1) = 2$, for all $z \in x_1 \circ x_2$ and $w \in x_{\sigma(1)} \circ x_{\sigma(2)}$, we have $z \alpha_2 w$, so $z \alpha^* w$ and $\alpha^*(x_1) \odot \alpha^*(x_2) = \alpha^*(z) = \alpha^*(x_2) \odot \alpha^*(x_1)$. $\quad \square$

Theorem 5.21. *The relation α^* is the smallest strongly regular relation on a semihypergroup H such that the quotient H/α^* is commutative semigroup.*

Proof. Suppose that ρ is a strongly regular relation such that H/ρ is a commutative semigroup and $\varphi : H \to H/\rho$ is the canonical projection. Then, φ is a good homomorphism. Moreover, if $x \alpha_n y$, then there exist

$(z_1, \ldots, z_n) \in H^n$ and $\sigma \in \mathbb{S}_n$ such that $x \in \prod_{i=1}^{n} z_i$ and $y \in \prod_{i=1}^{n} z_{\sigma(i)}$, whence $\varphi(x) = \varphi(z_1) \odot \ldots \odot \varphi(z_n)$ and $\varphi(y) = \varphi(z_{\sigma(1)}) \odot \ldots \odot \varphi(z_{\sigma(n)})$. By the commutativity of H/ρ, it follows that $\varphi(x) = \varphi(y)$ and $x\rho y$. Thus, $x\alpha_n y$ implies $x\rho y$, and obviously, $x\alpha y$ implies that $x\rho y$.

Finally, if $x\alpha^* y$, then there exist $m \in \mathbb{N}$ and
$$(w_0 = x, w_1, \ldots, w_{m-1}, w_m = y) \in H^{m+1}$$
such that $x = w_0 \alpha w_1 \alpha \ldots \alpha w_{m-1} \alpha w_m = y$. Therefore,
$$x = w_0 \rho w_1 \rho \ldots \rho w_{m-1} \rho w_m = y,$$
and transitivity of ρ implies that $x\rho y$. Therefore, $\alpha^* \subseteq \rho$. $\qquad\square$

Now, we want to determine some necessary and sufficient conditions so that the relation α is transitive.

Definition 5.8. Let M be a non-empty subset of H. We say that M is a α-*part* of H if for any non-zero natural number n, for all $(z_1, \ldots, z_n) \in H^n$ and for all $\sigma \in \mathbb{S}_n$, we have
$$M \cap \prod_{i=1}^{n} z_i \neq \emptyset \Rightarrow \prod_{i=1}^{n} z_{\sigma(i)} \subseteq M.$$

Lemma 5.2. *Let M be a non-empty subsets of H. Then, the following conditions are equivalent.*

(1) *M is an α-part of H;*
(2) *$x \in M$, $x\alpha y \Rightarrow y \in M$;*
(3) *$x \in M$, $x\alpha^* y \Rightarrow y \in M$.*

Proof. (1$\Rightarrow$2): If $(x, y) \in H^2$ is a pair such that $x \in M$ and $x\alpha y$, then there exist $n \in \mathbb{N}$, $(z_1, \ldots, z_n) \in H^n$ and $\sigma \in \mathbb{S}_n$ such that $x \in M \cap \prod_{i=1}^{n} z_i$ and $y \in \prod_{i=1}^{n} z_{\sigma(i)}$. Since M is an α-part of H, we have $\prod_{i=1}^{n} z_{\sigma(i)} \subseteq M$ and $y \in M$.

(2$\Rightarrow$3): Assume that $(x, y) \in H^2$ such that $x \in M$ and $x\alpha^* y$. Obviously, there exist $m \in \mathbb{N}$ and $(w_0 = x, w_1, \ldots, w_{m-1}, w_m = y) \in H^{m+1}$ such that $x = w_0 \alpha w_1 \alpha \ldots \alpha w_{m-1} \alpha w_m = y$. Since $x \in M$, applying (2) m times, we obtain $y \in M$.

(3$\Rightarrow$1): Suppose that $M \cap \prod_{i=1}^{n} x_i \neq \emptyset$ and $x \in M \cap \prod_{i=1}^{n} x_i$. For every $\sigma \in \mathbb{S}_n$ and for every $y \in \prod_{i=1}^{n} x_{\sigma(i)}$, we have $x\alpha y$. Thus, $x \in M$ and $x\alpha^* y$.

Finally, by (3), we obtain $y \in M$, whence $\prod_{i=1}^{n} x_{\sigma(i)} \subseteq M$. $\qquad\square$

Before proving the next theorem, we introduce the following notations: Let H be a semihypergroup. For all $x \in H$, we set

- $T_n(x) = \left\{ (x_1, \ldots, x_n) \in H^n \mid x \in \prod_{i=1}^{n} x_i \right\}$;

- $P_n(x) = \left\{ \prod_{i=1}^{n} x_{\sigma(i)} \mid \sigma \in \mathbb{S}_n, (x_1, \ldots, x_n) \in T_n(x) \right\}$;

- $P_\sigma(x) = \bigcup_{n \geq 1} P_n(x)$.

From the previous notations and definitions, it follows at once the following.

Lemma 5.3. *For every $x \in H$, $P_\sigma(x) = \{ y \in H \mid x\alpha y \}$.*

Proof. For all $x, y \in H$, we have

$$x\alpha y \Leftrightarrow \exists n \in \mathbb{N}, \exists (x_1, \ldots, x_n) \in H^n, \exists \sigma \in \mathbb{S}_n : x \in \prod_{i=1}^{n} x_i, \; y \in \prod_{i=1}^{n} x_{\sigma(i)}$$

$$\Leftrightarrow \exists n \in \mathbb{N} : y \in P_n(x)$$

$$\Leftrightarrow y \in P_\sigma(x). \qquad \square$$

Theorem 5.22. *Let H be a semihypergroup. Then, the following conditions are equivalent.*

(1) α is transitive;

(2) $\alpha^*(x) = P_\sigma(x)$, for all $x \in H$;

(3) $P_\sigma(x)$ is an α-part of H, for all $x \in H$.

Proof. $(1\Rightarrow2)$: By Lemma 5.3, for all $x, y \in H$, we have

$$y \in \alpha^*(x) \Leftrightarrow x\alpha^* y \Leftrightarrow x\alpha y \Leftrightarrow y \in P_\sigma(x).$$

$(2\Rightarrow3)$: By Lemma 5.2, if M is a non-empty subset of H, then M is an α-part of H if and only if it is union of equivalence classes modulo α^*. In particular, every equivalence class modulo α^* is an α-part of H.

$(3\Rightarrow1)$: If $x\alpha y$ and $y\alpha z$, then there exist $m, n \in \mathbb{N}$, $(x_1, \ldots, x_n) \in T_n(x)$, $(y_1, \ldots, y_m) \in T_m(y)$, $\sigma \in \mathbb{S}_n$ and $\tau \in \mathbb{S}_m$ such that $y \in \prod_{i=1}^{n} x_{\sigma(i)}$ and $z \in \prod_{i=1}^{m} y_{\tau(i)}$. Since $P_\sigma(x)$ is an α-part of H, we have

$$x \in \prod_{i=1}^{n} x_i \cap P_\sigma(x) \Rightarrow \prod_{i=1}^{n} x_{\sigma(i)} \subseteq P_\sigma(x) \Rightarrow y \in \prod_{i=1}^{m} y_i \cap P_\sigma(x)$$

$$\Rightarrow \prod_{i=1}^{m} y_{\tau(i)} \subseteq P_\sigma(x) \Rightarrow z \in P_\sigma(x)$$

$$\Rightarrow \exists k \in \mathbb{N} : z \in P_k(x) \Rightarrow z\alpha x.$$

Therefore, α is transitive. $\qquad \square$

5.4 Cyclic groups derived from hypergroups

In [134], Mousavi, Leoreanu-Fotea and Jafarpour introduced a strongly regular equivalence relation $\rho_{\mathcal{A}}^*$ on the hypergroup H such that in a particular case the quotient $H/\rho_{\mathcal{A}}^*$ is a cyclic group. In this section we study this relation. The main reference for this section is [134]. Throughout this section, H is a hypergroup and $\mathcal{X} = \bigcup_{a \in \mathcal{A}} H_a$, where $\emptyset \neq \mathcal{A} \subseteq H$, $H_a = \bigcup a^n$, $n \in \mathbb{N}$ and for all $a, b \in \mathcal{A}$ if $a \neq b$, then $H_a \cap H_b = \emptyset$. For all $n \geqslant 1$ define $\Re_n^{\mathcal{A}}$ as follows: $\Re_n^{\mathcal{A}} = \Im_n^{\mathcal{A}} \cup \wp_n^{\mathcal{A}} \cup \jmath_n^{\mathcal{A}}$, where

$$\Im_n^{\mathcal{A}} = \left\{ \left(\prod_{i=1}^n x_i, \prod_{i=1}^n y_i\right) \mid \exists \sigma \in S_n, \prod_{i=1}^n y_i = \prod_{i=1}^n x_{\sigma(i)} \right\},$$

$$\wp_n^{\mathcal{A}} = \left\{ \left(\prod_{i=1}^n x_i, \prod_{i=1}^n y_i\right) \mid \exists a \in \mathcal{A}, \{x_1, \ldots, x_n\} \cap H_a = \{y_1, \ldots, y_n\} \cap H_a \neq \emptyset, \right.$$
$$\left. \exists \sigma_a \in S_n, \ y_j \in \{y_1, \ldots, y_n\} \cap H_a \Rightarrow y_j = x_{\sigma_a(j)} \right\},$$

$$\jmath_n^{\mathcal{A}} = \left\{ \left(\prod_{i=1}^n x_i, \prod_{i=1}^n y_i\right) \mid \{x_1, \ldots, x_n\} \cap \mathcal{X} = \{y_1, \ldots, y_n\} \cap \mathcal{X} = \emptyset \right\}.$$

It is easy to see that $\Re_1^{\mathcal{A}} = \left\{ (\{x\}, \{y\}) \mid \{x, y\} \cap \mathcal{X} = \emptyset \text{ or } x = y \right\}$ and
$$\Re_2^{\mathcal{A}} = \{(x \circ z, y \circ z), (x \circ z, z \circ y), (z \circ x, z \circ y), (z \circ x, y \circ z) \mid (\{x\}, \{y\}) \in \Re_1^{\mathcal{A}}, z \in H\}.$$

For every $n \geqslant 1$, define the relation $\rho_{\mathcal{A}}^n$ on H by:

$$x \rho_{\mathcal{A}}^n y \Leftrightarrow \exists (A, B) \in \Re_n^{\mathcal{A}}, \ x \in A, \ y \in B.$$

Notice that for $n = 1$ we obtain

$$x \rho_{\mathcal{A}}^1 y \Leftrightarrow (\{x\}, \{y\}) \in \Re_1^{\mathcal{A}} \text{ or } x = y \in \mathcal{X}.$$

Obviously, for every $n \geqslant 1$, the relations $\rho_{\mathcal{A}}^n$ are symmetric, and the relation $\rho_{\mathcal{A}} = \bigcup_{n \geqslant 1} \rho_{\mathcal{A}}^n$ is reflexive and symmetric. Let $\rho_{\mathcal{A}}^*$ be the transitive closure of $\rho_{\mathcal{A}}$.

Theorem 5.23. *The relation $\rho_{\mathcal{A}}^*$ is a strongly regular relation.*

Proof.   Notice that $\rho_{\mathcal{A}}^*$ is an equivalence relation. In order to prove that it is strongly regular, we have to show first that:

$$x \rho_{\mathcal{A}} y \Rightarrow x \circ z \overline{\overline{\rho_{\mathcal{A}}}} y \circ z, \quad z \circ x \overline{\overline{\rho_{\mathcal{A}}}} z \circ y. \tag{5.1}$$

for every $z \in H$. Suppose that $x \rho_{\mathcal{A}} y$, thus there exists $n \geqslant 1$ such that $x \rho_{\mathcal{A}}^n y$. Therefore there exists $(B, C) \in \Re_n^{\mathcal{A}}$ such that $x \in B$ and $y \in C$. We obtain the following three cases:

Case 1. Suppose that $(B, C) \in \mathfrak{S}_n^{\mathcal{A}}$ so we have $B = \prod_{i=1}^{n} x_i$, $C = \prod_{i=1}^{n} y_i$ and there exist $\sigma \in S_n$ such that $\prod_{i=1}^{n} y_i = \prod_{i=1}^{n} x_{\sigma(i)}$. Set $z = y_{n+1} = x_{n+1}$ and let τ be the permutation of S_{n+1} such that:

$$\tau(i) = \begin{cases} \sigma(i) & \text{if } i \in \{1, 2, \ldots, n\}; \\ n+1 & \text{if } i = n+1. \end{cases}$$

So $C \circ z = \prod_{i=1}^{n+1} x_{\tau(i)}$ and hence $(B \circ z, C \circ z) \in \mathfrak{S}_{n+1}^{\mathcal{A}}$.

Case 2. Suppose that $(B, C) \in \wp_n^{\mathcal{A}}$ so we have $B = \prod_{i=1}^{n} x_i$, $C = \prod_{i=1}^{n} y_i$ and there exists $a \in \mathcal{A}$ such that $\{x_1, \ldots, x_n\} \cap H_a = \{y_1, \ldots, y_n\} \cap H_a \neq \emptyset$ and $\sigma_a \in S_n$ such that for all $1 \leqslant j \leqslant n$ if $y_j \in H_a$, then $y_j = x_{\sigma_a(j)}$. If $z \notin \mathcal{X}$, then it is easy to see that $(B \circ z, C \circ z) \in \wp_{n+1}^{\mathcal{A}}$. If $z \in \mathcal{X}$, then there exists $b \in \mathcal{A}$ such that $z \in H_b$. Set $z = y_{n+1} = x_{n+1}$ and let τ_a be the following permutation of S_{n+1}:

$$\tau_a(i) = \begin{cases} \sigma_a(i) & \text{if } i \in \{1, 2, \ldots, n\}; \\ n+1 & \text{if } i = n+1. \end{cases}$$

Hence, for $1 \leqslant j \leqslant n+1$ if $y_j \in H_a$, then $y_j = x_{\tau_a(j)}$. Therefore, we have $(B \circ z, C \circ z) \in \wp_{n+1}^{\mathcal{A}}$ and hence $(B \circ z, C \circ z) \in \Re_{n+1}^{\mathcal{A}}$.

Case 3. Suppose that $(B, C) \in \jmath_n^{\mathcal{A}}$ so we have $B = \prod_{i=1}^{n} x_i$, $C = \prod_{i=1}^{n} y_i$, $\{x_1, \ldots, x_n\} \cap \mathcal{X} = \{y_1, \ldots, y_n\} \cap \mathcal{X} = \emptyset$. If $z \notin \mathcal{X}$, then $(B \circ z, C \circ z) \in \jmath_{n+1}^{\mathcal{A}}$ and if $z \in \mathcal{X}$, then $(B \circ z, C \circ z) \in \wp_{n+1}^{\mathcal{A}}$.

Therefore, for all $v \in x \circ z$ and for all $y \in y \circ z$, we have $v \in x \circ z \subseteq B \circ z$ and $u \in y \circ z \subseteq C \circ z$. This yields that $v \rho_{\mathcal{A}}^m u$ for some $m \in \mathbb{N}$, because $(B \circ z, C \circ z) \in \Re_m^{\mathcal{A}}$ for some $m \in \mathbb{N}$ and hence $v \rho_{\mathcal{A}} u$. Thus $x \circ z \; \overline{\rho_{\mathcal{A}}} \; y \circ z$. In the same way we obtain that $x \rho_{\mathcal{A}} y \Rightarrow z \circ x \; \overline{\overline{\rho_{\mathcal{A}}}} \; z \circ y$. Moreover, if $x \rho_{\mathcal{A}}^* y$, then there exists $m \in \mathbb{N}$ and $(u_0 = x, u_1, \ldots, u_m = y) \in H^m$ such that $x = u_0 \rho_{\mathcal{A}} u_1 \rho_{\mathcal{A}} \cdots \rho_{\mathcal{A}} u_{m-1} \rho_{\mathcal{A}} u_m = y$, whence, by (6.4), we obtain that

$$x \circ z = u_0 \circ z \; \overline{\rho_{\mathcal{A}}} \; u_1 \circ z \; \overline{\rho_{\mathcal{A}}} \; u_2 \circ z \; \overline{\rho_{\mathcal{A}}} \cdots \overline{\rho_{\mathcal{A}}} \; u_{m-1} \; \overline{\rho_{\mathcal{A}}} \; u_m \circ z = y \circ z.$$

Finally, for all $v \in x \circ z = u_0 \circ z$ and for all $u \in u_m \circ z = y \circ z$, taking $z_1 \in u_1 \circ z$, $z_2 \in u_2 \circ z$, $\ldots$, $z_{m-1} \in u_{m-1} \circ z$, we have $v \rho_{\mathcal{A}} z_1 \rho_{\mathcal{A}} z_2 \rho_{\mathcal{A}} \cdots \rho_{\mathcal{A}} z_{m-1} \rho_{\mathcal{A}} u$, and so $v \rho_{\mathcal{A}}^* u$. Therefore, we obtain

$$x \rho_{\mathcal{A}}^* y \Rightarrow x \circ z \; \overline{\rho_{\mathcal{A}}^*} \; y \circ z.$$

Similarly, we can prove that $x \rho_{\mathcal{A}}^* y \Rightarrow z \circ x \; \overline{\rho_{\mathcal{A}}^*} \; z \circ y$, hence $\rho_{\mathcal{A}}^*$ is strongly regular. $\square$

Theorem 5.24. *The quotient $H/\rho_{\mathcal{A}}^*$ is an abelian group with the generators*

$$\{\rho_{\mathcal{A}}^*(x_0), \rho_{\mathcal{A}}^*(a) \mid x_0 \in (H \setminus \mathcal{X}), a \in \mathcal{A}\}.$$

Proof. Since $\rho_{\mathcal{A}}^*$ is a strongly regular equivalence, it follows that the quotient $H/\rho_{\mathcal{A}}^*$ is a group under the following operation:

$$\rho_{\mathcal{A}}^*(x) \odot \rho_{\mathcal{A}}^*(y) = \rho_{\mathcal{A}}^*(z), \quad \text{for all } z \in x \circ y.$$

Since $\gamma^* \subseteq \rho_{\mathcal{A}}^*$, it follows that $H/\rho_{\mathcal{A}}^*$ is an abelian group. For all $(x,y) \in (H \setminus \mathcal{X})^2$ since $\{x,y\} \cap \mathcal{X} = \emptyset$, we have $(\{x\}, \{y\}) \in \Re_1^{\mathcal{A}}$ and hence $x \rho_{\mathcal{A}}^* y$ so $\rho_{\mathcal{A}}^*(x) = \rho_{\mathcal{A}}^*(y)$. Now, assume that $\rho_{\mathcal{A}}^*(h)$ is given. If $h \in (H \setminus \mathcal{X})$, then $\rho_{\mathcal{A}}^*(h) = \rho_{\mathcal{A}}^*(x_0)$ and if $h \in \mathcal{X}$, then there exists $a \in \mathcal{A}$ such that $h \in H_a$. Therefore, we conclude that $\rho_{\mathcal{A}}^*(h) = [\rho_{\mathcal{A}}^*(a)]^k$ for some $k \in \mathbb{N}$. $\qquad\square$

Now, suppose that $\mathcal{A} = \{a\}$, so $\mathcal{X} = H_a$. Put $\eta_a^n = \rho_{\mathcal{A}}^n$, $\eta_a = \bigcup_{n \geqslant 1} \eta_a^n$ and $\eta_a^* = \rho_{\mathcal{A}}^*$. So we obtain the following corollary.

Corollary 5.4. *The quotient H/η_a^* is a cyclic group.*

Proof. By the proof of Theorem 5.24, we conclude that the equivalence classes determined by $\rho_{\mathcal{A}}^*$ of all elements of $(H \setminus \mathcal{X})$ coincide and the equivalence class of every element of $\mathcal{X}$ is a power of $\eta_a^*(a)$.

If $H \setminus \mathcal{X} \neq \emptyset$, and $x_0 \in H \setminus \mathcal{X}$ then there exists $u \in H \setminus \mathcal{X}$ such that $x_0 \in au$ since H is a hypergroup. We obtain $\eta_a^*(x_0) = \eta_a^*(a)\,\eta_a^*(u) = \eta_a^*(a)\,\eta_a^*(x_0)$ whence $\eta_a^*(a) = e_{H/\eta_a^*}$ the identity element of the group H/η_a^*. Hence $H/\eta_a^* = <\eta_a^*(x_0)>$. Moreover, since $a \in x_0 \circ v$ for some $v \in H \setminus \mathcal{X}$, we obtain that $\eta_a^*(x_0) = (\eta_a^*(x_0))^{-1}$ which means that the order of $\eta_a^*(x_0)$ is two, so in this case H/η_a^* is a cyclic group of order two.

Now if $H \setminus \mathcal{X} = \emptyset$, which means that H is a cyclic hypergroup, then $H/\eta_a^* = <\eta_a^*(a)>$.

Therefore, we conclude that H/η_a^* is a cyclic group, as desired. $\qquad\square$

Example 5.3. Let S_3 be the permutation group of order 3, i.e.,

$$S_3 = \{(1), (1\ 2), (1\ 3), (2\ 3), (1\ 2\ 3), (1\ 3\ 2)\}$$

and $a = (1\ 2\ 3)$ be a cycle of order 3 in S_3. Then $S_3/\eta_a^* \cong \mathbb{Z}_2$.

Proof. It is easy to see that $\Re_1^{\mathcal{A}} = \{(x,y) \mid x,y \text{ are odd permutations}\}$ and $\Re_2^{\mathcal{A}} = \{(z,t) \mid z,t \text{ are odd or } z,t \text{ are even permutations}\}$ and $\Re_n^{\mathcal{A}} = \Re_2^{\mathcal{A}} = \eta_a^* = \alpha^*$, so $S_3/\eta_a^* \cong \mathbb{Z}_2$. $\qquad\square$

Theorem 5.25. *The relation η_a^* is the smallest strongly regular equivalence such that the quotient H/η_a^* is a cyclic group and the equivalence classes of all elements of $(H \setminus \mathcal{X})$ are equal.*

Proof. Let ρ be a strongly regular equivalence such that H/ρ is a cyclic group and the equivalence classes of ρ of all elements of $(H \setminus \mathcal{X})$ are equal. Suppose that $\varphi : H \to H/\rho$ is the canonical projection. φ is a good homomorphism. We show that $\eta_a^* \subseteq \rho$.

Let $n \geqslant 1$ and $x \, \eta_a^n y$. So there exists $(B, C) \in \mathfrak{R}_n^{\mathcal{A}}$ such that $x \in B$ and $y \in C$. We have three cases:

Case 1. Suppose that $(B, C) \in \mathfrak{S}_n^{\mathcal{A}}$ so we have $B = \prod_{i=1}^{n} x_i$, $C = \prod_{i=1}^{n} y_i$, and there exist $\sigma \in S_n$ such that $\prod_{i=1}^{n} y_i = \prod_{i=1}^{n} x_{\sigma(i)}$. Since $\varphi(x) = \odot_{i=1}^n \varphi(x_i)$ and $\varphi(y) = \odot_{i=1}^n \varphi(y_i) = \odot_{i=1}^n \varphi(x_{\sigma(i)})$, it follows that $\varphi(x) = \varphi(y)$, because H/ρ is commutative. Thus, $x\rho y$.

Case 2. Suppose that $(B, C) \in \wp_n^{\mathcal{A}}$ so we have $B = \prod_{i=1}^{n} x_i$, $C = \prod_{i=1}^{n} y_i$, $\{x_1, \ldots, x_n\} \cap H_a = \{y_1, \ldots, y_n\} \cap H_a \neq \emptyset$ and there exists $\sigma \in S_n$ such that for all $1 \leqslant j \leqslant n$ if $y_j \in H_a$, then $y_j = x_{\sigma(j)}$. Renumber of the elements of the sets $\{x_1, \ldots, x_n\}$ and $\{y_1, \ldots, y_n\}$ such that $\{y_1, \ldots, y_m\} \subseteq \mathcal{X}$, where $1 \leqslant m \leqslant n$ and $x_k, y_k \notin \mathcal{X}$ for all $m + 1 \leqslant k \leqslant n$. So, we have $\varphi(x) = (\odot_{j=1}^m \varphi(x_j)) \odot (\odot_{k=m+1}^n \varphi(x_k))$ and $\varphi(y) = (\odot_{j=1}^m \varphi(x_{\sigma(j)})) \odot (\odot_{k=m+1}^n \varphi(y_k))$. For all $m + 1 \leqslant t, l \leqslant n$, $\varphi(x_t) = \varphi(y_l)$ and since H/ρ is an abelian group, we have $\varphi(x) = \varphi(y)$ and hence $x\rho y$.

Case 3. Suppose that $(B, C) \in \jmath_n^{\mathcal{A}}$ so we have $B = \prod_{i=1}^{n} x_i$, $C = \prod_{i=1}^{n} y_i$, $\{x_1, \ldots, x_n\} \cap \mathcal{X} = \{y_1, \ldots, y_n\} \cap \mathcal{X} = \emptyset$. So for all $1 \leqslant i, j \leqslant n$, $\varphi(x_i) = \varphi(y_j)$, thus $\varphi(x) = \varphi(y)$ and hence xRy.

In all cases we have $x\rho y$ and hence $x \, \eta_a y \Rightarrow x\rho y$ whence $x \, \eta_a^* y \Rightarrow x\rho y$ by transitivity of ρ. Therefore, we deduce that $\eta_a^* \subseteq \rho$. $\qquad\square$

Now, we introduce the concept of $\rho_{\mathcal{A}}^*$-*part* of a hypergroup H and we determine necessary and sufficient conditions such that the relation $\rho_{\mathcal{A}}$ to be transitive.

Definition 5.9. Let M be a non-empty subset of H and $\prod_{i=1}^{n} x_i \cap M \neq \emptyset$. We say that M is a $\rho_{\mathcal{A}}^*$-*part* of H if the following conditions hold:

(P1) For all $\sigma \in S_n$, $\prod_{i=1}^{n} x_{\sigma(i)} \subseteq M$;

(P2) If $\exists a \in \mathcal{A}$, $\{x_1, \ldots, x_n\} \cap H_a \neq \emptyset$ implies that for all $\prod_{i=1}^{n} y_i$ such that $\{y_1, \ldots, y_n\} \cap H_a = \{x_1, \ldots, x_n\} \cap H_a$ there exists $\sigma_a \in S_n$ for which for all $y_j \in \{y_1, \ldots, y_n\} \cap H_a$, $y_j = x_{\sigma_a(j)}$, then $\prod_{i=1}^{n} y_i \subseteq M$;

(P3) If $\{x_1, \ldots, x_n\} \cap \mathcal{X} = \emptyset$ then for all $\prod_{i=1}^{n} y_i$ such that $\{y_1, \ldots, y_n\} \cap$

$\mathcal{X} = \emptyset$, we have $\prod_{i=1}^{n} y_i \subseteq M$.

Theorem 5.26. *Let M be a non-empty subset of a hypergroup H. The following conditions are equivalent.*

    (1) M *is a $\rho_{\mathcal{A}}^*$-part;*
    (2) $x \in M, x\rho_{\mathcal{A}}y \Rightarrow y \in M$;
    (3) $x \in M, x\rho_{\mathcal{A}}^*y \Rightarrow y \in M$.

Proof.   (1$\Rightarrow$2): Suppose that the pair $(x, y) \in H^2$ is such that $x \in M$ and $x\rho_{\mathcal{A}}y$. Then, there exists $(B, C) \in \Re_n^{\mathcal{A}}$ such that $x \in B$ and $y \in C$. So, we have three cases.

Case 1. Let $(B, C) \in \Im_n^{\mathcal{A}}$ so we have $B = \prod_{i=1}^{n} x_i$, $C = \prod_{i=1}^{n} y_i$ and there exists $\sigma \in S_n$ such that $\prod_{i=1}^{n} y_i = \prod_{i=1}^{n} x_{\sigma(i)}$. Since $x \in \prod_{i=1}^{n} x_i \cap M$, by (P1) we have $\prod_{i=1}^{n} x_{\sigma(i)} \subseteq M$ and hence $y \in M$.

Case 2. Let $(B, C) \in \wp_n^{\mathcal{A}}$. So $B = \prod_{i=1}^{n} x_i$, $C = \prod_{i=1}^{n} y_i$, there exists $a \in \mathcal{A}$, such that $\{x_1, \ldots, x_n\} \cap H_a = \{y_1, \ldots, y_n\} \cap H_a \neq \emptyset$ and there exists $\sigma_a \in S_n$ such that for all $1 \leqslant j \leqslant n$ if $y_j \in H_a$, then $y_j = x_{\sigma_a(j)}$. Since $x \in \prod_{i=1}^{n} x_i \cap M$, by (P2) we have $\prod_{i=1}^{n} y_i \subseteq M$ and hence $y \in M$.

Case 3. Suppose that $(B, C) \in \jmath_n^{\mathcal{A}}$ so we have $B = \prod_{i=1}^{n} x_i$, $C = \prod_{i=1}^{n} y_i$, $\{x_1, \ldots, x_n\} \cap \mathcal{X} = \{y_1, \ldots, y_n\} \cap \mathcal{X} = \emptyset$. Since $x \in \prod_{i=1}^{n} x_i \cap M$, by (P3) we have $\prod_{i=1}^{n} y_i \subseteq M$ and hence $y \in M$.

(2$\Rightarrow$3): Suppose that $(x, y) \in H^2$ is such that $x \in M$ and $x\rho_{\mathcal{A}}^*y$. So there exist $m \in \mathbb{N}$ and $(w_0 = x, w_1, \ldots, w_{m-1}, w_m = y) \in H^m$ such that $x = w_0\rho_{\mathcal{A}}w_1\rho_{\mathcal{A}}w_2 \ldots \rho_{\mathcal{A}}w_{m-1}\rho_{\mathcal{A}}w_m = y$. Since $x \in M$, applying (2) m times, we obtain $y \in M$.

(3$\Rightarrow$1): Let $x \in \prod_{i=1}^{n} x_i \cap M$. We shall check the conditions (P1), (P2) and (P3).

Let $\sigma \in S_n$ be given, thus $(\prod_{i=1}^{n} x_i, \prod_{i=1}^{n} x_{\sigma(i)}) \in \mathfrak{S}_n^{\mathcal{A}}$. So for all $y \in \prod_{i=1}^{n} x_{\sigma(i)}$ we have $x\rho_{\mathcal{A}}^{*}y$ and hence $y \in M$. Therefore $\prod_{i=1}^{n} x_{\sigma(i)} \subseteq M$. This means that (P1) holds.

Let $\{x_1, \ldots, x_n\} \cap H_a \neq \emptyset$ for some $a \in \mathcal{A}$. Now suppose that for all $\prod_{i=1}^{n} y_i$ if $\{y_1, \ldots, y_n\} \cap H_a = \{x_1, \ldots, x_n\} \cap H_a$ there exists $\sigma_a \in S_n$ such that for every $y_j \in \{y_1, \ldots, y_n\} \cap H_a$, $y_j = x_{\sigma_a(j)}$. Thus $(\prod_{i=1}^{n} x_i, \prod_{i=1}^{n} y_i) \in \wp_n^{\mathcal{A}}$ and hence for all $y \in \prod_{i=1}^{n} y_i$ we have $x\rho_{\mathcal{A}}^{*}y$. Consequently, $y \in M$. Hence, $\prod_{i=1}^{n} y_i \subseteq M$, and so (P2) holds.

Let $\{x_1, \ldots, x_n\} \cap \mathcal{X} = \emptyset$ and $\prod_{i=1}^{n} y_i$ be such that $\{y_1, \ldots, y_n\} \cap \mathcal{X} = \emptyset$. Thus $(\prod_{i=1}^{n} x_i, \prod_{i=1}^{n} y_i) \in \jmath_n^{\mathcal{A}}$ and hence for all $y \in \prod_{i=1}^{n} y_i$ we have $x\rho_{\mathcal{A}}^{*}y$. Therefore $y \in M$ and so $\prod_{i=1}^{n} y_i \subseteq M$. This yields that (P3) holds. $\square$

Notation 5.1. Let x be an arbitrary element of a hypergroup H.
For $n \geqslant 1$, set $P_n(x) = P_{\mathfrak{S}_n}(x) \cup P_{\jmath_n}(x) \cup P_{\wp_n}(x)$, where

$$P_{\mathfrak{S}_n}(x) = \bigcup \{\prod_{i=1}^{n} x_{\sigma(i)} \mid x \in \prod_{i=1}^{n} x_i\};$$

$$P_{\wp_n}(x) = \bigcup \{\prod_{i=1}^{n} y_i \mid x \in \prod_{i=1}^{n} x_i, \{y_1, \ldots, y_n\} \cap H_a = \{x_1, \ldots, x_n\} \cap H_a,$$
$$\exists \sigma_a \in S_n, y_j \in \{y_1, \ldots, y_n\} \cap H_a \Rightarrow y_j = x_{\sigma_a(j)}\};$$

$$P_{\jmath_n}(x) = \bigcup \{\prod_{i=1}^{n} y_i \mid x \in \prod_{i=1}^{n} x_i, \{y_1, \ldots, y_n\} \cap \mathcal{X} = \{x_1, \ldots, x_n\} \cap \mathcal{X} = \emptyset\}.$$

Denote $P(x) = \bigcup_{n \geqslant 1} P_n(x)$.

Notice that if $x \notin \mathcal{C}$, then $P_1(x) = H \setminus \mathcal{X}$ and

$$P_2(x) = \bigcup_{z \in H} \{dz, zd \mid d \in (H \setminus \mathcal{X}), \exists b \in (H \setminus \mathcal{X}), x \in bz \cup zb\},$$

while if $x \in a^n \subseteq \mathcal{C}$ for some $n \in \mathbb{N}$, then $P_m(x) = a^n$ for all $m \geqslant 1$.

Theorem 5.27. *For every $x \in H$, $P(x) = \{y \in H \mid x\rho_{\mathcal{A}}y\}$.*

Proof. Suppose that $x \in H$ and $y \in P(x)$ are given. Then there exist B and C such that $x \in B$, $y \in C$ and

(1) $y \in P_{\mathfrak{I}_n}(x) \Rightarrow (B, C) \in \mathfrak{I}_n^{\mathcal{A}};$
(2) $y \in P_{\wp_n}(x) \Rightarrow (B, C) \in \wp_n^{\mathcal{A}};$
(3) $y \in P_{\mathfrak{I}_n}(x) \Rightarrow (B, C) \in \mathfrak{I}_n^{\mathcal{A}}.$

Therefore $x\rho_{\mathcal{A}}y$ and so $P(x) \subseteq \{y \in H \mid x\rho_{\mathcal{A}}y\}$. The proof of the converse of the inclusion is similar to the above one. □

Lemma 5.4. *Suppose that H is a hypergroup and M is a $\rho_{\mathcal{A}}^*$-part of H. If $x \in M$, then $P(x) \subseteq M$.*

Proof. It follows by Definition 6.8. □

Theorem 5.28. *Let H be a hypergroup. Then, the following conditions are equivalent:*

(1) *$\rho_{\mathcal{A}}$ is transitive;*
(2) *for every $x \in H, \rho_{\mathcal{A}}^*(x) = P(x);$*
(3) *for every $x \in H, P(x)$ is a $\rho_{\mathcal{A}}^*$-part of H.*

Proof. (1⇒2): By Theorem 6.4, for all pair $(x, y) \in H^2$ we have:

$$y \in \rho_{\mathcal{A}}^*(x) \Leftrightarrow x\rho_{\mathcal{A}}^*y \Leftrightarrow x\rho_{\mathcal{A}}y \Leftrightarrow y \in P(x).$$

(2⇒3): By Theorem 6.3, if M is a non-empty subset of H, then M is a $\rho_{\mathcal{A}}^*$-part of H if and only if it is a union of equivalence classes modulo $\rho_{\mathcal{A}}^*$. Particularly, every equivalence class modulo $\rho_{\mathcal{A}}^*$ is a $\rho_{\mathcal{A}}^*$-part of H.

(3⇒1): Suppose that $x\rho_{\mathcal{A}}y$ and $y\rho_{\mathcal{A}}z$ so $x \in P(y)$ and $y \in P(z)$. By Lemma 6.2, we have $P(y) \subseteq P(z)$ and hence $x \in P(z)$. By Theorem 6.4 it follows that $x\rho_{\mathcal{A}}z$ and the proof is complete. □

Let H be a hypergroup and A be an invertible subhypergroup of H. For all $x \in H$, there exists $y \in H$ such that $x \in Ay$. Therefore the relation $_AE$ on H defined by:

$$x \; _AE \; y \Leftrightarrow x \in Ay$$

is a regular equivalence and a congruence. Moreover, the quotient $H/_AE = \{Ax \mid x \in H\}$ is a hypergroup with respect to the hyperoperation

$$Ax \odot Ay = \{Az \mid z \in xAy\}.$$

and the canonical projection $\pi : H \to H/_AE$ is an almost strong homomorphism. This means that if $\pi(x) = \pi(a) \odot \pi(t)$, then there exist $a', t' \in H$ such that $\pi(a) = \pi(a'), \pi(t) = \pi(t')$ and $x \in a' \circ t'$.

Let $\mathcal{A} = \{a\}$ and $\eta_a^*(a) \neq e_{H/\eta_a^*}$.

Theorem 5.29. *For all $x \in (H \setminus \mathcal{X})$, $\eta_a^*(x) = e_{H/\eta_a^*}$.*

Proof. Suppose that $\eta_a^*(x) \neq e_{H/\eta_a^*}$. By Theorem 5.25, $H/\eta_a^* = < \eta_a^*(a) >$, so $\eta_a^*(x) = \eta_a^*(t)$ such that $t \in a^n$ for some $n \in \mathbb{N}$. Therefore $x\,\eta^*t$ and hence there exist the elements $x = x_1, \ldots, x_{n-2}, x_{n-1}, x_n = t$ in H such that

$$x = x_1\,\eta_a\,\cdots\,x_{n-2}\,\eta_a\,x_{n-1}\,\eta_a\,x_n = t.$$

Thus, we have $x_{n-1}\,\eta_a\,t$ and by Theorem 6.4 it follows that $x_{n-1} \in P(t)$. Since $t \in a^n$, $P(t) = a^n$ and hence $x_{n-1} \in a^n$. Similarly, from $x_{n-2}\,\eta_a\,x_{n-1}$ we obtain $x_{n-2} \in a^n$. After $(n-1)$ steps, we conclude that $x = x_1 \in a^n \subseteq C$ and it is a contradiction. $\qquad\square$

5.5 Solvable groups derived from hypergroups

In [85], Jafarpour, Aghabozorgi and Davvaz, introduced and analyzed a new strongly regular relation on a hypergroup H, is named τ_n^*, such that H/τ_n^* is a solvable group. In this section we study this relation. The main reference is [85].

Definition 5.10. Let H be a hypergroup. We define

(1) $H^{(0)} = H$ and
(2) $H^{(k+1)} = \left\{ h \in H^{(k)} \mid x \circ y \cap h \circ y \circ x \neq \emptyset \text{ for some } x, y \in H^{(k)} \right\},$

for all $k \geq 0$. Suppose that $n \in \mathbb{N}$ and $\tau_n = \bigcup_{m>1} \tau_{m,n}$, where $\tau_{1,n}$ is the diagonal relation and for every integer $m > 1$, $\tau_{m,n}$ is the relation defined as follows:

$$x\,\tau_{m,n}\,y \Leftrightarrow \exists (z_1, \ldots, z_m) \in H^m, \exists \sigma \in \mathbb{S}_m : \sigma(i) = i \text{ if } z_i \notin H^{(n)} \text{ such that}$$
$$x \in \prod_{i=1}^{m} z_i \quad \text{and} \quad y \in \prod_{i=1}^{m} z_{\sigma(i)}.$$

Obviously, for every $n \geq 1$, the relation τ_n is reflexive and symmetric. Now, let τ_n^* be the transitive closure of τ_n.

Corollary 5.5. *For all $n \in \mathbb{N}$, we have $\beta^* \subseteq \tau_n^* \subseteq \alpha^*$.*

Notation 5.2. We denote $\bar{x}$ as the equivalence class of x and instead of $\bar{x} \odot \bar{y}$ we write $\bar{x}\bar{y}$.

Theorem 5.30. *For all $n \in \mathbb{N}$, the relation τ_n^* is a strongly regular relation.*

Proof. Suppose that $n \in \mathbb{N}$. Clearly, τ_n^* is an equivalence relation. In order to prove that it is strongly regular, first we have to show that:

$$x\tau_n y \Rightarrow x \circ z \; \overline{\overline{\tau_n^*}} \; y \circ z, \quad z \circ x \; \overline{\overline{\tau_n^*}} \; z \circ y, \tag{5.2}$$

for every $z \in H$. Suppose that $x\tau_n y$. Then, there exists $m \in \mathbb{N}$ such that $x\tau_{m,n} y$. Hence, there exist $(z_1, \ldots, z_m) \in H^m$, $\sigma \in \mathbb{S}_m$ with $\sigma(i) = i$ if $z_i \notin H^{(n)}$, such that $x \in \prod_{i=1}^{m} z_i$ and $y \in \prod_{i=1}^{m} z_{\sigma(i)}$. Suppose that $z \in H$. We have $x \circ z \subseteq (\prod_{i=1}^{m} z_i) \circ z$, $y \circ z \subseteq (\prod_{i=1}^{m} z_{\sigma(i)}) \circ z$ and $\sigma(i) = i$ if $z_i \notin H^{(n)}$. Now, suppose that $z_{i+1} = z$ and we define the permutation $\sigma' \in \mathbb{S}_{m+1}$ as follows:

$$\sigma'(i) = \sigma(i), \text{ for all } 1 \leq i \leq m \text{ and } \sigma'(m+1) = m+1.$$

Thus, $x \circ z \subseteq \prod_{i=1}^{m+1} z_i$ and $y \circ z \subseteq \prod_{i=1}^{m+1} z_{\sigma'(i)}$ such that $\sigma'(i) = i$ if $z_i \notin H^{(n)}$. Therefore, $x \circ z \; \overline{\overline{\tau_n^*}} \; y \circ z$. Similarly, we have $z \circ x \; \overline{\overline{\tau_n^*}} \; z \circ y$. Now, if $x\tau_n^* y$ then there exists $k \in \mathbb{N}$ and $(x = u_0, u_1, \ldots, u_k = y) \in H^{k+1}$ such that $x = u_0 \tau_n u_1 \tau_n \ldots \tau_n u_{k-1} \tau_n u_k = y$. Hence, by the above results, we obtain

$$x \circ z = u_0 \circ z \; \overline{\overline{\tau_n^*}} \; u_1 \circ z \; \overline{\overline{\tau_n^*}} \; u_2 \circ z \; \overline{\overline{\tau_n^*}} \ldots \overline{\overline{\tau_n^*}} \; u_{k-1} \circ z \; \overline{\overline{\tau_n^*}} \; u_k \circ z = y \circ z$$

and so $x \circ z \; \overline{\overline{\tau_n^*}} \; y \circ z$.

Similarly, we can prove that $z \circ x \; \overline{\overline{\tau_n^*}} \; z \circ y$. Therefore, τ_n^* is a strongly regular relation on H. $\qquad\square$

Theorem 5.31. *For all $n \in \mathbb{N}$, we have $\tau_{n+1}^* \subseteq \tau_n^*$.*

Corollary 5.6. *If H is a commutative hypergroup, then $\beta^* = \tau_n^* = \alpha^*$ for all $n \in \mathbb{N}$.*

Theorem 5.32. *H/τ_n^* is a solvable group of length at most $n+1$.*

Proof. Since τ_n^* is a strongly regular relation, it follows that $G = H/\tau_n^*$ is a group and so it is enough to prove that G is solvable of length at most $n+1$. First, we prove that

$$G^{(k)} = \langle \bar{t} \mid t \in H^{(k)} \rangle,$$

for all $k \in \mathbb{N}$. We do this by induction on k. For $k = 0$, we have $G = G^{(0)} = \langle \bar{t} \mid t \in H^{(0)} = H \rangle$. Now, suppose that $\bar{a} \in \langle \bar{t} \mid t \in H^{(k+1)} \rangle$. Then, $a \in H^{(k+1)}$ and so there exist $x, y \in H^{(k)}$ such that $x \circ y \cap a \circ y \circ x \neq \emptyset$.

Thus, $\bar{x}\bar{y} = \bar{a}\bar{y}\bar{x}$. By hypothesis of induction we conclude that $\bar{a} \in (G^{(k)})' = G^{(k+1)}$. Conversely, let $\bar{a} \in G^{(k+1)}$. Without loss of generality, suppose that $\bar{a} = \bar{x}\bar{y}\bar{x}^{-1}\bar{y}^{-1}$, where $\bar{x}, \bar{y} \in G^{(k)}$ which implies that $\bar{x}\bar{y} = \bar{a}\bar{y}\bar{x}$. Thus, there exist $c \in x \circ y$ and $d \in a \circ y \circ x$ such that $c\tau_n^* d$ which means $\bar{c} = \bar{d}$. Since H is a hypergroup, it follows that there exists $u \in H$ such that $c \in x \circ y \cap u \circ y \circ x$. From $\bar{x}, \bar{y} \in G^{(k)}$ and hypothesis of induction we have $x, y \in H^{(k)}$. Thus, $u \in H^{(k+1)}$ and

$$\bar{a}\bar{y}\bar{x} = \bar{d} = \bar{c} = \bar{x}\bar{y} = \bar{u}\bar{y}\bar{x}.$$

So, $\bar{a} = \bar{u} \in \langle \bar{t} \mid t \in H^{(k)} \rangle$. Consequently, $G^{(n)}$ is an abelian group and $G^{(n+1)} = (G^{(n)})' = \{\bar{0}_G\}$. $\square$

The relation τ_n^* defined in Definition 5.10, can be also introduced, in a natural way, for groups. Note that every group is a hypergroup, too.

Example 5.4. Suppose that $G = \mathbb{S}_4$. It is easy to see that $G^{(1)} = \mathbb{A}_4$ (alternating subgroup of $\mathbb{S}_4$) and $G^{(2)} = V = \{e, (12)(34), (13)(24), (14)(23)\}$. Now we have

$$\tau_2^*(e) = \{e, (13)(24), (12)(34), (14)(23)\};$$
$$\tau_2^*(12) = (12)\tau_2^*(e);$$
$$\tau_2^*(13) = (13)\tau_2^*(e);$$
$$\tau_2^*(23) = (23)\tau_2^*(e);$$
$$\tau_2^*(123) = (123)\tau_2^*(e);$$
$$\tau_2^*(124) = (124)\tau_2^*(e).$$

Therefore, $G/\tau_2^* \cong \mathbb{S}_3$, which is solvable group of length 2.

Now, we introduce the smallest strongly relation τ^* on a finite hypergroup H such that H/τ^* is a solvable group.

Definition 5.11. Let H be a finite hypergroup. Then, we define the relation τ^* on H as follows:

$$\tau^* = \bigcap_{n \geq 1} \tau_n^*.$$

Theorem 5.33. *The relation τ^* is a strongly regular relation on a finite hypergroup H such that H/τ^* is a solvable group.*

Proof. It easy to see that τ^* is a strongly regular relation on H. Now, we conclude that there exists $k \in \mathbb{N}$ such that $\tau_{k+1}^* = \tau_k^*$ and so $\tau^* = \tau_m^*$, for some $m \in \mathbb{N}$. $\square$

Theorem 5.34. *The relation τ^* is the smallest strongly regular relation on a finite hypergroup H such that H/τ^* is a solvable group.*

Proof. Suppose that ρ is a strongly regular relation on H such that $L = H/\rho$ is a solvable group of length n. We can see our claim by proving the fact that $\langle \bar{a} \mid a \in H^{(k)} \rangle \subseteq L^{(k)}$ for every $k \in \mathbb{N}$, which we prove it by induction on k. It holds for $k = 0$ easily. From $\bar{a} \in \langle \bar{a} \mid a \in H^{(k+1)} \rangle$ we have $a \in H^{(k+1)}$ so there exist $x, y \in H^{(k)}$ such that $x \circ y \cap a \circ y \circ x \neq \emptyset$ thus $\bar{x}\bar{y} = \bar{a}\bar{y}\bar{x}$. By hypothesis of induction, we conclude that $\bar{a} \in (L^{(k)})' = L^{(k+1)}$.

Suppose that $x\tau^* y$. Then, $x\tau_n^* y$ and so there exists $m \in \mathbb{N}$ such that

$$x \; \tau_{m,n} \; y \Leftrightarrow \exists (z_1, \ldots, z_m) \in H^m, \exists \sigma \in \mathbb{S}_m : \sigma(i) = i \text{ if } z_i \notin H^{(n)} \text{ such that}$$
$$x \in \prod_{i=1}^{m} z_i \quad \text{and} \quad y \in \prod_{i=1}^{m} z_{\sigma(i)},$$

and so we have

$$\bar{x} = \prod_{i=1}^{m} \bar{z}_i \quad \text{and} \quad \bar{y} = \prod_{i=1}^{m} \bar{z}_{\sigma(i)}.$$

Since $\bar{z}_i = 0$, for all $z_i \in H^{(n)}$ we conclude that $\bar{x} = \bar{y}$ and so $x\rho y$ as we need. $\qquad\square$

Example 5.5. If $G = \mathbb{S}_4$, then we can see that $\beta^* = \tau^* = \tau_3^*$. Therefore, $G/\tau^* \cong G$.

Definition 5.12. Let M be a non-empty subset of H. Then, we say that M is a τ-*part* of H if for every $k \in \mathbb{N}$ and $(z_1, \ldots, z_k) \in H^k$ and for every $\sigma \in \mathbb{S}_k$ such that $\sigma(i) = i$ if $z_i \notin \bigcup_{n \geq 1} H^{(n)}$, then

$$\prod_{i=1}^{k} z_i \cap M \neq \emptyset \Rightarrow \prod_{i=1}^{k} z_{\sigma(i)} \subseteq M.$$

Lemma 5.5. *Let M be a non-empty subset of a hypergroup H. Then, the following conditions are equivalent:*

(1) *M is a τ-part of H;*
(2) *$x \in M, x\tau y \Rightarrow y \in M$;*
(3) *$x \in M, x\tau^* y \Rightarrow y \in M$.*

Proof. (1$\Rightarrow$2): If $(x, y) \in H^2$ is a pair such that $x \in M$ and $x\tau y$, then there exists $(z_1, \ldots z_k) \in H^k$ such that $x \in \prod_{i=1}^{k} z_i \cap M$, $y \in \prod_{i=1}^{k} z_{\sigma(i)}$ and

$\sigma(i) = i$ if $z_i \notin \bigcup_{n \geq 1} H^n$. Since M is a τ-part of H, we have $\prod_{i=1}^{k} z_{\sigma(i)} \subseteq M$ and so $y \in M$.

$(2 \Rightarrow 3)$: Suppose that $(x, y) \in H^2$ is a pair such that $x \in M$ and $x\tau^*y$. Then, there exists $(z_1, \ldots, z_k) \in H^k$ such that $x = z_0\tau z_1\tau \ldots \tau z_{k-1}\tau z_k = y$. Now, by using (2) k times we obtain $y \in M$.

$(3 \Rightarrow 1)$: Suppose that $x \in \prod_{i=1}^{k} z_i \cap M$ and $\sigma \in \mathbb{S}_k$ such that $\sigma(i) = i$, if $z_i \notin \bigcup_{n \geq 1} H^n$. Let $y \in \prod_{i=1}^{k} z_{\sigma(i)}$. Since $x\tau y$, by (3) we have $y \in M$.

Consequently, $\prod_{i=1}^{k} z_{\sigma(i)} \subseteq M$ and so M is a τ-part. $\square$

Theorem 5.35. *The following conditions are equivalent:*

(1) *for every $a \in H$, $\tau(a)$ is a τ-part of H;*

(2) *τ is transitive.*

Proof. $(1 \Rightarrow 2)$: Suppose that $x\tau^*y$. Then, there exists $(z_1, \ldots, z_k) \in H^k$ such that $x = z_0\tau z_1\tau \ldots \tau z_{k-1}\tau z_k = y$. Since $\tau(z_i)$, for all $0 \leq i \leq k$, is a τ-part, we have $z_i \in \tau(z_{i-1})$, for all $1 \leq i \leq k$. Thus, $y \in \tau(x)$ which means that $x\tau y$.

$(2 \Rightarrow 1)$: Suppose that $x \in H$, $z \in \tau(x)$ and $z\tau y$. By transitivity of τ, we have $y \in \tau(x)$. Now, according to Lemma 5.5, $\tau(x)$ is a τ-part of H. $\square$

Definition 5.13. The intersection of all τ-parts which contain A is called τ-*closure*, of A in H and it will be denoted by $K(A)$.

In what follows, we determine the set $T(A)$, where A is a non-empty subset of H. We set

(1) $T_1(A) = A$ and

(2) $T_{n+1}(A) = \{x \in H \mid \exists (z_1, \ldots, z_k) \in H^k, x \in \prod_{i=1}^{k} z_i$ and $\exists \sigma \in \mathbb{S}_k$ such that

$\sigma(i) = i$, if, $z_i \notin \bigcup_{s \geq 1} H^s$, and $\prod_{i=1}^{k} z_{\sigma(i)} \cap T_n(A) \neq \emptyset\}$.

We denote $T(A) = \bigcup_{n \geq 1} T_n(A)$.

Theorem 5.36. *For any non-empty subset of H the following statements hold:*

(1) $T(A) = K(A)$;

(2) $K(A) = \bigcup_{a \in A} K(a)$.

Proof.   (1) It is enough to prove:

(a) $T(A)$ is a τ-part;

(b) if $A \subseteq B$ and B is a τ-part, then $T(A) \subseteq B$.

In order to prove (a), suppose that $\prod_{i=1}^{k} z_i \cap T(A) \neq \emptyset$ and $\sigma \in \mathbb{S}_k$ such that

$\sigma(i) = i$ if $z_i \notin \bigcup_{s \geq 1} H^s$ Therefore, there exists $n \in \mathbb{N}$ such that $\prod_{i=1}^{k} z_i \cap$

$T_n(A) \neq \emptyset$ whence it follows that $\prod_{i=1}^{k} z_{\sigma(i)} \subseteq T_{n+1}(A) \subseteq T(A)$. Now,

we prove (b) by induction on n. We have $T_1(A) = A \subseteq B$. Suppose

that $T_n(A) \subseteq B$. We prove that $T_{n+1}(A) \subseteq B$. If $z \in T_{n+1}(A)$, then

$z \in \prod_{i=1}^{k} z_i$ and there exists $\sigma \in \mathbb{S}_k$ such that $\sigma(i) = i$, if $z_i \notin \bigcup_{s \geq 1} H^s$ and also

$\prod_{i=1}^{k} z_{\sigma(i)} \cap T_n(A) \neq \emptyset$. Therefore, $\prod_{i=1}^{k} z_{\sigma(i)} \cap B \neq \emptyset$ and hence $z \in \prod_{i=1}^{k} z_i \subseteq B$.

(2) It is clear that for all $a \in A$, $K(a) \subseteq K(A)$. By part (1), we have

$K(A) = \bigcup_{n \geq 1} T_n(A)$ and $T_1(A) = A = \bigcup_{a \in A} \{a\}$. It is enough to prove that

$T_n(A) = \bigcup_{a \in A} T_n(a)$, for all $n \in \mathbb{N}$. We follow the assertion by induction on

n. Suppose that it true for n. We prove that $T_{n+1}(A) = \bigcup_{a \in A} T_{n+1}(a)$. If

$z \in T_{n+1}(A)$, then $z \in \prod_{i=1}^{k} z_i$ and there exists $\sigma \in S_k$ such that $\sigma(i) = i$, if

$z_i \notin \bigcup_{s \geq 1} H^s$ and also $\prod_{i=1}^{k} z_{\sigma(i)} \cap T_n(A) \neq \emptyset$. By the hypothesis of induction

$\prod_{i=1}^{k} z_{\sigma(i)} \cap T_n(a') \neq \emptyset$, for some $a' \in A$. Therefore $z \in T_{n+1}(a')$ and so

$T_{n+1}(A) \subseteq \bigcup_{a \in A} T_{n+1}(a)$. Hence, $K(A) = \bigcup_{a \in A} K(a)$. $\square$

Theorem 5.37. *The following relation is an equivalence relation on H,*

$$xTy \Leftrightarrow x \in T(y),$$

for every $(x, y) \in H^2$, where $T(y) = T(\{y\})$.

Proof.   It is easy to see that T is reflexive and transitive. We prove that T is symmetric. To this end, we check that:

(1) for all $n \geq 2$ and $x \in H$, $T_n(T_2(x)) = T_{n+1}(x)$;

(2) $x \in T_n(y)$ if and only if $y \in T_n(x)$.

We prove (1) by induction on n. Suppose that $z \in T_2(T_2(x))$. Then, $z \in \prod_{i=1}^{k} z_i$ and there exists $\sigma \in \mathbb{S}_k$ such that $\sigma(i) = i$, if $z_i \notin \bigcup_{s \geq 1} H^s$ and also $\prod_{i=1}^{k} z_{\sigma(i)} \cap T_2(x) \neq \emptyset$. Thus, $z \in T_3(x)$. If $z \in T_{n+1}(T_2(x))$, then $z \in \prod_{i=1}^{k} z_i$ and there exists $\sigma \in \mathbb{S}_k$ such that $\sigma(i) = i$, if $z_i \notin \bigcup_{s \geq 1} H^s$ and also $\prod_{i=1}^{k} z_{\sigma(i)} \cap T_n(T_2(x)) \neq \emptyset$. By hypothesis of induction, we have $\prod_{i=1}^{k} z_{\sigma(i)} \cap T_{n+1}(x) \neq \emptyset$ and so $z \in T_{n+2}(x)$. Now, we prove (2) by induction on n, too. It is clear that $x \in T_2(y)$ if and only if $y \in T_2(x)$. Suppose that $x \in T_n(y)$ if and only if $y \in T_n(x)$. Let $x \in T_{n+1}(y)$. Then, $x \in \prod_{i=1}^{k} z_i$ and there exists $\sigma \in \mathbb{S}_k$ such that $\sigma(i) = i$, if $z_i \notin \bigcup_{s \geq 1} H^s$ and also $\prod_{i=1}^{k} z_{\sigma(i)} \cap T_n(y) \neq \emptyset$. Suppose that $b \in \prod_{i=1}^{k} z_{\sigma(i)} \cap T_n(y)$. Then, we have $y \in T_n(b)$. From $x \in \prod_{i=1}^{k} z_i \cap T_1(x)$ and $b \in \prod_{i=1}^{k} z_{\sigma(i)}$ we conclude that $b \in T_2(x)$. Therefore, $y \in T_n(T_2(x)) = T_{n+1}(x)$. $\square$

Definition 5.14. Let H be a hypergroup and $\varphi : H \to H/\tau$ be the canonical projection. We denote by 1 the identity of the group H/τ. The set $\varphi^{-1}(1)$ is called the τ-heart of H and it is denoted by ω_τ.

Theorem 5.38. ω_τ is the smallest subhypergroup of H, which is also a τ-part of H.

Proof. First, we check that ω_τ is a subhypergroup of H. If $x, y \in \omega_\tau$ and $z \in x \circ y$, then $\tau(z) = \tau(x)\tau(y) = 1$, hence $z \in \omega_\tau$. On the other hand, there exists $u \in H$, such that $x \in u \circ y$ and so $1 = \tau(x) = \tau(u)\tau(y) = \tau(u)$. Therefore, $u \in \omega_\tau$. This means that $\omega_\tau y = \omega_\tau$, for all $y \in H$. Similarly, we obtain that $y\omega_\tau = \omega_\tau$ which follows that ω_τ is a subhypergroup of H. Now, we prove that $K(x) = \varphi^{-1}\varphi(x) = \omega_\tau x = x\omega_\tau$, for every $x \in H$. Indeed, we have $z \in \varphi^{-1}\varphi(x)$ if and only if $\varphi(z) = \varphi(x)$ which means that $z\tau x$. Therefore, $z \in T(x) = K(x)$. Hence, $K(x) = \varphi^{-1}\varphi(x)$. It is easy to see that $\varphi^{-1}\varphi(x) = \omega_\tau x = x\omega_\tau$. From the above, we obtain that if $x \in \omega_\tau$, then $K(x) = \omega_\tau$ which means that ω_τ is a τ-part of H. Moreover, if L is a subhypergroup of H which is also a τ-part, then $L = K(L) = \bigcup_{a \in L} K(a) = \bigcup_{a \in L} \omega_\tau a = \omega_\tau L$. Thus, $\omega_\tau \subseteq L$. $\square$

5.6 The relation δ^n and multisemi-direct hyperproducts of hypergroups

In this section we generalize the relations β and α, by introducing a new relation that we denote by δ^n. If H is a hypergroup, then δ^{n*} is the smallest equivalence relation, such that the quotient H/δ^{n*} is a group, for which for all $\bar{x} \in H/\delta^{n*}$, $\bar{x}^n = 1$, where 1 is the identity of H/δ^{n*}. In particular, using the equivalence relation δ^{n*}, we obtain p-elementary abelian groups and Burnside groups as quotient structures. Then, using the relation δ^{n*} and the notion of a generalized automorphism, we introduce and analyze multisemi-direct hyperproducts of hypergroups. The results of this section are contained in [107].

First, let us recall the next definition.

Definition 5.15. Let $(H, *)$ and $(H', *')$ be (semi)hypergroups. A function $f : H \to P^*(H')$ is called *a good multihomomorphism* if and only if

$$\forall (x, y) \in H^2, \ f(x * y) = f(x) *' f(y).$$

If $(H, *) = (H', *')$ and $\bigcup_{h \in H} f(h) = H$, then f is called *a generalized automorphism*. Moreover, we denote by $GAut(H)$ the set of all generalized automorphisms of $(H, *)$.

Example 5.6. Let $(H, *)$ be the commutative hypergroup given by the following table:

$*$	x	y	z
x	x	y	z
y	y	x	z
z	z	z	$\{x, y, z\}$

The map $f : H \to P^*(H)$ defined by $f(x) = x$, $f(y) = x$, $f(z) = \{x, y, z\}$ is a generalized automorphism which is not an automorphism.

Theorem 5.39. $(GAut(H), \circ)$ *is a monoid, where $\circ$ is defined as follows:*

$$\forall (f, g) \in GAut(H)^2, \ \forall h \in H, \ (f \circ g)(h) = \bigcup_{a \in g(h)} f(a).$$

Moreover, Aut(H) (i.e., the group of automorphism of H) is a subgroup of GAut(H).

Now, we introduce and analyze a new relation on a semihypergroup $(H, \cdot)$, which generalizes the relations β and α.

For any natural number n, we introduce a strongly equivalence relation δ^n on H as follows: $\delta^n = \bigcup_{m \geq 1} \delta^n_m$, where for every integer $m \geq 1$, δ^n_m is the relation defined as follows:

$$x \, \delta^n_m \, y \Leftrightarrow \exists (x_1, ..., x_m) \in H^m, \exists \tau \in S_m,$$

$$x \in \prod_{i=1}^{m} x_i, \; y \in \prod_{i=1}^{m} x_{\tau(i)}^{j_{\tau(i)}} \quad \text{or} \quad x \in \prod_{i=1}^{m} x_i^{j_i}, \; y \in \prod_{i=1}^{m} x_{\tau(i)},$$

where for all $i \in \{1, 2..., m\}$, $j_i \in \{1, n+1\}$ and $x_i^{j_i} = x_i \cdot x_i \cdot ... \cdot x_i (j_i \; times)$. Clearly, $\beta \subseteq \delta^n$ and $\alpha \subseteq \delta^n$. Moreover, $\alpha = \delta^n$ if for all $x \in H, x^{n+1} = x$.

Theorem 5.40. *δ^{n*} is a strongly regular equivalence relation on H.*

Proof. Notice that δ^{n*} is an equivalence relation. We check that:

$$x \delta^n y \Rightarrow xz \, \overline{\overline{\delta^n}} \, yz, \quad zx \, \overline{\overline{\delta^n}} \, zy. \tag{5.3}$$

for every $z \in H$. If $x\delta^n y$, then there exists $m \geq 1$ such that $x\delta^n_m y$ whence there exists $(x_1, ..., x_m) \in H^m$ and $\tau \in S_m$, such that

(1) $\quad x \in \prod_{i=1}^{m} x_i, \; y \in \prod_{i=1}^{m} x_{\tau(i)}^{j_{\tau(i)}}$ or

(2) $\quad x \in \prod_{i=1}^{m} x_i^{j_i}, \; y \in \prod_{i=1}^{m} x_{\tau(i)}$, where $j_i \in \{1, n+1\}$.

Set $z \in H$. If (1) holds, then we have $x \cdot z \subseteq \prod_{i=1}^{m} x_i \cdot z$ and $y \cdot z \subseteq \prod_{i=1}^{m} x_{\tau(i)}^{j_{\tau(i)}} \cdot z$. Set $x_{m+1} = z$ and define $\sigma \in S_{m+1}$ as follows:

$$\sigma(i) = \tau(i), \quad 1 \leq i \leq m \quad and \quad \sigma(m+1) = m+1.$$

Moreover, set $j_{m+1} = 1$. Thus $x \cdot z \subseteq \prod_{i=1}^{m+1} x_i$ and $y \cdot z \subseteq \prod_{i=1}^{n+1} x_{\sigma(i)}^{j_{\sigma(i)}}$. Similarly we have $z \cdot x \subseteq \prod_{i=1}^{m+1} x_i$ and $z \cdot y \subseteq \prod_{i=1}^{m+1} x_{\sigma'(i)}^{j_{\sigma'(i)}}$ for some $\sigma' \in S_{m+1}$. Hence, $xz \, \overline{\overline{\delta^n}} \, yz, \; zx \, \overline{\overline{\delta^n}} \, zy.$

Now, if $x\delta^{n*}y$, then there exists $m \in \mathbb{N}$ and $(u_0 = x, u_1, ..., u_m) \in H^{m+1}$ such that $x = u_0\delta^n u_1\delta^n...\delta^n u_{m-1}\delta^n u_m = y$, whence, by the above results, we obtain that

$$xz = u_0 z \, \overline{\overline{\delta^n}} \, u_1 z \, \overline{\overline{\delta^n}} \, u_2 z \, \overline{\overline{\delta^n}} \, ... \, \overline{\overline{\delta^n}} \, u_{m-1} z \, \overline{\overline{\delta^n}} \, u_m z.$$

Hence, for all $a \in xz = u_0 z$ and for all $b \in u_m z = yz$, we obtain that $a \ \delta^n \ t_1 \ \delta^n \ t_2 \ \delta^n ... \delta^n \ t_{m-1} \ \delta^n \ b$, for some $t_1 \in u_1 z, t_2 \in u_2 z, ..., t_{m-1} \in u_{m-1} z$. By the transitivity of δ^{n*}, it follows that $a \delta^{n*} b$ whence $xz \ \overline{\overline{\delta^{n*}}} \ yz$.

Similarly we can prove that $x \delta^{n*} y \Rightarrow zx \ \overline{\overline{\delta^{n*}}} \ zy$, hence δ^{n*} is a strongly regular equivalence relation on H. $\qquad \square$

Theorem 5.41. *The quotient H/δ^{n*} is an abelian group, for which $[\delta^{n*}(x)]^n = e$, the identity element of H/δ^{n*}.*

Moreover, δ^{n*} is the smallest strongly regular equivalence relation on H such that the quotient structure is an abelian group, for which $[\delta^{n*}(x)]^n = e$.

Proof. By Theorem 5.40, δ^{n*} is a strongly regular equivalence relation, hence the quotient structure H/δ^{n*} is a group with respect to the following operation

$$\delta^{n*}(x) \otimes \delta^{n*}(y) = \delta^{n*}(z), \quad \text{for all } z \in xy.$$

Since $\alpha \subseteq \delta^n$, we conclude that H/δ^{n*} is an abelian group. Moreover, for all $x \in H$, $\delta^n(x) = [\delta^n(x)]^{n+1}$ holds in the abelian group H/δ^n, thus $[\delta^{n*}(x)]^n = e$, the identity element of H/δ^{n*}.

Let us consider now R a strongly regular equivalence relation on H, such that H/R is an abelian group, such that $[\delta^{n*}(x)]^n = e$. Denote by $\pi : H \to H/R$ the canonical projection. Set $x \ \delta^n_m \ y$. Suppose that $x \in \prod_{i=1}^{m} z_i$ and $y \in \prod_{i=1}^{m} z_{\tau(i)}^{j_{\tau(i)}}$, for some $z_1, z_2, ..., z_m \in H$ and $\tau \in S_m$. It follows that

$$\pi(x) = \otimes_{i=1}^m \pi(z_i) = \otimes_{i=1}^m \pi(z_{\tau(i)}^{j_{\tau(i)}}) = \pi(y),$$

since $j_{\tau(i)} \in \{1, n+1\}$ and H/R is an abelian group.

This means that $\delta^n_m \subseteq R$ for all natural nonzero number m, whence $\delta^{n*} \subseteq R$, which means that δ^{n*} is the smallest strongly regular equivalence relation on H, such that $[\delta^{n*}(x)]^n = e$. $\qquad \square$

Corollary 5.7. *If $n = p$, a prime number, then the quotient H/δ^{p*} is an p-elementary abelian group, (i.e., $[\delta^{p*}(x)]^p = e$, for all $x \in H$).*

Moreover, δ^{p*} is the smallest strongly regular equivalence relation on H such that the quotient structure is an p-elementary abelian group.

Corollary 5.8. *If $|H/\delta^{p*}| = p^n$, where p is a prime natural number, then $Aut(H/\delta^{p*}) \cong GL(n,p)$, where $GL(n,p)$ is the generalized linear group.*

Example 5.7. Let $p = 2$ and $H = S_3 \times S_3$, where S_3 be the permutation group of order 3, i.e.,

$$S_3 = \{(1), (12), (13), (23), (123), (132)\}.$$

Then $H/\delta^{p*} \cong \mathbb{Z}_2 \times \mathbb{Z}_2$.

Proof. It is easy to see that:

$$\delta_1^p = \{< (x, y), (x', y') > \mid xx' \;\; and \;\; yy' \;\; \text{are even permutations}\}$$

therefore $\delta^p((x, y)) = \{(a, b) \mid xa, yb \;\; \text{are even}\}$, thus $H/\delta^{p*} \cong \mathbb{Z}_2 \times \mathbb{Z}_2$. $\square$

Finally, we consider the following relation on H, which is included in δ_m^n :

$$x \rho_m^n y \Leftrightarrow \exists (x_1, ..., x_m) \in H^m \text{ such that}$$

$$x \in \prod_{i=1}^m x_i, \; y \in \prod_{i=1}^m x_i^{j_i} \;\; \text{or} \;\; x \in \prod_{i=1}^m x_i^{j_i}, \; y \in \prod_{i=1}^m x_i,$$

where for all $i \in \{1, 2, ..., m\}$, $j_i \in \{1, n+1\}$ and $x_i^{j_i} = x_i \cdot x_i \cdot ... \cdot x_i$ (j_i times). Set $\rho^n = \cup_{m \in \mathbb{N}^*} \rho_m^n$ and let ρ^{n*} be the transitive closure of ρ^n. Clearly, $\beta \subseteq \rho^{n*}$.

Similarly to the proof of Theorem 5.40, we show that

Theorem 5.42. ρ^{n*} *is a strongly regular equivalence relation on* H.

Hence H/ρ^{n*} is a group. Moreover, for all $x \in H$, $\rho^{n*}(x) = [\rho^{n*}(x)]^{n+1}$ holds, which means that $[\rho^{n*}(x)]^n = e$, the identity of the group H/ρ^{n*}.

Let $(H, \cdot)$ be a hypergroup and $X = \{x_1, x_2, ..., x_r\}$ a finite subset of H. Set

$$G = \{\prod_{i=1}^m y_i \mid m \in \mathbb{N}^*, \forall i \in \{1, 2, ..., m\}, \exists j_i \in \{1, 2, ..., r\} : y_i = x_{j_i}\}.$$

We say that X *generates* H if for all $g \in H$ there exists $P \in G$ such that $g \in P$. Recall that a *Burnside group* of order n and r generators is a group $B(n, r)$ such that the following properties hold:

(1) $B(n, r) = [x_1, x_2, ..., x_r]$;
(2) $x^n = 1$, for all $x \in B(n, r)$.

Theorem 5.43. *If* X *generates* H, *then* H/ρ^{n*} *is a Burnside group.*

Proof. We check that $\rho^{n*}(X) = \{\rho^{n*}(x_1), \rho^{n*}(x_2), ..., \rho^{n*}(x_r)\}$ generates the group H/ρ^{n*}. Let $\rho^{n*}(h)$ be an arbitrary element of H/ρ^{n*}. Since X generates H, it follows that there exists $P = \prod_{k=1}^{t} x_k \in G$ such that for all k, $x_k \in X$ and $h \in P$. Hence $\rho^{n*}(h) = \prod_{k=1}^{t} \rho^{n*}(x_k)$. Moreover, for all $\rho^{n*}(h)$ of H/ρ^{n*}, $[\rho^{n*}(h)]^n = e$, hence H/ρ^{n*} is a Burnside group of order n and r generators. $\qquad \square$

By Theorem 5.41 it follows that:

Theorem 5.44. ρ^{n*} *is the smallest strongly regular equivalence relation on* H, *such that the quotient structure is a Burnside group of order* n *and* r *generators.*

Theorem 5.45.
$$\delta_m^n \subseteq \rho_m^n \circ \alpha_m \cap \alpha_m \circ \rho_m^n.$$

Proof. If $u \, \delta_m^n \, v$, then there exists $z_1, z_2, ..., z_m \in H$ and $\tau \in S_m$, such that $u \in \prod_{i=1}^{m} x_i$, $v \in \prod_{i=1}^{m} x_{\tau(i)}^{j_{\tau(i)}}$ or $u \in \prod_{i=1}^{m} x_i^{j_i}$, $v \in \prod_{i=1}^{m} x_{\tau(i)}$, where for all $i \in \{1, 2..., m\}$, $j_i \in \{1, n+1\}$.

Suppose that $u \in \prod_{i=1}^{m} x_i$, $v \in \prod_{i=1}^{m} x_{\tau(i)}^{j_{\tau(i)}}$. Then for all $t \in \prod_{i=1}^{m} x_{\tau(i)}$, we have $u \alpha_m t$ and $t \rho_m^n v$, whence $u \, [\rho_m^n \circ \alpha_m] \, v$. Hence $\delta_m^n \subseteq \rho_m^n \circ \alpha_m$.

On the other hand, from $u \in \prod_{i=1}^{m} x_i$, $v \in \prod_{i=1}^{m} x_{\tau(i)}^{j_{\tau(i)}}$, it follows that for all $s \in \prod_{i=1}^{m} x_i^{j_i}$ we have $u \rho_m^n s$, $s \alpha_m v$, whence $u \, [\alpha_m \circ \rho_m^n] \, v$. Hence $\delta_m^n \subseteq \alpha_m \circ \rho_m^n$. $\qquad \square$

From here, it follows that $\delta^n \subseteq \rho^n \circ \alpha \cap \alpha \circ \rho^n$.

5.7 Multisemi-direct hyperproduct of hypergroups

Definition 5.16. Let $(K, \cdot)$ and $(H, *)$ be two hypergroups. We consider the monoid $GAut(K)$ and the group H/δ^{n*}. Let:
$$\varphi: \{ H/\delta^{n*} \to GAut(K),$$
$$\delta^{n*}(x) \to \varphi_{\delta^{n*}(x)}$$
be a homomorphism. Then we define a hyperoperation in $K \times H$ as follows:
$$(x_1, y_1) \circ (x_2, y_2) = \{(x, y) \mid x \in x_1 \cdot \varphi_{\delta^{n*}(y_1)}(x_2), y \in y_1 * y_2\}.$$
We call it a *multisemi-direct hyperproduct* of hypergroups K and H through φ.

Theorem 5.46. $K \times H$ *equipped with the multisemi-hyperproduct is a hypergroup.*

Proof. Let (x_1, y_1), (x_2, y_2) and (x_3, y_3) are elements in $K \times H$. If $(s, t) \in [(x_1, y_1) \circ (x_2, y_2)] \circ (x_3, y_3)$ then we have:
$(s, t) \in (u, v) \circ (x_3, y_3)$ for some $(u, v) \in (x_1, y_1) \circ (x_2, y_2)$. Therefore $s \in u \cdot \varphi_{\delta^{n*}(v)}(x_3), t \in v * y_3$ and $u \in x_1 \cdot \varphi_{\delta^{n*}(y_1)}(x_2), v \in y_1 * y_2$. Thus, $s \in (x_1 \cdot \varphi_{\delta^{n*}(y_1)}(x_2)) \cdot \varphi_{\delta^{n*}(v)}(x_3)$ and $t \in (y_1 * y_2) * y_3$. By the associativity of $\cdot$ and $*$ we have: $s \in x_1 \cdot (\varphi_{\delta^{n*}(y_1)}(x_2) \cdot \varphi_{\delta^{n*}(v)}(x_3))$ and $t \in y_1 * (y_2 * y_3)$. Since $v \in y_1 * y_2$ we conclude that $s \in x_1 \cdot (\varphi_{\delta^{n*}(y_1)}(x_2) \cdot \varphi_{\delta^{n*}(y_1)\delta^{n*}(y_2)}(x_3))$. On the other hand, if $(s, t) \in (x_1, y_1) \circ [(x_2, y_2) \circ (x_3, y_3)]$, then we have: $(s', t') \in (x_1, y_1) \circ (u', v')$ for some $(u', v') \in (x_2, y_2) \circ (x_3, y_3)$. Therefore $s' \in x_1 \cdot \varphi_{\delta^{n*}(y_1)}(u'), t' \in y_1 * v'$ and $u' \in x_2 \cdot \varphi_{\delta^{n*}(y_2)}(x_3), v' \in y_2 * y_3$. Thus, $s' \in x_1 \cdot \varphi_{\delta^{n*}(y_1)}(x_2 \cdot \varphi_{\delta^{n*}(y_2)}(x_3)) = x_1 \cdot (\varphi_{\delta^{n*}(y_1)}(x_2) \cdot \varphi_{\delta^{n*}(y_1)\delta^{n*}(y_2)}(x_3))$ and $t' \in y_1 * (y_2 * y_3)$. By the above results, we conclude the associativity of $\circ$.

Now let $(a, x) \in K^2$ and $(b, y) \in H^2$. Since K and H are hypergroups, there exists $(u, w) \in K \times H$ such that $a \in x \cdot u$ and $b \in y * w$. Since $u \in K = \varphi_{\delta^{n*}(y)}(K)$, we conclude that there exists $t \in H$ such that $u \in \varphi_{\delta^{n*}(y)}(t)$ and so we have $(a, b) \in (x, y) \circ (t, w)$. Similarly there exists $(t', w') \in K \times H$ such that $(a, b) \in (t', w') \circ (x, y)$. $\square$

From now on, $K \times_\varphi H$ denotes the multisemi-direct hyperproduct hypergroup of K and H through φ.

Example 5.8. Let $(H, *)$ be the hypergroup, defined by the following table:

*	x	y	z
x	x	y	z
y	y	x	z
z	z	z	$\{x, y\}$

If we consider the homomorphism $\varphi : Z_2 \to GAut(H)$ by $\varphi([0]) = \varphi([1]) = \sigma$, where $\sigma(x) = \{x, y\} = \sigma(y)$, $\sigma(z) = H$, then the multisemi-direct hyperproduct through φ hypergroup $(H \times Z_2, \circ)$ is the following one:

$\circ$	e	a	b	c	d	f
e	$\{e, a\}$	$\{e, a\}$	$\{e, a, b\}$	$\{c, d\}$	$\{c, d\}$	$\{c, d, f\}$
a	$\{e, a\}$	$\{e, a\}$	$\{e, a, b\}$	$\{c, d\}$	$\{c, d\}$	$\{c, d, f\}$
b	b	b	$\{e, a, b\}$	f	f	$\{c, d, f\}$
c	$\{c, d\}$	$\{c, d\}$	$\{c, d, f\}$	$\{e, a\}$	$\{e, a\}$	$\{e, a, b\}$
d	$\{c, d\}$	$\{c, d\}$	$\{c, d, f\}$	$\{e, a\}$	$\{e, a\}$	$\{e, a, b\}$
f	f	f	$\{c, d, f\}$	b	b	$\{e, a, b\}$

where $e = (x, [0]), a = (y, [0]), b = (z, [0]), c = (x, [1]), d = (y, [1]), f = (z, [1])$.

Theorem 5.47. *Let $(K, \cdot)$ and $(H, *)$ be two hypergroups and n is a natural number. Now suppose that δ_K^*, δ_H^* are the δ^{n*}-strongly regular equivalence relations on K and H respectively. If*

$$\varphi: \ H/\delta_H^* \to GAut(K),$$

$$\delta_H^*(x) \to \varphi_{\delta_H^*(x)}$$

is a homomorphism which satisfies the condition

$$\forall a \in K, \quad \exists b \in \varphi_{\delta_H^*(x)}(a) \text{ such that } \varphi_{\delta_H^*(x)}(a) = b^s \text{ for some } s \in N,$$

then

$$\forall (x, y) \in K \times H, \quad \varphi_{\delta_H^*(y)}(\delta_K^*(x)) \subset \delta_K^*(\varphi_{\delta_H^*(y)}(x)).$$

Proof.   The proof consists into two steps.

(1) Let $z \in \varphi_{\delta_H^*(y)}(\delta_K(x))$, so there exists $t \in \delta_K(x)$ such that $z \in \varphi_{\delta_H^*(y)}(t)$. Consequently, there exist $(x_1, ..., x_m) \in H^m$ and $\tau \in S_m$, such that $[x \in \prod_{i=1}^m x_i, \ t \in \prod_{i=1}^m x_{\tau(i)}^{j_{\tau(i)}}]$ or $[x \in \prod_{i=1}^m x_i^{j_i}, \ t \in \prod_{i=1}^m x_{\tau(i)}]$, where $j_i \in \{1, n+1\}$. Now suppose that $a_i^{t_i} = \varphi_{\delta_H^*(y)}(x_i)$ therefore $[\varphi_{\delta_H^*(y)}(x) \subset \prod_{i=1}^n a_i^{t_i}, \ \varphi_{\delta_H^*(y)}(t) \subset \prod_{i=1}^n (a_{\tau(i)}^{j_{\tau(i)}})^{t_{\tau(i)}}]$ or $[\varphi_{\delta_H^*(y)}(x) \subset \prod_{i=1}^n (a_i^{t_i})^{j_i}, \ \varphi_{\delta_H^*(y)}(t) \subset \prod_{i=1}^n (a_{\tau(i)})^{t_{\tau(i)}}]$.

Thus $z \in \delta_K^*(\varphi_{\delta_H^*(y)}(x))$.

(2) Let $z \in \varphi_{\delta_H^*(y)}(\delta_K^*(x))$, so $t \in \delta_K^*(x)$ exists, such that $z \in \varphi_{\delta_H^*(y)}(t)$ thus we have $(t = t_1, t_2, ..., t_m = x) \in K^m$ such that $t_i \in \delta_K(t_{i+1})$, $0 < i < m$, hence $\varphi_{\delta_H^*(y)}(t_i) \subset \varphi_{\delta_H^*(y)}(\delta_K(t_{i+1}))$. By (i), we conclude that $\varphi_{\delta_H^*(y)}(t_i) \subset \delta_K^*(\varphi_{\delta_H^*(y)}(t_{i+1}))$ and so $z \in \delta_K^*(\varphi_{\delta_H^*(y)}(x))$ follows. $\square$

Theorem 5.48. *Let $(K, \cdot)$ and $(H, *)$ be two finite hypergroups and n be a natural number. Let δ_K^*, δ_H^* be the δ^{n*}- strongly regular equivalence relation on K and H respectively. If*

$$\varphi: \ H/\delta_H^* \to GAut(K),$$

$$\delta_H^*(x) \to \varphi_{\delta_H^*(x)}$$

is a homomorphism, then $\varphi_{\delta_H^*(e)} \in Aut(K)$ if and only if $\varphi_{\delta_H^*(y)} \in Aut(K)$ for all $y \in H$, where $\delta_H^*(e)$ is the identity element of H/δ_H^*.

Proof. Let $\varphi_{\delta_H^*(e)} \in Aut(K)$ and $y \in H$. Since $\bigcup_{x \in K} \varphi_{\delta_H^*(y)}(x) = K$ we have $\bigcup_{x \in K} \varphi_{\delta_H^*(y)^m}(x) = K$ for all $m \geq 1$. Now if $b \in K$, then there exists $a \in K$ such that $b \in \varphi_{\delta_H^*(y)^{n-1}}(a)$ and therefore $\varphi_{\delta_H^*(y)}(b) \in \varphi_{\delta_H^*(y)^n}(a) = \varphi_{\delta_H^*(e)}(a) = a$ hence $\varphi_{\delta_H^*(y)}(b) = a$, that is $\varphi_{\delta_H^*(y)} \in Aut(K)$. $\square$

Corollary 5.9. *Let $(K, \cdot)$ and $(H, *)$ be two finite hypergroups with scalar identities e and e' respectively. In the hypothesis of previous theorem, (e, e') is the scalar identity of $(K \times H, \circ)$ if and only if $Im(\varphi)$ is a subgroup of $Aut(K)$.*

Theorem 5.49. *Let $(K, \cdot)$ and $(H, *)$ be two hypergroups, n be a natural number and*

$$\varphi', \varphi : H/\delta_H^* \to GAut(K)$$

be two homomorphisms. If there exists $\alpha \in Aut(K)$ such that

$$\varphi'_{\delta_H^*(y)} = \alpha \varphi_{\delta_H^*(y)} \alpha^{-1}$$

for all $y \in H$, then $K \times_\varphi H \cong K \times_{\varphi'} H$.

Proof. Consider the map

$$\psi : K \times_\varphi H \to K \times_{\varphi'} H, \quad (a, b) \mapsto (\alpha(a), b).$$

Then

$$
\begin{aligned}
\psi(a,b)\psi(c,d) &= (\alpha(a), b)(\alpha(c), d) \\
&= \{(x, y) | x \in \alpha(a)\varphi'_{\delta_H^*(b)}(\alpha(c)), y \in b * d\} \\
&= \{(x, y) | x \in \alpha(a)\alpha(\varphi_{\delta_H^*(b)}(c)), y \in b * d\} \\
&= \{(x, y) | x \in \alpha(a\varphi_{\delta_H^*(b)}(c)), y \in b * d\}.
\end{aligned}
$$

On the other hand

$$
\begin{aligned}
\psi((a,b)(c,d)) &= \{\psi(x, y) | x \in a\varphi_{\delta_H^*(b)}(c), y \in b * d\} \\
&= \{(\alpha(x), y) | x \in a\varphi_{\delta_H^*(b)}(c), y \in b * d\}.
\end{aligned}
$$

Therefore ψ is a good homomorphism. It is easy to see that ψ is an isomorphism. $\square$

Theorem 5.50. *Let $(K, \cdot)$ and $(H, *)$ be two hypergroups, n be a natural number and*

$$\varphi', \varphi : H/\delta_H^* \to GAut(K)$$

be two homomorphisms. If there exists $\alpha \in Aut(H)$ such that

$$\varphi' \circ \bar{\alpha} = \varphi,$$

where $\bar{\alpha}(\delta_H^*(x)) = \delta_H^*(\alpha(x))$ for all $x \in H$, then $K \times_\varphi H \cong K \times_{\varphi'} H$.

Proof.   The map $(a, b) \mapsto (a, \alpha(b))$ is an isomorphism $K \times_\varphi H \to K \times_{\varphi'} H$.
$\square$

Definition 5.17. Let $(H, \cdot)$ be a polygroup, n is a natural number and K be a normal subhypergroup and a complete part of H. Then an n-subgroup Q of H is an *n-complement* of K in H if $K \cap Q \subset \omega_H$ and $H = K \cdot Q$.

Example 5.9. Consider the polygroup $(H, \cdot)$ defined as follows:

$\cdot$	e	a	b	c
e	e	a	b	c
a	a	e	c	b
b	b	c	$\{e, b, c\}$	$\{a, b, c\}$
c	c	b	$\{a, b, c\}$	$\{e, b, c\}$

Then the subgroup $Q = \{e, a\}$ is a 2-complement of H in H.

Theorem 5.51. *If Q is an n-complement of K in a polygroup $(H, \cdot)$, then the following statements hold:*

(1) *If $h \in a \cdot x \cap b \cdot y$, then $x \cdot \omega_H = y \cdot \omega_H$ and $a \cdot \omega_H = b \cdot \omega_H$, where $(a, b) \in K^2$ and $(x, y) \in Q^2$;*

(2) *There is a homomorphism*

$$\Theta : Q \to GAut(K), \quad \Theta(x) = \Theta_x,$$

defined by $\Theta_x(a) = x \cdot a \cdot x^{-1} \cdot \omega_H$ for all $x \in Q$ and $a \in K$. Moreover, for all $x, y \in Q$ and all $a \in K$, $a = \Theta_e(a)$ and $\Theta_x(\Theta_y(a)) = \Theta_{x \cdot y}(a)$, where e is the scalar identity element of H;

(3) *The map $(a, x) \mapsto a \cdot x \cdot \omega_H$ is a generalized automorphism $K \times_\Theta Q \to H$. Moreover, if H is a group then $K \times_\Theta Q \cong H$.*

Proof.   (1) Suppose that $h \in a \cdot x \cap b \cdot y$, where $(a, b) \in K^2$ and $(x, y) \in Q^2$. Then $y \in b^{-1} \cdot h \subset b^{-1} \cdot a \cdot x$, hence there exists $t \in K$, such that $y \in t \cdot x$ consequently $t \in y \cdot x^{-1}$. So $t \in K \cap Q \subset \omega_H$ and therefore $y \cdot x^{-1} \subset \omega_H$. Thus $x \cdot \omega_H = y \cdot \omega_H$. Similarly we have $a \cdot \omega_H = b \cdot \omega_H$.

(2) Let $(a, b) \in K^2$, $x \in Q$. Then $\Theta_x(a \cdot b) = \bigcup_{t \in a \cdot b} \Theta_x(t) = \bigcup_{t \in a \cdot b} x \cdot t \cdot$
$x^{-1} \cdot \omega_H = x \cdot a \cdot b \cdot x^{-1} \cdot \omega_H = x \cdot a \cdot x^{-1} \cdot x \cdot b \cdot x^{-1} \cdot \omega_H = \Theta_x(a) \cdot \Theta_x(b)$.
Now let $(x, y) \in Q^2$ and $a \in K$. Then $\Theta_{x.y}(a) = x \cdot y \cdot a \cdot y^{-1} \cdot x^{-1} \cdot \omega_H$.

On the other hand

$$
\begin{aligned}
\Theta_x(\Theta_y(a)) &= \Theta_x(y \cdot a \cdot y^{-1} \cdot \omega_H) \\
&= \bigcup_{u \in y \cdot a \cdot y^{-1} \cdot \omega_H} \Theta_x(u) \\
&= \bigcup_{u \in y \cdot a \cdot y^{-1} \cdot \omega_H} x \cdot u \cdot x^{-1} \cdot \omega_H \\
&= x \cdot (y \cdot a \cdot y^{-1} \cdot \omega_H) \cdot x^{-1} \cdot \omega_H \\
&= x \cdot y \cdot a \cdot y^{-1} \cdot x^{-1} \cdot \omega_H.
\end{aligned}
$$

Thus $\Theta_{x.y}(a) = \Theta_x(\Theta_y(a))$.

(3) The proof is easy. $\qquad\qquad\qquad\qquad\qquad\qquad\qquad\qquad$ $\square$

Chapter 6

More about the Corresponding Quotient Structures

In this chapter we introduce and analyze the relation ζ_e, for which the transitive closure leads to a monoid as a quotient structure. We analyze when the relation ζ_e is transitive and we give some necessary and sufficient conditions. Using the relation ζ_e^*, we characterize the derived ζ_e^*-strong semihypergroup. Then we analyze a new relation, denoted by τ_e^* which extends ζ_e^* and for which the quotient structure is a commutative monoid. Finally, we generalize the notion of complete parts and α-parts by introducing the notion of $\Re$-parts. The results of this chapter are contained in [134; 133].

6.1 The relation ζ_e

Let $(S, \cdot)$ be a semihypergroup and e be a whichever element of S.

Definition 6.1. We say that a pair $(\prod_{i=1}^{n} x_i, \prod_{i=1}^{m} y_i)$ *satisfies the condition* $\mathfrak{P}$ if $m > n$ and there exist k and l in $\mathbb{N}$, such that $1 \leqslant k \leqslant n$, $k \leqslant l \leqslant m$, $m = n + (l - k + 1)$ and

$$
y_t = \begin{cases}
x_t, & \text{if } 1 \leqslant t < k; \\
e, & \text{if } k \leqslant t \leqslant l; \\
x_{t+k-l-1}, & \text{if } l < t \leqslant m.
\end{cases} \tag{6.1}
$$

We denote:

$$
\Re^{\mathfrak{P},e} := \Big\{ (\prod_{i=1}^{n} x_i, \prod_{i=1}^{m} y_i) \mid (\prod_{i=1}^{n} x_i, \prod_{i=1}^{m} y_i) \text{ satisfies the condition } \mathfrak{P} \Big\},
$$

$$\Re_u \mathfrak{P}e := \Re^{\mathfrak{P},e} \cup (\Re^{\mathfrak{P},e})^{-1},$$

$$\mathfrak{I} := \left\{ \left(\prod_{i=1}^{n} x_i, \prod_{i=1}^{n} x_i \right) \mid \forall i \in \{1, 2, ..., n\}, x_i \in S, \ n \in \mathbb{N}, \ n \geqslant 1 \right\}$$

and $\Re_u e := \Re_u \mathfrak{P}e \cup \mathfrak{I}$.

Definition 6.2. We define the relation ζ_e on $(S, \cdot)$ as follows:

$$x \ \zeta_e \ y \Leftrightarrow \exists (A, B) \in \Re_u e, \ x \in A, \ y \in B.$$

The relation ζ_e is reflexive and symmetric and $\beta \subseteq \zeta_e$. Let ζ_e^* be the transitive closure of ζ_e.

Lemma 6.1. ζ_e^* *is a strongly regular relation.*

Proof. We can see that ζ_e^* is an equivalence relation. In order to prove that it is strongly regular, we have to show first that

$$x \ \zeta_e \ y \Rightarrow xz \ \overset{=}{\zeta_e} \ yz, \quad zx \ \overset{=}{\zeta_e} \ zy. \tag{6.2}$$

for all $z \in S$. Since $x \ \zeta_e \ y$, it follows that there exists $(A, B) \in \Re_u e$ such that $x \in A$ and $y \in B$. We distinguish the following situations:

Case 1: There exists a pair $\left(\prod_{i=1}^{n} x_i, \prod_{i=1}^{m} y_i \right)$ which satisfies the condition $\mathfrak{P}$ and $A = \prod_{i=1}^{n} x_i$ and $B = \prod_{i=1}^{m} y_i$. We obtain

$$xz \subseteq \left(\prod_{i=1}^{n} x_i \right) z \quad \text{and} \quad yz \subseteq \left(\prod_{i=1}^{m} y_i \right) z.$$

For all $1 \leqslant i \leqslant n$ set $x_i' = x_i$ and $x_{n+1}' = z$ and for all $1 \leqslant t \leqslant m$ set $y_t' = y_t$ and $y_{m+1}' = z$. Thus

$$xz \subseteq \prod_{i=1}^{n+1} x_i' \quad \text{and} \quad yz \subseteq \prod_{t=1}^{m+1} y_t'.$$

It is easy to see that the pair $\left(\prod_{i=1}^{n+1} x_i', \prod_{t=1}^{m+1} y_t' \right)$ satisfies the condition $\mathfrak{P}$ and hence this pair belongs to $\Re_u \mathfrak{P}e \subseteq \Re_u e$. Therefore for all $v \in xz$ and for all $u \in yz$, we have $v \in xz \subseteq Az$ and $u \in yz \subseteq Bz$, so $v \ \zeta_e \ u$, because $(Az, Bz) \in \Re_u e$. Thus $xz \ \overset{=}{\zeta_e} \ yz$.

Case 2: If there exists a pair $\left(\prod_{i=1}^{n} x_i, \prod_{i=1}^{m} y_i \right)$ which satisfies the condition $\mathfrak{P}$ and $x \in \prod_{i=1}^{m} y_i = B, y \in \prod_{i=1}^{n} x_i = A$, then according to Case 1, $(Az, Bz) \in \Re_u e$ and so $(Bz, Az) \in \Re_u e$. Thus $xz \ \overset{=}{\zeta_e} \ yz$.

Case 3: There exists $n \geqslant 1$, such that $x, y \in A = B = \prod_{i=1}^{n} x_i$. Thus $xz \cup yz \subseteq (\prod_{i=1}^{n} x_i)z$. It follows that $xz \overset{=}{\zeta_e} yz$.

In the same way, we can show that $x \zeta_e y \Rightarrow zx \overset{=}{\zeta_e} zy$.

Moreover, if $x \zeta_e^* y$, then there exist $m \in \mathbb{N}$ and $(u_0 = x, u_1, ..., u_m = y) \in S^m$ such that $x = u_0 \zeta_e u_1 \zeta_e \cdots \zeta_e u_{m-1} \zeta_e u_m = y$, whence, by (6.4), we obtain

$$xz = u_0 z \overset{=}{\zeta_e} u_1 z \overset{=}{\zeta_e} u_2 z \overset{=}{\zeta_e} \cdots \overset{=}{\zeta_e} u_{m-1} z \overset{=}{\zeta_e} u_m z = yz.$$

Hence, for all $v \in xz = u_0 z$ and for all $u \in u_m z = yz$, taking $z_1 \in u_1 z, z_2 \in u_2 z, ..., z_{m-1} \in u_{m-1} z$, we have $v \zeta_e z_1 \zeta_e z_2 \zeta_e \cdots \zeta_e z_{m-1} \zeta_e u$, and so $v \zeta_e^* u$. Therefore

$$x \zeta_e^* y \Rightarrow xz \overset{=}{\zeta_e^*} yz.$$

Similarly we can prove that $x \zeta_e^* y \Rightarrow zx \overset{=}{\zeta_e^*} zy$, hence ζ_e^* is strongly regular. $\square$

Theorem 6.1. *The quotient* S/ζ_e^* *is a monoid with the identity* $\zeta_e^* (e)$.

Proof. By Lemma 6.1, ζ_e^* is a strongly regular equivalence, so the quotient S/ζ_e^* is a semigroup under the following operation:

$$\zeta_e^* (x) \otimes \zeta_e^* (y) = \zeta_e^* (z), \quad \text{for all } z \in xy.$$

Moreover, for all $x \in S$ we have $(x, ex) \in \Re_u \mathfrak{P}e$ and $(x, xe) \in \Re_u \mathfrak{P}e$. So for all $y \in xe \cup ex$, $x \zeta_e y$ and hence $\zeta_e^* (x) = \zeta_e^* (y)$. Therefore $\zeta_e^* (x) \otimes \zeta_e^* (e) = \zeta_e^* (x) = \zeta_e^* (e) \otimes \zeta_e^* (x)$ and the proof is complete. $\square$

Example 6.1. Consider the semihypergroup $(S, *)$, where $*$ is defined on S as follows:

$*$	e	a	b	c
e	a	$\{e, a\}$	b	c
a	$\{e, a\}$	$\{e, a\}$	b	c
b	b	b	b	b
c	c	c	b	c

We can see that the monoid S/ζ_e^* is as follows:

$\star$	$\zeta_e^* (e)$	$\zeta_e^* (b)$	$\zeta_e^* (c)$
$\zeta_e^* (e)$	$\zeta_e^* (e)$	$\zeta_e^* (b)$	$\zeta_e^* (c)$
$\zeta_e^* (b)$	$\zeta_e^* (b)$	$\zeta_e^* (b)$	$\zeta_e^* (b)$
$\zeta_e^* (c)$	$\zeta_e^* (c)$	$\zeta_e^* (b)$	$\zeta_e^* (c)$

Note that $\zeta_e^* (e) = \zeta_e^* (a)$.

Theorem 6.2. *The relation ζ_e^* is the smallest equivalence relation such that the quotient S/ζ_e^* is a monoid with the identity $\zeta_e^*(e)$.*

Proof. Let R be a strongly regular equivalence such that S/R is a monoid with the identity $R(e)$. Let $\phi_e : S \to S/R$ be the canonical projection, so ϕ_e is a good homomorphism.

Let $x \, \zeta_e \, y$. Suppose that there exists a pair $(\prod_{i=1}^{n} x_i, \prod_{i=1}^{m} y_i)$ which satisfies the condition $\mathfrak{P}$ and $x \in A = \prod_{i=1}^{n} x_i$, $y \in B = \prod_{i=1}^{m} y_i$. Therefore $\phi_e(x) = \phi_e(z)$ and $\phi_e(y) = \phi_e(r)$ for all $z \in \prod_{i=1}^{n} x_i$ and $r \in \prod_{i=1}^{m} y_i$. Since $\phi_e(e) = R(e)$ is the identity element of the monoid S/R, by (6.3) we obtain $\bigotimes_{i=1}^{n} \phi_e(x_i) = \bigotimes_{t=1}^{m} \phi_e(y_t)$ and hence $\phi_e(z) = \phi_e(r)$. Therefore $\phi_e(x) = \phi_e(y)$ and so xRy.

Similarly, if (A, B) satisfies the condition $\mathfrak{P}$ and $x \in B, y \in A$, we obtain xRy. If $x, y \in \prod_{i=1}^{n} x_i$ then $\phi_e(x) = \phi_e(y)$, hence xRy. Thus $x \, \zeta_e \, y$ implies that xRy.

Finally, let $x \, \zeta_e^* \, y$. Since R is transitively closed, we obtain
$$x \in \zeta_e^*(y) \Rightarrow x \in R(y).$$
Therefore $\zeta_e^* \subseteq R$.   □

6.2   Transitivity condition of ζ_e

In this section we determine some necessary and sufficient conditions such that the relation ζ_e is transitive.

Let M be a non-empty subset of a semihypergroup $(S, \cdot)$.

Definition 6.3. We say that M is a ζ_e^*-*part* of S if $\prod_{i=1}^{n} x_i \cap M \neq \emptyset$ implies that

(P1) $\prod_{i=1}^{m} y_i \subseteq M$ for all $\prod_{i=1}^{m} y_i$ such that one of the pairs $(\prod_{i=1}^{n} x_i, \prod_{i=1}^{m} y_i)$, $(\prod_{i=1}^{m} y_i, \prod_{i=1}^{n} x_i)$ satisfies the condition $\mathfrak{P}$ and

(P2) $\prod_{i=1}^{n} x_i \subseteq M$.

Theorem 6.3. *The following conditions are equivalent:*

(1) *M is a ζ_e^*-part;*

(2) $x \in M, x \; \zeta_e \; y \Rightarrow y \in M;$

(3) $x \in M, x \; \zeta_e^* \; y \Rightarrow y \in M.$

Proof. $(1 \Rightarrow 2)$: Let $(x, y) \in S^2$ be such that $x \in M$ and $x \; \zeta_e \; y$. Suppose that there exists a pair $(\prod_{i=1}^{n} x_i, \prod_{i=1}^{m} y_i)$, which satisfies the condition $\mathfrak{P}$ and such that $x \in A = \prod_{i=1}^{n} x_i$, $y \in B = \prod_{i=1}^{m} y_i$. Since $x \in M$, $\prod_{i=1}^{n} x_i \cap M \neq \emptyset$ and so we obtain $\prod_{i=1}^{m} y_i \subseteq M$ by Definition 6.3 (P1). Thus $y \in M$.

Similarly, if $(A = \prod_{i=1}^{n} x_i, B = \prod_{i=1}^{m} y_i)$ satisfies the condition $\mathfrak{P}$ and $x \in B, y \in A$, then $\prod_{i=1}^{m} y_i \cap M \neq \emptyset$ and so we obtain $\prod_{i=1}^{m} x_i \subseteq M$ by Definition 6.3 (P1). Thus $y \in M$.

On the other hand, if $x, y \in \prod_{i=1}^{n} x_i$, then $y \in \prod_{i=1}^{m} y_i \subseteq M$, by Definition 6.3 (P2).

$(2 \Rightarrow 3)$: Let $(x, y) \in S^2$ be such that $x \in M$ and $x \; \zeta_e^* \; y$. So there exist $m \in \mathbb{N}$ and $(w_0 = x, w_1, ..., w_{m-1}, w_m = y) \in S^m$ such that $x = w_0 \; \zeta_e \; w_1 \; \zeta_e \; w_2 \cdots \; \zeta_e \; w_{m-1} \; \zeta_e \; w_m = y$. Since $x \in M$, applying (2) m times, we obtain $y \in M$.

$(3 \Rightarrow 1)$: Let $\prod_{i=1}^{n} x_i \cap M \neq \emptyset$ and $\prod_{i=1}^{m} y_i$ be such that $(\prod_{i=1}^{n} x_i, \prod_{i=1}^{m} y_i)$ or $(\prod_{i=1}^{m} y_i, \prod_{i=1}^{n} x_i)$ satisfies the condition $\mathfrak{P}$. Suppose that $(\prod_{i=1}^{n} x_i, \prod_{i=1}^{m} y_i)$ satisfies the condition $\mathfrak{P}$, the other case is similar. Since $\prod_{i=1}^{n} x_i \cap M \neq \emptyset$, it follows that there exists $x \in \prod_{i=1}^{n} x_i \cap M$. Let y be an arbitrary element of $\prod_{i=1}^{m} y_i$. From $(\prod_{i=1}^{n} x_i, \prod_{i=1}^{m} y_i) \in \Re_u \mathfrak{P}e$ it follows that $x \; \zeta_e^* \; y$. Thus by (3) we have $y \in M$ and so $\prod_{i=1}^{m} y_i \subseteq M$. Finally, let $x \in M \cap \prod_{i=1}^{n} x_i$. Since for all $y \in \prod_{i=1}^{n} x_i, x \; \zeta_e^* \; y$, from (3) it follows that $y \in M$, so $\prod_{i=1}^{n} x_i \subseteq M$. $\square$

Before proving the next theorem, we introduce the following notations.

Notation 6.1. Let x be an arbitrary element of a semihypergroup $(S, \cdot)$. For all $n \geqslant 1$, set:

(N1) $P_{\wp_n}(x) = \bigcup \left\{ \prod_{i=1}^{m} y_i \mid x \in \prod_{i=1}^{n} x_i, \left(\prod_{i=1}^{n} x_i, \prod_{i=1}^{m} y_i \right) \text{ or } \left(\prod_{i=1}^{m} y_i, \prod_{i=1}^{n} x_i \right) \right.$

satisfies the condition $\mathfrak{P}$ $\Big\}$;

(N2) $P_{\Im_n}(x) = \bigcup \left\{ \prod_{i=1}^{n} x_i \mid x \in \prod_{i=1}^{n} x_i \right\}$;

(N3) $P(x) = \bigcup_{n \geqslant 1} (P_{\wp_n}(x) \cup P_{\Im_n}(x))$.

From the above notations and definitions, we obtain:

Theorem 6.4. *For all $x \in S$, $P(x) = \{y \in S \mid x \ \zeta_e \ y\}$.*

Proof. Let $x \in S$ and $y \in P(x)$. So there exists $n \geqslant 1$, such that $y \in P_{\wp_n}(x) \cup P_{\Im_n}(x)$. If $y \in P_{\wp_n}(x)$, then there exists $\prod_{i=1}^{m} y_i$, such that one of the pairs $\left(\prod_{i=1}^{n} x_i, \prod_{i=1}^{m} y_i \right)$, $\left(\prod_{i=1}^{m} y_i, \prod_{i=1}^{n} x_i \right)$ satisfies the condition $\mathfrak{P}$ and $x \in \prod_{i=1}^{n} x_i$ and $y \in \prod_{i=1}^{m} y_i$. Therefore $\left(\prod_{i=1}^{n} x_i, \prod_{i=1}^{m} y_i \right) \in \Re_u \mathfrak{P} e$ and hence $x \ \zeta_e \ y$. If $y \in P_{\Im_n}(x)$, then the proof is immediate. Thus $P(x) \subseteq \{y \in S \mid x \ \zeta_e \ y\}$. The proof of the reverse of the inclusion is obvious. $\square$

Lemma 6.2. *Let $(S, \cdot)$ be a semihypergroup and M be a ζ_e^*-part of S. If $x \in M$, then $P(x) \subseteq M$.*

Proof. If $y \in P(x)$, then $x \ \zeta_e \ y$. First, suppose that there exists a pair $\left(\prod_{i=1}^{n} x_i, \prod_{i=1}^{m} y_i \right)$ which satisfies the condition $\mathfrak{P}$ and such that $x \in A = \prod_{i=1}^{n} x_i$, $y \in B = \prod_{i=1}^{m} y_i$. Since $x \in \prod_{i=1}^{n} x_i \cap M$ and M is a ζ_e^*-part, it follows that $\prod_{i=1}^{m} y_i \subseteq M$, by Definition 6.3 (P1) and so $y \in M$. Similarly, if $\left(\prod_{i=1}^{m} y_i, \prod_{i=1}^{n} x_i \right)$ satisfies the condition $\mathfrak{P}$, $x \in \prod_{i=1}^{n} x_i$, $y \in \prod_{i=1}^{m} y_i$, then by Definition 6.3 (P1) we obtain $y \in M$. Finally, if $x, y \in \prod_{i=1}^{n} x_i$, then $y \in M$, by Definition 6.3 (P2). $\square$

Theorem 6.5. *Let S be a semihypergroup. The following conditions are equivalent:*

(1) *ζ_e is transitive;*

(2) *for every $x \in S$, $\zeta_e^*(x) = P(x)$;*

(3) *for every $x \in S$, $P(x)$ is a ζ_e^*-part of S.*

Proof. (1⇒2): By Theorem 6.4, for all pair $(x, y) \in S^2$ we have:

$$y \in \zeta_e^*(x) \Leftrightarrow x \zeta_e^* y \Leftrightarrow x \zeta_e y \Leftrightarrow y \in P(x).$$

(2⇒3): By Theorem 6.3, if M is a non-empty subset of S, then M is a ζ_e^*-part of S if and only if it is a union of equivalence classes modulo ζ_e^*. Particularly, every equivalence class modulo ζ_e^* is a ζ_e^*-part of S.

(3⇒1): Let $x \zeta_e y$ and $y \zeta_e z$, so $x \in P(y)$ and $y \in P(z)$ by Theorem 6.4. Since $P(z)$ is a ζ_e^*-part, by Lemma 6.2, we have $P(y) \subseteq P(z)$ and hence $x \in P(z)$. Therefore, $x \zeta_e z$ by Theorem 6.4 and the proof is complete. □

6.3 ζ_e^*-strong semihypergroup and a characterization of a derived ζ_e^*-strong semihypergroup

In this section, using the relation ζ_e, we characterize a derived ζ_e^*-strong semihypergroup.

Definition 6.4. Let $(S, \cdot)$ be a semihypergroup and ϕ_e be the canonical projection $\phi_e : S \to S/\zeta_e^*$. Define $D_e(S)$ as the *kernel* of the canonical projection ϕ_e, i.e., $D_e(S) = \phi_e^{-1}(e_{S/\zeta_e^*})$.

Theorem 6.6. *For a non-empty subset M of a semihypergroup S we have $D_e(S)M \cup MD_e(S) \subseteq \phi_e^{-1}(\phi_e(M))$.*

Proof. For all $x \in D_e(S)M$, there exists a pair $(d, m) \in D_e(S) \times M$ such that $x \in dm$, so $\phi_e(x) = \phi_e(d) \otimes \phi_e(m) = \phi_e(m)$. Therefore $\phi_e(x) = \phi_e(m)$ and hence $\phi_e(x) \in \phi_e(M)$. Thus $x \in \phi_e^{-1}(\phi_e(M))$. Similarly, $MD_e(S) \subseteq \phi_e^{-1}(\phi_e(M))$. □

Definition 6.5. A semihypergroup S is called ζ_e^*-*strong* whenever

(1) for all $x, y \in S$ if $x \zeta_e^* y$, then $xe \cap ye \neq \emptyset$ and $ex \cap ey \neq \emptyset$ and
(2) $\{e\}$ is invertible.

Remark 6.1. At the Example 6.1 the semihypergroup $(S, *)$ is ζ_e^*-*strong* semihypergroup.

Theorem 6.7. *For a non-empty subset M of a ζ_e^*-strong semihypergroup S we have*

(1) $MD_e(S) = D_e(S)M = \phi_e^{-1}(\phi_e(M));$

(2) *M is an ζ_e^*-part if and only if $\phi_e^{-1}(\phi_e(M)) = M$.*

Proof.   (1) By Theorem 6.6 it is enough prove that

$$\phi_e^{-1}(\phi_e(M)) \subseteq D_e(S)M \cap MD_e(S).$$

For all $x \in \phi_e^{-1}(\phi_e(M))$, an element $m \in M$ exists such that $\phi_e(x) = \phi_e(m)$. Since S is a ζ_e^*-strong semihypergroup, it follows that $xe \cap me \neq \emptyset$. So there exists $z \in xe \cap me$. Since $\{e\}$ is invertible, we have $x \in ze$ and hence $x \in mee$. Therefore $x \in MD_e(S)$, because $ee \subseteq D_e(S)$. Similarly we can prove $\phi_e^{-1}(\phi_e(M)) \subseteq D_e(S)M$.

(2) Let M be an ζ_e^*-part and set $x \in \phi_e^{-1}(\phi_e(M))$. Thus there exists $m \in M$ such that $\phi_e(x) = \phi_e(m)$ and hence $m \, \zeta_e^* \, x$, so by Theorem 6.3 we have $x \in M$. Therefore $\phi_e^{-1}(\phi_e(M)) \subseteq M$. Since $M \subseteq \phi_e^{-1}(\phi_e(M))$ it follows that $\phi_e^{-1}(\phi_e(M)) = M$. For the proof of the sufficiency suppose that $m \, \zeta_e^* \, x$ and $m \in M$. Thus $\phi_e(x) = \phi_e(m) \in \phi_e(M)$ and so $x \in \phi_e^{-1}(\phi_e(M)) = M$. Therefore by Theorem 6.3 it follows that M is an ζ_e^*-part of S. $\square$

Theorem 6.8. *If S is a ζ_e^*-strong semihypergroup, then ζ_e is transitive.*

Proof.   By Theorem 6.5, it is enough to show that for all $x \in S$, $P(x)$ is an ζ_e^*-part of H. According to Theorem 6.7, we have to check that $\phi_e^{-1}(\phi_e(P(x))) = P(x)$.

Let $z \in \phi_e^{-1}(\phi_e(P(x)))$, so there exists $k \in P(x)$ such that $\phi_e(z) = \phi_e(k)$ and hence $\zeta_e^*(z) = \zeta_e^*(k)$. Since $k \in P(x)$, $x \, \zeta_e \, k$ by Theorem 6.4. Thus $\zeta_e^*(k) = \zeta_e^*(x)$ and so $\zeta_e^*(z) = \zeta_e^*(x)$. Since S is a ζ_e^*-strong semihypergroup, we have $xe \cap ze \neq \emptyset$ and hence there exists $s \in xe \cap ze$. Therefore $x \in zee$ and $z \in xee$, because $\{e\}$ is an invertible and so $z \in zeeee$. Since $(zee, zeeee) \in \Re_u \mathfrak{P} e$, we have $x \, \zeta_e \, z$ and hence $z \in P(x)$. So we prove that $\phi_e^{-1}(\phi_e(P(x))) \subseteq P(x)$; it is obvious that $P(x) \subseteq \phi_e^{-1}(\phi_e(P(x)))$. Therefore $\phi_e^{-1}(\phi_e(P(x))) = P(x)$ and the proof is complete. $\square$

6.4   The relation τ_e^*

Let $(S, \cdot)$ be a semihypergroup and e be a whichever element of S. Denote by S_n the group of all permutations of the set $\{1, 2, ..., n\}$.

Definition 6.6. We say that a pair $(\prod_{i=1}^{n} x_i, \prod_{i=1}^{m} y_i)$ *satisfies the condition* $\mathfrak{T}$ if $m > n$ and there exist k and l in $\mathbb{N}$, such that $1 \leqslant k \leqslant n$, $k \leqslant l \leqslant m$,

$m = n + (l - k + 1)$ and $\exists \sigma \in S_n$, for which

$$
y_t = \begin{cases}
x_{\sigma(t)}, & \text{if } 1 \leqslant t < k; \\
e, & \text{if } k \leqslant t \leqslant l; \\
x_{\sigma(t+k-l-1)}, & \text{if } l < t \leqslant m.
\end{cases} \tag{6.3}
$$

Similarly as above, we denote

$$
\mathfrak{U}^{\mathfrak{T},e} := \left\{ \left(\prod_{i=1}^{n} x_i, \prod_{i=1}^{m} y_i \right) \mid \left(\prod_{i=1}^{n} x_i, \prod_{i=1}^{m} y_i \right) \text{ satisfies the condition } \mathfrak{T} \right\},
$$

$$
\mathfrak{U}_{\mathfrak{T}}^{e} := \mathfrak{U}^{\mathfrak{T},e} \cup (\mathfrak{U}^{\mathfrak{T},e})^{-1},
$$

$$
\mathfrak{J} := \left\{ \left(\prod_{i=1}^{n} x_i, \prod_{i=1}^{n} x_{\sigma(i)} \right) \mid \forall i \in \{1,2,...,n\}, x_i \in S, \ n \in \mathbb{N}, \ n \geqslant 1, \sigma \in S_n \right\}
$$

and $\mathfrak{U}^e := \mathfrak{U}_{\mathfrak{T}}^{e} \cup \mathfrak{J}$.

Definition 6.7. We define the relation τ_e on $(S, \cdot)$ as follows:

$$
x \tau_e y \Leftrightarrow \exists (A, B) \in \mathfrak{U}^e, \ x \in A, \ y \in B.
$$

Remark 6.2. The relation τ_e is reflexive and symmetric.

Let τ_e^* be the transitive closure of τ_e.

Theorem 6.9. τ_e^* *is a strongly regular relation and the quotient S/τ_e^* is a commutative monoid with the identity $\tau_e^*(e)$.*

Proof. Clearly, τ_e^* is an equivalence relation. First, we check that

$$
x \tau_e y \Rightarrow xz \overline{\overline{\tau_e}} yz, \quad zx \overline{\overline{\tau_e}} zy. \tag{6.4}
$$

for all $z \in S$. From $x \tau_e y$, it follows that there exists $(A, B) \in \mathfrak{U}^e$ such that $x \in A$ and $y \in B$.

If $(A, B) \in \mathfrak{U}^{\mathfrak{T},e}$, then $A = \prod_{i=1}^{n} x_i$, $B = \prod_{i=1}^{m} y_i$ and (A, B) satisfies the condition $\mathfrak{T}$ and the corresponding permutation of this pair is $\sigma \in S_n$. Then $xz \subseteq (\prod_{i=1}^{n} x_i)z$ and $yz \subseteq (\prod_{i=1}^{m} y_i)z$. Similarly as in Lemma 6.1, for all $1 \leqslant i \leqslant n$ set $x_i' = x_i$ and $x_{n+1}' = z$ and for all $1 \leqslant t \leqslant m$ set $y_t' = y_t$ and $y_{m+1}' = z$. Thus

$$
xz \subseteq \prod_{i=1}^{n+1} x_i' \quad \text{and} \quad yz \subseteq \prod_{t=1}^{m+1} y_t'.
$$

Define $\sigma' \in S_{n+1}$ as follows: $\sigma'(i) = \sigma(i)$ for all $i \in \{1,2,...,n\}$ and $\sigma'(n+1) = \sigma(n+1)$. Hence the pair $\left(\prod_{i=1}^{n+1} x_i', \prod_{t=1}^{m+1} y_t' \right)$ satisfies the condition

$\mathfrak{T}$ and the corresponding permutation of this pair is σ' and hence this pair belongs to $\mathfrak{U}^{\mathfrak{T},e} \subseteq \mathfrak{U}^e$. Thus $xz \;\overline{\overline{\tau_e}}\; yz$.

If $(A, B) \in (\mathfrak{U}^{\mathfrak{T},e})^{-1}$ then $(Bz, Az) \in \mathfrak{U}^{\mathfrak{T},e}$ and so we obtain again that $xz \;\overline{\overline{\tau_e}}\; yz$.

If $(A, B) \in \mathfrak{J}$ and the corresponding permutation of this pair is $\sigma \in S_n$, then we define σ' as above and it follows that $(Az, Bz) \in \mathfrak{J}$, whence $xz \;\overline{\overline{\tau_e}}\; yz$.

In a similar way, we can show that $x\tau_e y \Rightarrow zx \;\overline{\overline{\tau_e}}\; zy$. Now, if $x\tau_e^* y$, then there exist $m \in \mathbb{N}$ and $(u_0 = x, u_1, ..., u_m = y) \in S^m$ such that $x = u_0 \tau_e u_1 \tau_e ... \tau_e u_{m-1} \tau_e u_m = y$, whence, by (6.4), we obtain

$$xz = u_0 z \;\overline{\overline{\tau_e}}\; u_1 z \;\overline{\overline{\tau_e}}\; u_2 z \;\overline{\overline{\tau_e}}\; ... \;\overline{\overline{\tau_e}}\; u_{m-1} \;\overline{\overline{\tau_e}}\; u_m z = yz.$$

Hence $x\tau_e^* y \Rightarrow xz \;\overline{\overline{\tau_e^*}}\; yz$. Similarly we can prove that $x\tau_e^* y \Rightarrow zx \;\overline{\overline{\tau_e^*}}\; zy$, thus τ_e^* is strongly regular. Therefore the quotient S/τ_e^* is a semigroup under the following operation:

$$\tau_e^*(x) \otimes \tau_e^*(y) = \tau_e^*(z), \quad \text{for all } z \in xy.$$

On the other hand, since $(xy, yx) \in \mathfrak{J}$, it follows that $\tau_e^*(x) \otimes \tau_e^*(y) = \tau_e^*(y) \otimes \tau_e^*(x)$. Moreover, for all $x \in S$, $(x, xe) \in \mathfrak{U}^e$, so $\tau_e^*(x) \otimes \tau_e^*(e) = \tau_e^*(x))$, hence S/τ_e^* is a commutative monoid.   □

Theorem 6.10. τ_e^* *is the smallest equivalence relation such that the quotient is a commutative monoid.*

Proof. Let R be a strongly regular equivalence such that $(S/R, \otimes)$ is a commutative monoid with the identity $R(e)$. Let $\phi_e : S \to S/R$ be the canonical projection, which is a good homomorphism. We check that $x\tau_e y$ implies that xRy.

If there exists a pair $(A = \prod_{i=1}^{n} x_i, B = \prod_{i=1}^{m} y_i)$ which satisfies the condition $\mathfrak{T}$ and $x \in A = \prod_{i=1}^{n} x_i, y \in B = \prod_{i=1}^{m} y_i$, then we obtain $\phi_e(x) = \bigotimes_{i=1}^{n} \phi_e(x_i) = \bigotimes_{t=1}^{m} \phi_e(y_t) = \phi_e(y)$. Hence xRy. Similarly, if (A, B) satisfies the condition $\mathfrak{T}$ and $x \in B, y \in A$, we obtain xRy. Finally, if $x \in \prod_{i=1}^{n} x_i, y \in \prod_{i=1}^{n} x_{\sigma(i)}$, where $\sigma \in S_n$, then $\phi_e(x) = \bigotimes_{i=1}^{n} \phi_e(x_i) = \bigotimes_{i=1}^{n} \phi_e(x_{\sigma(i)}) = \phi_e(y)$, whence xRy. Since R is transitively closed, we obtain

$$x \in \tau_e^*(y) \Rightarrow x \in R(y).$$

Therefore $\tau_e^* \subseteq R$.   □

Corollary 6.1. *If $(S, \cdot)$ is a hypergroup and e is an element of S, then S/τ_e^* is a commutative group with the identity $\tau_e^*(e)$. Moreover, τ_e^* is the smallest equivalence relation such that the quotient is a commutative group.*

Similarly as for the relation ζ_e, we introduce the following notations, in order to characterize the transitivity of the relation τ_e.

Notation 6.2. Let x be an arbitrary element of a semihypergroup $(S, \cdot)$. For all $n \geqslant 1$, set:

$$P_{\mathcal{T}}(x) = \bigcup \left\{ \prod_{i=1}^{m} y_i \mid x \in \prod_{i=1}^{n} x_i, \left(\prod_{i=1}^{n} x_i, \prod_{i=1}^{m} y_i \right) \text{ or } \left(\prod_{i=1}^{m} y_i, \prod_{i=1}^{n} x_i \right) \right.$$
$$\left. \text{satisfies the condition } \mathfrak{T} \right\};$$

$$P_{\mathcal{I}_\backslash}(x) = \bigcup \left\{ \prod_{i=1}^{n} x_{\sigma(i)} \mid x \in \prod_{i=1}^{n} x_i, \sigma \in S_n \right\};$$

$$T(x) = \bigcup_{n \geqslant 1} \left(P_{\mathcal{T}}(x) \cup P_{\mathcal{I}_\backslash}(x) \right).$$

The next results have a similar proof to Theorem 6.4, Lemma 6.2 and Theorem 6.5.

Theorem 6.11. *For all $x \in S$, $T(x) = \{ y \in S \mid x \tau_e y \}$.*

If M be a non-empty subset of a semihypergroup $(S, \cdot)$, then we give the following definition

Definition 6.8. We say that M is a τ_e^*-*part* of S if $\prod_{i=1}^{n} x_i \cap M \neq \emptyset$ implies that

(T1) $\prod_{i=1}^{m} y_i \subseteq M$ for all $\prod_{i=1}^{m} y_i$ such that one of the pairs $\left(\prod_{i=1}^{n} x_i, \prod_{i=1}^{m} y_i \right)$, $\left(\prod_{i=1}^{m} y_i, \prod_{i=1}^{n} x_i \right)$ satisfies the condition $\mathfrak{T}$ and

(T2) $\prod_{i=1}^{n} x_{\sigma(i)} \subseteq M$, for all $\sigma \in S_n$.

Lemma 6.3. *Let $(S, \cdot)$ be a semihypergroup and M be a τ_e^*-part of S. If $x \in M$, then $T(x) \subseteq M$.*

Theorem 6.12. *Let S be a semihypergroup. The following conditions are equivalent:*

(1) *τ_e is transitive;*
(2) *for all $x \in S$, $\tau_e^*(x) = T(x)$;*
(3) *for all $x \in S$, $T(x)$ is a τ_e^*-part of S.*

6.5 $\Re$-parts in (semi)hypergroups

In continue of this chapter, first we generalize the notion of complete parts and α-parts by introducing the notion of $\Re$-parts and then we study $\Re$-closures in semihypergroups. The main reference is [135].

Let H be a semihypergroup, $\mathcal{U}$ be the set of finite products of elements of H and $\Re$ be a relation on $\mathcal{U}$.

Definition 6.9. For a non-empty subset A of H we say that:

(1) A is a *left $\Re$-part* of H with respect to $\mathcal{U}$ (or briefly is $\mathcal{L}\Re_{\mathcal{U}}$-*part*) if for all $\prod_{i=1}^{n} x_i$ and $\prod_{i=1}^{m} y_i$ in $\mathcal{U}$ the following implication is valid:

$$[\prod_{i=1}^{n} x_i \cap A \neq \emptyset \quad \text{and} \quad \prod_{i=1}^{m} y_i \ \Re \ \prod_{i=1}^{n} x_i] \Rightarrow \prod_{i=1}^{m} y_i \subseteq A;$$

(2) A is a *right $\Re$-part* of H with respect to $\mathcal{U}$ (or briefly is $\mathcal{R}\Re_{\mathcal{U}}$-*part*) if for all $\prod_{i=1}^{n} x_i$ and $\prod_{i=1}^{n} y_i$ in $\mathcal{U}$ the following implication is valid:

$$[\prod_{i=1}^{n} x_i \cap A \neq \emptyset \quad \text{and} \quad \prod_{i=1}^{n} x_i \ \Re \ \prod_{i=1}^{m} y_i] \Rightarrow \prod_{i=1}^{m} y_i \subseteq A;$$

(3) A is a *$\Re$-part* of H with respect to $\mathcal{U}$ (or briefly is $\Re_{\mathcal{U}}$-*part*) if it is $\mathcal{L}\Re_{\mathcal{U}}$-part and $\mathcal{R}\Re_{\mathcal{U}}$-part.

Notice that if $\Re$ is the diagonal relation on H, then $\Re$-parts are complete parts of H.

Remark 6.3. For a non-empty subset A of H,

(1) A is a $\mathcal{L}\Re_{\mathcal{U}}^{-1}$-part if and only if A is a $\mathcal{R}\Re_{\mathcal{U}}$-part;
(2) A is a $\mathcal{R}\Re_{\mathcal{U}}^{-1}$-part if and only if A is a $\mathcal{L}\Re_{\mathcal{U}}$-part.

First of all, we establish a connection between $\mathcal{L}\Re_{\mathcal{U}}$-parts, $\mathcal{R}\Re_{\mathcal{U}}$-parts of a semihypergroup H and $\mathcal{L}\Re_{\mathcal{U}}$-parts, $\mathcal{R}\Re_{\mathcal{U}}$-parts of a semihypergroup K_H, generated by H.

We recall that a K_H-semihypergroup is a semihypergroup constructed from a semihypergroup $(H, \circ)$ and a family $\{A(x)\}_{x \in H}$ of non-empty and mutually disjoint subsets of H. Set $K_H = \bigcup_{x \in H} A(x)$ and define the hyperoperation $*$ on K_H as follows:

$$\forall (a, b) \in K_H^2; \quad a \in A(x), b \in A(y), \quad a * b = \bigcup_{z \in x \circ y} A(z).$$

$(H, \circ)$ is a hypergroup if and only if $(K_H, *)$ is a hypergroup (see [60]). For all $P \in P^*(H)$, set $A(P) = \bigcup_{x \in P} A(x)$.

Theorem 6.13. *If $\Re$ is a relation on $\mathcal{U}$, then P is a $L\Re_u$-part of H if and only if $A(P)$ is a $L\widehat{\Re}_u$-part of K_H, where the relation $\widehat{\Re}$ is defined as follows:*

$$\bigcup_{v \in \prod_{i=1}^{n} x_i} A(v) \ \widehat{\Re} \ \bigcup_{u \in \prod_{i=1}^{m} y_i} A(u) \ \Leftrightarrow \ \prod_{i=1}^{n} x_i \ \Re \ \prod_{i=1}^{m} y_i.$$

Proof. Suppose that $A(P)$ is a $L\widehat{\Re}_u$-part of K_H and $(\prod_{i=1}^{n} x_i, \prod_{i=1}^{m} y_i) \in \Re$ is such that $\prod_{i=1}^{m} y_i \cap P \neq \emptyset$. So $\bigcup_{v \in \prod_{i=1}^{n} x_i} A(v) \ \widehat{\Re} \ \bigcup_{u \in \prod_{i=1}^{m} y_i} A(u)$ and

$$\prod_{i=1}^{m} y_i \cap P \neq \emptyset \Rightarrow \exists p \in P, \text{ such that } p \in \prod_{i=1}^{m} y_i$$

$$\Rightarrow \exists p \in P, \text{ such that } A(p) \subseteq \bigcup_{u \in \prod_{i=1}^{m} y_i} A(u)$$

$$\Rightarrow \bigcup_{u \in \prod_{i=1}^{m} y_i} A(u) \cap A(P) \neq \emptyset$$

$$\Rightarrow \bigcup_{v \in \prod_{i=1}^{n} x_i} A(v) \subseteq A(P), \text{ because } A(P) \text{ is a } L\widehat{\Re}_u\text{-part.}$$

For all $t \in \prod_{i=1}^{n} x_i$, $A(t) \subseteq A(P)$, so there exists $q \in P$ such that $A(t) \cap A(q) \neq \emptyset$. Therefore $t = q$ and hence $t \in P$, thus $\prod_{i=1}^{n} x_i \subseteq P$.

Conversely, let $(z_1, ..., z_m) \in K_H^m$ be such that $* \prod_{i=1}^{m} z_i \cap A(P) \neq \emptyset$, where $* \prod$ denotes a hyperproduct of elements in K_H. Suppose that $* \prod_{i=1}^{s} t_i \ \widehat{\Re} \ * \prod_{i=1}^{m} z_i$. There exists $(x_1, ..., x_m) \in H^m$ such that for all $1 \leqslant i \leqslant m$, $z_i \in A(x_i)$. Suppose that $u \in \bigcup_{y \in \prod_{i=1}^{m} x_i} A(y) \cap A(P)$, thus $u \in A(y_0)$ for some $y_0 \in \prod_{i=1}^{m} x_i$. Since $u \in A(P)$, there exists $y_1 \in P$ such that $u \in A(y_1)$. Therefore $A(y_0) \cap A(y_1) \neq \emptyset$, which implies that $y_0 = y_1 \in \prod_{i=1}^{m} x_i \cap P$. Since

P is $\mathcal{LR}_{\mathcal{U}}$-part of H and $\prod_{i=1}^{s} x_i' \, \mathfrak{R} \, \prod_{i=1}^{m} x_i$, where $t_i \in A(x_i')$ for all $1 \leqslant i \leqslant s$,

it follows that $\prod_{i=1}^{s} x_i' \subseteq P$. So

$$\prod_{i=1}^{s} t_i = \bigcup_{w \in \prod_{i=1}^{s} x_i'} A(w) \subseteq \bigcup_{u \in P} A(u) = A(P).$$

This completes the proof. $\qquad\square$

Definition 6.10. The intersection of all $\mathcal{LR}_{\mathcal{U}}$-parts (or $\mathcal{RR}_{\mathcal{U}}$-parts, $\mathfrak{R}$-parts) which is contain A is called $\mathcal{LR}_{\mathcal{U}}$-*closure* (or $\mathcal{RR}_{\mathcal{U}}$-*closure*, $\mathfrak{R}$-*closure*) of A in H and it will be denoted by $\overline{\mathcal{LR}_{\mathcal{U}}}(A)$ (or $\overline{\mathcal{RR}_{\mathcal{U}}}(A)$, $\overline{\mathfrak{R}_{\mathcal{U}}}(A)$).

Theorem 6.14. *For any non-empty subset A of H,*

(1) $\overline{\mathcal{LR}_{\mathcal{U}}^{-1}}(A) = \overline{\mathcal{RR}_{\mathcal{U}}}(A)$;

(2) $\overline{\mathcal{RR}_{\mathcal{U}}^{-1}}(A) = \overline{\mathcal{LR}_{\mathcal{U}}}(A)$.

Proof. It follows from Remark 6.3. $\qquad\square$

Lemma 6.4. *For a non-empty subset A of H define:*

$$_A{\sum}^{\mathcal{U}} \overset{def}{:=} \{\mathfrak{R} \subseteq \mathcal{U} \times \mathcal{U} \mid \overline{\mathcal{LR}_{\mathcal{U}}}(A) = A\} \ and \ {\sum}_A^{\mathcal{U}} \overset{def}{:=} \{\mathfrak{R} \subseteq \mathcal{U} \times \mathcal{U} \mid \overline{\mathcal{RR}_{\mathcal{U}}}(A) = A\}$$

If $_A{\sum}^{\mathcal{U}} \neq \emptyset$ *(or* ${\sum}_A^{\mathcal{U}} \neq \emptyset$*), then* $(_A{\sum}^{\mathcal{U}}, \circ)$ *(or* $({\sum}_A^{\mathcal{U}}, \circ)$*) is a semigroup, where $\circ$ is the relation composition.*

Proof. Suppose that $\mathfrak{R}, \mathfrak{R}' \in \ _A{\sum}^{\mathcal{U}}$ and $(\prod_{i=1}^{n} x_i, \prod_{i=1}^{m} y_i) \in \mathcal{U} \times \mathcal{U}$ are given. Also let $\prod_{i=1}^{n} x_i \cap A \neq \emptyset$ and $\prod_{i=1}^{m} y_i \, \mathfrak{R} \circ \mathfrak{R}' \, \prod_{i=1}^{n} x_i$. So there exists $(z_1, ..., z_k) \in H^n$ such that $\prod_{i=1}^{k} z_i \, \mathfrak{R} \, \prod_{i=1}^{n} x_i$ and $\prod_{i=1}^{m} y_i \, \mathfrak{R}' \, \prod_{i=1}^{k} z_i$. From $\prod_{i=1}^{k} z_i \, \mathfrak{R} \, \prod_{i=1}^{n} x_i$ and $\mathfrak{R} \in \ _A{\sum}^{\mathcal{U}}$ it follows that $\prod_{i=1}^{k} z_i \subseteq A$ and since $\mathfrak{R}' \in \ _A{\sum}^{\mathcal{U}}$ and $\prod_{i=1}^{m} y_i \, \mathfrak{R}' \, \prod_{i=1}^{k} z_i$, we obtain that $\prod_{i=1}^{m} y_i \subseteq A$. Hence $\mathfrak{R} \circ \mathfrak{R}' \in \ _A{\sum}^{\mathcal{U}}$ and so $(_A{\sum}^{\mathcal{U}}, \circ)$ is a semigroup. $\qquad\square$

Theorem 6.15. *If $\mathfrak{R}$ is a permutation of finite order in $S_{\mathcal{U}}$, then the following assertions are equivalent:*

(1) A is $\mathcal{L}\Re_{\mathcal{U}}$-part;

(2) A is $\mathcal{R}\Re_{\mathcal{U}}$-part;

(3) A is $\Re_{\mathcal{U}}$-part.

Proof. It is enough to prove (1⇒2). For this reason suppose that A is a $\mathcal{L}\Re_{\mathcal{U}}$-part. So $\overline{\mathcal{L}\Re_{\mathcal{U}}}(A) = A$ and hence $\Re \in {}_A\sum^{\mathcal{U}}$. Since $\Re$ is a permutation of finite order in $S_{\mathcal{U}}$, $\langle \Re \rangle = \{\Re^n \mid n \in \mathbb{N}\}$ is a subgroup of ${}_A\sum^{\mathcal{U}}$ and so $\Re^{-1} \in {}_A\sum^{\mathcal{U}}$. By Theorem 6.14, $A = \overline{\mathcal{L}\Re_{\mathcal{U}}^{-1}(A))} = \overline{\mathcal{R}\Re_{\mathcal{U}}}(A)$, thus $\Re \in {}_A\sum^{\mathcal{U}}$ and hence A is a $\mathcal{R}\Re_{\mathcal{U}}$-part. $\square$

In what follows, we determine the sets $\overline{\mathcal{L}\Re_{\mathcal{U}}}(A)$, $\overline{\mathcal{R}\Re_{\mathcal{U}}}(A)$ and $\overline{\Re_{\mathcal{U}}}(A)$, where $\Re \subseteq \mathcal{U} \times \mathcal{U}$ and A is a non-empty subset of H.

Set $K_{1,\Re}^{\mathcal{L}}(A) = A$ and

$$K_{t+1,\Re}^{\mathcal{L}}(A) = \{x \in H \mid \exists(\prod_{i=1}^{n} x_i, \prod_{i=1}^{m} y_i) \in \Re, x \in \prod_{i=1}^{n} x_i, \prod_{i=1}^{m} y_i \cap K_{t,\Re}^{\mathcal{L}}(A) \neq \emptyset\}.$$

Denote $K_{\Re}^{\mathcal{L}}(A) = \bigcup_{n \geq 1} K_{n,\Re}^{\mathcal{L}}(A)$. Similarly, set $K_{1,\Re}^{\mathcal{R}}(A) = A$ and

$$K_{t+1,\Re}^{\mathcal{R}}(A) = \{x \in H \mid \exists(\prod_{i=1}^{m} y_i, \prod_{i=1}^{n} x_i) \in \Re, x \in \prod_{i=1}^{n} x_i, \prod_{i=1}^{m} y_i \cap K_{t,\Re}^{\mathcal{R}}(A) \neq \emptyset\}.$$

Denote $K_{\Re}^{\mathcal{R}}(A) = \bigcup_{n \geq 1} K_{n,\Re}^{\mathcal{R}}(A)$. Finally, set $K_{1,\Re}(A) = A$ and

$$K_{t+1,\Re}(A) = \{x \in H \mid \exists(\prod_{i=1}^{n} x_i, \prod_{i=1}^{m} y_i) \in \Re \cup \Re^{-1},$$
$$x \in \prod_{i=1}^{n} x_i, \prod_{i=1}^{m} y_i \cap K_{t,\Re}(A) \neq \emptyset\}.$$

Denote $K_{\Re}(A) = \bigcup_{n \geq 1} K_{n,\Re}(A)$.

Theorem 6.16. *The following statements hold:*

(1) $K_{\Re}^{\mathcal{L}}(A) = \overline{\mathcal{L}\Re_{\mathcal{U}}}(A)$;

(2) $K_{\Re}^{\mathcal{R}}(A) = \overline{\mathcal{R}\Re_{\mathcal{U}}}(A)$;

(3) $K_{\Re}(A) = \overline{\Re_{\mathcal{U}}}(A)$.

Proof. (1) It is necessary to prove:

(a) $K_{\Re}^{\mathcal{L}}(A)$ is a $\mathcal{L}\Re_{\mathcal{U}}$-part;

(b) if $A \subseteq B$ and B is a $\mathcal{L}\Re_{\mathcal{U}}$-part, then $K_{\Re}^{\mathcal{L}}(A) \subseteq B$.

In order to prove (a), suppose that $\prod_{i=1}^{n} x_i \cap K_{\Re}^{\mathcal{L}}(A) \neq \emptyset$ and $\prod_{i=1}^{m} y_i \Re \prod_{i=1}^{n} x_i$. Therefore there exists $t \in \mathbb{N}$ such that $\prod_{i=1}^{n} x_i \cap K_{t,\Re}^{\mathcal{L}}(A) \neq$

$\emptyset$, whence it follows that $\prod_{i=1}^{m} y_i \subseteq K^{\mathcal{L}}_{t+1,\Re}(A) \subseteq K^{\mathcal{L}}_{\Re}(A)$. Now we prove (b) by induction on t. We have $K^{\mathcal{L}}_{1,\Re}(A) = A \subseteq B$. Suppose that $K^{\mathcal{L}}_{t,\Re}(A) \subseteq B$. We prove that $K^{\mathcal{L}}_{t+1,\Re}(A) \subseteq B$. If $z \in K^{\mathcal{L}}_{t+1,\Re}(A)$, then there exists $(\prod_{i=1}^{n} x_i, \prod_{i=1}^{m} y_i) \in \mathcal{U} \times \mathcal{U}$ such that $z \in \prod_{i=1}^{n} x_i$, $\prod_{i=1}^{n} x_i \,\Re\, \prod_{i=1}^{m} y_i$ and $\prod_{i=1}^{m} y_i \cap K^{\mathcal{L}}_{t,\Re}(A) \neq \emptyset$. Therefore $\prod_{i=1}^{m} y_i \cap B \neq \emptyset$ and hence $z \in \prod_{i=1}^{n} x_i \subseteq B$. So $K^{\mathcal{L}}_{t+1,\Re}(A) \subseteq B$.

(2) We have

$$K^{\mathcal{R}}_{\Re}(A) = K^{\mathcal{L}}_{\Re^{-1}}(A)$$
$$= \overline{\mathcal{L}\Re_{\mathcal{U}}^{-1}}(A), \quad \text{by part (1)}$$
$$= \overline{\mathcal{R}\Re_{\mathcal{U}}}(A), \quad \text{by Theorem 6.14.}$$

(3) It follows from (1) and (2). $\qquad\qquad\qquad\qquad\qquad\qquad\square$

Theorem 6.17. *Let A be a non-empty subset of H and $\Re$ be a relation on $\mathcal{U}$. Then, we have*

(1) $\overline{\mathcal{L}\Re_{\mathcal{U}}}(A) = \bigcup_{a \in A} \overline{\mathcal{L}\Re_{\mathcal{U}}}(a);$

(2) $\overline{\mathcal{R}\Re_{\mathcal{U}}}(A) = \bigcup_{a \in A} \overline{\mathcal{R}\Re_{\mathcal{U}}}(a);$

(3) $\overline{\Re_{\mathcal{U}}}(A) = \bigcup_{a \in A} \overline{\Re_{\mathcal{U}}}(a).$

Proof. It is clear that for all $a \in A$, $\overline{\mathcal{L}\Re_{\mathcal{U}}}(a) \subseteq \overline{\mathcal{L}\Re_{\mathcal{U}}}(A)$. By Theorem 6.16(1) we have $\overline{\mathcal{L}\Re_{\mathcal{U}}}(A) = \bigcup_{n \geq 1} K^{\mathcal{L}}_{n,\Re}(A)$ and $K^{\mathcal{L}}_{1,\Re}(A) = A = \bigcup_{a \in A} \{a\} = \bigcup_{a \in A} K^{\mathcal{L}}_{1,\Re}(a)$. We prove the Theorem by induction on n. Supposing it true for n, we prove that $K^{\mathcal{L}}_{n+1,\Re}(A) \subseteq \bigcup_{a \in A} K^{\mathcal{L}}_{n+1,\Re}(a)$. If $z \in K^{\mathcal{L}}_{n+1,\Re}(A)$, then there exists $(\prod_{i=1}^{n} x_i, \prod_{i=1}^{m} y_i) \in \Re$ such that $z \in \prod_{i=1}^{n} x_i$ and $\prod_{i=1}^{m} y_i \cap K^{\mathcal{L}}_{n,\Re}(A) \neq \emptyset$. By the hypothesis of induction, $\prod_{i=1}^{m} y_i \cap (\bigcup_{a \in A} K^{\mathcal{L}}_{n,\Re}(a)) \neq \emptyset$ and therefore there exists $a' \in A$ exists such that $\prod_{i=1}^{m} y_i \cap K^{\mathcal{L}}_{n,\Re}(a') \neq \emptyset$, and so $z \in K^{\mathcal{L}}_{n+1,\Re}(a')$, whence $\overline{\mathcal{L}\Re_{\mathcal{U}}}(A) \subseteq \bigcup_{a \in A} \overline{\mathcal{L}\Re_{\mathcal{U}}}(a)$. $\qquad\square$

Theorem 6.18. *If $\Re \subseteq \mathcal{U} \times \mathcal{U}$, then the following relation $K^{\mathcal{L}}_{\Re}$ ($K^{\mathcal{R}}_{\Re}$ respectively) on H:*

$$x \, K^{\mathcal{L}}_{\underset{\mathfrak{R}}{}} \, y \Leftrightarrow x \in K^{\mathcal{L}}_{\underset{\mathfrak{R}}{}}(y) \quad (x \, K^{\mathcal{R}}_{\underset{\mathfrak{R}}{}} \, y \Leftrightarrow x \in K^{\mathcal{R}}_{\underset{\mathfrak{R}}{}}(y)),$$

where $K^{\mathcal{L}}_{\underset{\mathfrak{R}}{}}(y) = K^{\mathcal{L}}_{\underset{\mathfrak{R}}{}}(\{y\})$ $(K^{\mathcal{R}}_{\underset{\mathfrak{R}}{}}(y) = K^{\mathcal{R}}_{\underset{\mathfrak{R}}{}}(\{y\}))$ is a preorder. Furthermore, if $\mathfrak{R}$ is symmetric, then $K^{\mathcal{L}}_{\underset{\mathfrak{R}}{}}$ $(K^{\mathcal{R}}_{\underset{\mathfrak{R}}{}}$ respectively) is an equivalence relation.

Proof. It is easy to see that $K^{\mathcal{L}}_{\underset{\mathfrak{R}}{}}$ is reflexive. Now suppose that $x \, K^{\mathcal{L}}_{\underset{\mathfrak{R}}{}} \, y$ and $y \, K^{\mathcal{L}}_{\underset{\mathfrak{R}}{}} \, z$. So $x \in K^{\mathcal{L}}_{\underset{\mathfrak{R}}{}}(y)$ and $y \in K^{\mathcal{L}}_{\underset{\mathfrak{R}}{}}(z)$. By Theorem 6.16(1), $K^{\mathcal{L}}_{\underset{\mathfrak{R}}{}}(z)$ is $\mathcal{L}\mathfrak{R}_{\mathcal{U}}$-part thus $K^{\mathcal{L}}_{\underset{\mathfrak{R}}{}}(y){\subseteq}K^{\mathcal{L}}_{\underset{\mathfrak{R}}{}}(z)$ and hence $x \in K^{\mathcal{L}}_{\underset{\mathfrak{R}}{}}(z)$. Therefore $K^{\mathcal{L}}_{\underset{\mathfrak{R}}{}}$ is preorder. Now if $\mathfrak{R}$ be symmetric, then we prove that $K^{\mathcal{L}}_{\underset{\mathfrak{R}}{}}$ is symmetric as well. To this end, we check that:

(1) for all $t \geq 2$ and $x \in H$, $K^{\mathcal{L}}_{t,\mathfrak{R}}(K^{\mathcal{L}}_{2,\mathfrak{R}}(x)) = K^{\mathcal{L}}_{t+1,\mathfrak{R}}(x)$;

(2) $x \in K^{\mathcal{L}}_{n,\mathfrak{R}}(y)$ if and only if $y \in K^{\mathcal{L}}_{n,\mathfrak{R}}(x)$.

We prove (1) by induction on n. Suppose that $z \in K^{\mathcal{L}}_{2,\mathfrak{R}}(K^{\mathcal{L}}_{2,\mathfrak{R}}(x))$ so there exists $(\prod_{i=1}^{n} x_i, \prod_{i=1}^{m} y_i) \in \mathfrak{R}$ such that $z \in \prod_{i=1}^{n} x_i$ and $\prod_{i=1}^{m} y_i \cap K^{\mathcal{L}}_{2,\mathfrak{R}}(x) \neq \emptyset$. Thus $z \in K^{\mathcal{L}}_{3,\mathfrak{R}}(x)$. If $K^{\mathcal{L}}_{t,\mathfrak{R}}(K^{\mathcal{L}}_{2,\mathfrak{R}}(x)) = K^{\mathcal{L}}_{t+1,\mathfrak{R}}(x)$ then:

$$z \in K^{\mathcal{L}}_{t+1,\mathfrak{R}}(K^{\mathcal{L}}_{2,\mathfrak{R}}(x))$$

$$\Leftrightarrow \exists (\prod_{i=1}^{n} x_i, \prod_{i=1}^{m} y_i) \in \mathfrak{R}, z \in \prod_{i=1}^{n} x_i, \prod_{i=1}^{m} y_i \cap K^{\mathcal{L}}_{t,\mathfrak{R}}(K^{\mathcal{L}}_{2,\mathfrak{R}}(x)) \neq \emptyset$$

$$\Leftrightarrow \exists (\prod_{i=1}^{n} x_i, \prod_{i=1}^{m} y_i) \in \mathfrak{R}, z \in \prod_{i=1}^{n} x_i, \prod_{i=1}^{m} y_i \cap K^{\mathcal{L}}_{t+1,\mathfrak{R}}(x) \neq \emptyset$$

$$\Leftrightarrow z \in K^{\mathcal{L}}_{t+2,\mathfrak{R}}(x)$$

Hence, for all $t \geq 2$ and $x \in H$, $K^{\mathcal{L}}_{t,\mathfrak{R}}(K^{\mathcal{L}}_{2,\mathfrak{R}}(x)) = K^{\mathcal{L}}_{t+1,\mathfrak{R}}(x)$.

We prove (2) by induction on n, too. It is clear that $x \in K^{\mathcal{L}}_{2,\mathfrak{R}}(y)$ if and only if $y \in K^{\mathcal{L}}_{2,\mathfrak{R}}(x)$. Suppose that $x \in K^{\mathcal{L}}_{t,\mathfrak{R}}(y)$ if and only if $y \in K^{\mathcal{L}}_{t,\mathfrak{R}}(x)$. If $x \in K^{\mathcal{L}}_{t+1,\mathfrak{R}}(y)$, then there exists $(\prod_{i=1}^{n} x_i, \prod_{i=1}^{m} y_i) \in \mathfrak{R}$ such that $x \in \prod_{i=1}^{n} x_i$ and $\prod_{i=1}^{m} y_i \cap K^{\mathcal{L}}_{t,\mathfrak{R}}(y) \neq \emptyset$. Therefore there exists $b \in \prod_{i=1}^{m} y_i \cap K^{\mathcal{L}}_{t,\mathfrak{R}}(y)$ and hence $y \in K^{\mathcal{L}}_{t,\mathfrak{R}}(b)$. Since $\mathfrak{R}$ is symmetric, $(\prod_{i=1}^{m} y_i, \prod_{i=1}^{n} x_i) \in \mathfrak{R}$. From $b \in \prod_{i=1}^{m} y_i$ and $x \in \prod_{i=1}^{n} x_i \cap K^{\mathcal{L}}_{1,\mathfrak{R}}(x)$ it follows that $b \in K^{\mathcal{L}}_{2,\mathfrak{R}}(x)$ and hence $y \in K^{\mathcal{L}}_{t,\mathfrak{R}}(K^{\mathcal{L}}_{2,\mathfrak{R}}(x)) = K^{\mathcal{L}}_{t+1,\mathfrak{R}}(x)$. Similarly we can show if $y \in K^{\mathcal{L}}_{t+1,\mathfrak{R}}(x)$, then $x \in K^{\mathcal{L}}_{t+1,\mathfrak{R}}(x)$. $\square$

Theorem 6.19. *Let $\mathfrak{R}$ be a relation on $\mathcal{U}$ and A be a non-empty subset of a hypergroup H. The following conditions are equivalent:*

(1) A is a $\mathcal{LR}_{\mathcal{U}}$-part ($\mathcal{RR}_{\mathcal{U}}$-part) of H;

(2) $x \in A$, $z\, K^{\mathcal{L}}_{\mathcal{R}}\, x \Rightarrow z \in A$ ($x\, K^{\mathcal{L}}_{\mathcal{R}}\, z \Rightarrow z \in A$ respectively).

Proof. (1$\Rightarrow$2): If $x \in A$ and $z \in H$ is such that $z\, K^{\mathcal{L}}_{\mathcal{R}}\, x$, then there exists $(\prod_{i=1}^{n} x_i, \prod_{i=1}^{m} y_i) \in \mathfrak{R}$ such that $z \in \prod_{i=1}^{n} x_i$ and $\prod_{i=1}^{m} y_i \cap K^{\mathcal{L}}_{t,\mathfrak{R}}(A) \neq \emptyset$ for some $t \in \mathbb{N}$. Since A is a $\mathcal{LR}_{\mathcal{U}}$-part, $K^{\mathcal{L}}_{t,\mathfrak{R}}(A) \subseteq A$ according to Theorem 6.16 and so $\prod_{i=1}^{m} y_i \cap A \neq \emptyset$. Therefore $\prod_{i=1}^{n} x_i \subseteq A$ and hence $z \in A$.

(2$\Rightarrow$1): Let $\prod_{i=1}^{n} x_i \cap A \neq \emptyset$ and $\prod_{i=1}^{m} y_i\, \mathfrak{R} \prod_{i=1}^{n} x_i$. Hence, there exists $x \in A \cap \prod_{i=1}^{n} x_i$ and so $\prod_{i=1}^{n} x_i \cap K^{\mathcal{L}}_{1,\mathfrak{R}}(x) \neq \emptyset$. Set $z \in \prod_{i=1}^{m} y_i$. So, we have

$$\prod_{i=1}^{m} y_i\, \mathfrak{R} \prod_{i=1}^{n} x_i \Rightarrow z \in K^{\mathcal{L}}_{2,\mathfrak{R}}(x), \quad \Rightarrow z\, K^{\mathcal{L}}_{\mathcal{R}}\, x \Rightarrow z \in A, \text{ (since } x \in A\text{).}$$

Therefore, we conclude that $\prod_{i=1}^{m} y_i \subseteq A$ and so A is a $\mathcal{LR}_{\mathcal{U}}$-part of H. $\square$

6.6 Groups derived from strongly $\mathcal{U}$-regular relations

We introduce now the notion of a strongly $\mathcal{U}$-regular relation and we investigate some properties of it.

Definition 6.11. Let $\mathfrak{R} \subseteq \mathcal{U} \times \mathcal{U}$.

(1) For all $(x,y) \in H^2$ define the relation $\rho_{\mathcal{L},\mathfrak{R}}$, as follows:

$$x\, \rho_{\mathcal{L},\mathfrak{R}}\, y \Leftrightarrow x = y \text{ or}$$

$$\exists (\prod_{i=1}^{n} x_i, \prod_{i=1}^{m} y_i) \in \mathfrak{R} \text{ such that } x \in \prod_{i=1}^{n} x_i \text{ and } y \in \prod_{i=1}^{m} y_i$$

and $\rho^{*}_{\mathcal{L},\mathfrak{R}}$ is the transitive closure of $\rho_{\mathcal{L},\mathfrak{R}}$;

(2) For all $(x,y) \in H^2$ define the relation $\rho_{\mathcal{R},\mathfrak{R}}$, as follows:

$$x\, \rho_{\mathcal{R},\mathfrak{R}}\, y \Leftrightarrow x = y \text{ or}$$

$$\exists (\prod_{i=1}^{m} y_i, \prod_{i=1}^{n} x_i) \in \mathfrak{R} \text{ such that } x \in \prod_{i=1}^{n} x_i \text{ and } y \in \prod_{i=1}^{m} y_i$$

and $\rho^{*}_{\mathcal{R},\mathfrak{R}}$ is the transitive closure of $\rho_{\mathcal{R},\mathfrak{R}}$;

(3) For all $(x,y) \in H^2$ define the relation $\rho_{\mathfrak{R}}$, as follows:

$$x\, \rho_{\mathfrak{R}}\, y \Leftrightarrow x = y \text{ or}$$

$$\exists (\prod_{i=1}^{n} x_i, \prod_{i=1}^{m} y_i) \in \mathfrak{R} \cup \mathfrak{R}^{-1} \text{ such that } x \in \prod_{i=1}^{n} x_i \text{ and } y \in \prod_{i=1}^{m} y_i$$

and $\rho^{*}_{\mathfrak{R}}$ is the transitive closure of $\rho_{\mathfrak{R}}$.

Theorem 6.20. *If* $\Re \subseteq \mathcal{U} \times \mathcal{U}$, *then for all* $(x, y) \in H^2$ *we have*

(1) $x \, K_{\Re}^{\mathcal{L}} \, y$ *if and only if* $x \, \rho_{\mathcal{L}, \Re}^* \, y$;

(2) $x \, K_{\Re}^{\mathcal{R}} \, y$ *if and only if* $x \, \rho_{\mathcal{R}, \Re}^* \, y$;

(3) $x K_{\Re} y$ *if and only if* $x \, \rho_{\Re}^* \, y$.

Proof. (1) It is easy to see that $\rho_{\mathcal{L}, \Re}^* \subseteq K_{\Re}^{\mathcal{L}}$. On the converse suppose that $x \, K_{\Re}^{\mathcal{L}} \, y$ so by Theorem 6.18, $x \in K_{t+1, \Re}^{\mathcal{L}}(y)$ for some $t \in \mathbb{N}$. So there exists $(\prod_{i=1}^{n_1} x_{1,i}, \prod_{i=1}^{m_1} y_{1,i}) \in \Re$ such that $x \in \prod_{i=1}^{n_1} x_{1,i}$ and $\prod_{i=1}^{m_1} y_{1,i} \cap K_{t, \Re}^{\mathcal{L}}(y) \neq \emptyset$ thus there exists $x_1 \in \prod_{i=1}^{m_1} y_{1,i} \cap K_{t, \Re}^{\mathcal{L}}(y)$ and hence $x \, \rho_{\mathcal{L}, \Re} \, x_1$. Since $x_1 \in K_{t, \Re}^{\mathcal{L}}(y)$, there exists $(\prod_{i=1}^{n_2} x_{2,i}, \prod_{i=1}^{m_2} y_{2,i}) \in \Re$ such that $x_1 \in \prod_{i=1}^{n_2} x_{2,i}$ and $\prod_{i=1}^{m_2} y_{2,i} \cap K_{t-1, \Re}^{\mathcal{L}}(y) \neq \emptyset$. Therefore $x_1 \, \rho_{\mathcal{L}, \Re} \, x_2$, where $x_2 \in \prod_{i=1}^{m_2} y_{2,i} \cap K_{t-1, \Re}^{\mathcal{L}}(y)$. After t steps we obtain that $x_t \in \prod_{i=1}^{m_t} y_{t,i} \cap K_{t-(t-1), \Re}^{\mathcal{L}}(y)$ exists such that $x_{t-1} \, \rho_{\mathcal{L}, \Re} \, x_t$. Thus we have:

$$x \, \rho_{\mathcal{L}, \Re} \, x_1 \, \rho_{\mathcal{L}, \Re} \, x_2 \, \cdots \, x_t \, \rho_{\mathcal{L}, \Re} \, y$$

and from this it follows that $K_{\Re}^{\mathcal{L}} \subseteq \rho_{\mathcal{L}, \Re}^*$, so the proof is complete.

Similarly we obtain (2) and (3). $\square$

Theorem 6.21. *If* $\Re$ *is a permutation of finite order in* $S_{\mathcal{U}}$, *then* $\rho_{\mathcal{L}, \Re}^* = \rho_{\mathcal{R}, \Re}^*$.

Proof. By Theorem 6.16, $K_{\Re}^{\mathcal{L}}(y)$ is $\mathcal{L}\Re_{\mathcal{U}}$-part so by Theorem 6.15, $K_{\Re}^{\mathcal{L}}(y)$ is a $\mathcal{R}\Re_{\mathcal{U}}$-part and hence $K_{\Re}^{\mathcal{R}}(y) \subseteq K_{\Re}^{\mathcal{L}}(y)$. Analogously $K_{\Re}^{\mathcal{L}}(y) \subseteq K_{\Re}^{\mathcal{R}}(y)$ and the proof is complete. $\square$

Definition 6.12. Let $(H, \circ)$ be a semihypergroup. A relation $\Re$ on $\mathcal{U}$ is called

(1) *compatible on the left (on the right)* with$\circ$ if for all $P_1, P_2, P \in \mathcal{U}$ from $P_1 \, \Re \, P_2$ it follows $P \circ P_1 \, \Re \, P \circ P_2$ ($P_1 \circ P \, \Re \, P_2 \circ P$). $\Re$ is *compatible* with $\circ$ if it is compatible on the left and on the right with $\circ$.

(2) *regular* if for all $x \in H$, one has $K_{\Re}^{\mathcal{L}}(x) = K_{\Re}^{\mathcal{R}}(x)$;

(3) A regular relation $\Re$ on $\mathcal{U}$ is called *strongly regular on the left (on the right)* if $\rho_{\mathcal{L}, \Re}^*$ ($\rho_{\mathcal{R}, \Re}^*$) is strongly regular on the left (on the right, respectively).

(4) A regular relation $\Re$ on $\mathcal{U}$ is called *strongly regular* if $\rho^*_{\Re}$ is strongly regular.

Theorem 6.22. *If $\Re$ is a regular relation on $\mathcal{U}$, then*

(1) $\Re^{-1}$ *is regular;*

(2) $\rho^*_{\mathcal{L},\Re} = \rho^*_{\Re,\Re} = \rho^*_{\Re}$ *and $\rho^*_{\Re}$ is an equivalence relation.*

Proof.   The proof follows from Theorem 6.14 and Theorem 6.20.   □

Let $\Re \subseteq \mathcal{U} \times \mathcal{U}$. For every element x of a hypergroup H, set:

$$P^n_{\mathcal{L},\Re}(x) = \bigcup \Big\{ \prod_{i=1}^m y_i \ \Big| \ \prod_{i=1}^m y_i \ \Re \ \prod_{i=1}^n x_i, \ x \in \prod_{i=1}^n x_i \Big\};$$

$$P_{\mathcal{L},\Re}(x) = \bigcup_{n \geqslant 1} P^n_{\mathcal{L},\Re}(x) \cup \{x\};$$

$$\rho^*_{\mathcal{L},\Re}(x) = \{ y \in H \ | \ y \, \rho^*_{\mathcal{L},\Re} \, x \}.$$

Theorem 6.23. *Let $\Re$ be a relation on $\mathcal{U}$ and H be a semihypergroup. The following conditions are equivalent:*

(1) $\rho_{\mathcal{L},\Re}$ *is transitive;*

(2) *for every $x \in H$, $\rho^*_{\mathcal{L},\Re}(x) = P_{\mathcal{L},\Re}(x)$;*

(3) *for every $x \in H$, $P_{\mathcal{L},\Re}(x)$ is a $\mathcal{L}\Re_{\mathcal{U}}$-part of H.*

Proof.   (1⇒2): For every pair (x,y) of elements of H we have:

$$y \in \rho^*_{\mathcal{L},\Re}(x) \Leftrightarrow y \, \rho^*_{\mathcal{L},\Re} \, x \Leftrightarrow y \, \rho_{\mathcal{L},\Re} \, x \Leftrightarrow y \in P_{\mathcal{L},\Re}(x).$$

(2⇒3): Let $\left(\prod_{i=1}^m y_i, \prod_{i=1}^n x_i \right) \in \Re$ be such that $\prod_{i=1}^n x_i \cap P_{\mathcal{L},\Re}(x) \neq \emptyset$. So $\prod_{i=1}^n x_i \cap \rho^*_{\mathcal{L},\Re}(x) \neq \emptyset$ and hence there exists $z \in \prod_{i=1}^n x_i$ and $z \in \rho^*_{\mathcal{L},\Re}(x)$, thus $z \in K^{\mathcal{L}}_{\Re}(x)$, by Theorem 6.20. On the other hand, $z \in K^{\mathcal{L}}_{\Re}(z)$, so $\prod_{i=1}^n x_i \cap K^{\mathcal{L}}_{\Re}(z) \neq \emptyset$ and hence $\prod_{i=1}^m y_i \subseteq K^{\mathcal{L}}_{\Re}(z)$, because $\prod_{i=1}^m y_i \ \Re \ \prod_{i=1}^n x_i$ and $K^{\mathcal{L}}_{\Re}(z)$ is a $\mathcal{L}\Re_{\mathcal{U}}$-part of H, by Theorem 6.16. Now suppose that $t \in \prod_{i=1}^m y_i$ is an arbitrary element, thus $t \in K^{\mathcal{L}}_{\Re}(x)$ and hence $t \, \rho^*_{\mathcal{L},\Re} \, x$. Therefore $t \in \rho^*_{\mathcal{L},\Re}(x) = P_{\mathcal{L},\Re}(x)$ and so $\prod_{i=1}^m y_i \subseteq P_{\mathcal{L},\Re}(x)$.

(3⇒1): Let x,y and z be in H such that $x \, \rho_{\mathcal{L},\Re} \, y$ and $y \, \rho_{\mathcal{L},\Re} \, z$. Since $x \, \rho_{\mathcal{L},\Re} \, y$, there exists $\left(\prod_{i=1}^n x_i, \prod_{i=1}^m y_i \right) \in \Re$ such that $x \in \prod_{i=1}^n x_i$ and

$y \in \prod\limits_{i=1}^{m} y_i$. Therefore $\prod\limits_{i=1}^{m} y_i \cap P_{\mathcal{L},\Re}(y) \neq \emptyset$ and since $P_{\mathcal{L},\Re}(y)$ is a $\mathcal{L}\Re_{\mathcal{U}}$-part, $\prod\limits_{i=1}^{n} x_i \subseteq P_{\mathcal{L},\Re}(y)$, whence $x \in P_{\mathcal{L},\Re}(y)$. We can easily check that $P_{\mathcal{L},\Re}(y) \subseteq P_{\mathcal{L},\Re}(z)$. Indeed, similarly as above, from $y \, \rho_{\mathcal{L},\Re} \, z$ we obtain $y \in P_{\mathcal{L},\Re}(z)$, then we use that $P_{\mathcal{L},\Re}(z)$ is a $\mathcal{L}\Re_{\mathcal{U}}$-part of H. Therefore $x \in P_{\mathcal{L},\Re}(z)$ and hence $x \, \rho_{\mathcal{L},\Re} \, z$. $\square$

Theorem 6.24. *If $(H, \circ)$ is a regular hypergroup and $\Re$ is a compatible relation with $\circ$ on $\mathcal{U}$, then $\rho_{\mathcal{L},\Re}$ is transitive.*

Proof. According to the above theorem, it is sufficient to check that for every $x \in H$, $P_{\mathcal{L},\Re}(x)$ is a $\mathcal{L}\Re_{\mathcal{U}}$-part of H.

Let $\prod\limits_{i=1}^{n} x_i$, $\prod\limits_{i=1}^{m} y_i$ be such that $\prod\limits_{i=1}^{n} x_i \cap P_{\mathcal{L},\Re}(x) \neq \emptyset$ and $\prod\limits_{i=1}^{m} y_i \, \Re \, \prod\limits_{i=1}^{n} x_i$. We check that $\prod\limits_{i=1}^{m} y_i \subseteq P_{\mathcal{L},\Re}(x)$. Since H is a regular hypergroup, there exists an identity e of H. Moreover, there exist $u, v \in H$ such that $e \in u \circ x$ and $x \in t \circ v$, where $t \in \prod\limits_{i=1}^{n} x_i \cap P_{\mathcal{L},\Re}(x)$. Hence there exist $P_1, P_2 \in \mathcal{U}$ such that $t \in P_1, x \in P_2$ and $P_1 \, \Re \, P_2$. We obtain

$$x \in t \circ v \subseteq \prod_{i=1}^{n} x_i \circ v \subseteq \prod_{i=1}^{n} x_i \circ e \circ v \subseteq \prod_{i=1}^{n} x_i \circ u \circ x \circ v$$

$$\subseteq \prod_{i=1}^{n} x_i \circ u \circ P_2 \circ v = P_3,$$

$$\prod_{i=1}^{m} y_i \subseteq \prod_{i=1}^{m} y_i \circ e \subseteq \prod_{i=1}^{m} y_i \circ u \circ t \circ v \subseteq \prod_{i=1}^{m} y_i \circ u \circ P_1 \circ v = P_4.$$

Since $\prod\limits_{i=1}^{n} x_i \, \Re \, \prod\limits_{i=1}^{m} y_i$, $P_1 \, \Re \, P_2$ and $\Re$ is regular, it follows that $P_3 \, \Re \, P_4$. Therefore, $\prod\limits_{i=1}^{m} y_i \subseteq P_{\mathcal{L},\Re}(x)$ and so, $\rho_{\mathcal{L},\Re}$ is transitive. $\square$

Similarly, we can prove that if $(H, \circ)$ is a regular hypergroup and $\Re$ is a compatible relation with $\circ$ on $\mathcal{U}$, then $\rho_{\Re}$ is transitive.

Theorem 6.25. *Let $(H, \circ)$ be a hypergroup and $K = \bigcup\limits_{n \geqslant 1} A_n$, where A_n is the alternating subgroup of the symmetric group S_n of order n or $K = \{\varepsilon\}$, the identity of S_n. If we define the relation $\Re^K$ on $\mathcal{U}$, as follows: for all $(\prod\limits_{i=1}^{n} x_i, \prod\limits_{i=1}^{m} y_i) \in \mathcal{U}^2$,*

$$\prod_{i=1}^{n} x_i \, \Re^K \, \prod_{i=1}^{m} y_i \Leftrightarrow \exists \tau \in K, \prod_{i=1}^{m} y_i = \prod_{i=1}^{n} x_{\tau(i)},$$

then

$$\rho_{\Re k} = \rho^*_{\Re K} = \begin{cases} \beta^* & \text{if } K = \{\varepsilon\} \\ \alpha^* & \text{if } K = \bigcup_{n \geqslant 1} A_n. \end{cases}$$

Proof. If $K = \{\varepsilon\}$ the proof is obvious, see [22]. Now suppose that $K = \bigcup_{n \geqslant 1} A_n$. So $\rho^*_{\Re K} \subseteq \alpha^*$. For the converse inclusion, we prove that $H/\rho^*_{\Re K}$ is an abelian group. Let $x_1, x_2 \in H$. From $H = x_2 \circ H$, it follows that there exists $x_3 \in H$ such that $x_2 \in x_2 \circ x_3$ so we have $x_1 \circ x_2 \subseteq x_1 \circ x_2 \circ x_3$ and $x_2 \circ x_1 \subseteq x_2 \circ x_3 \circ x_1$. Since $\prod_{i=1}^{3} x_i \, \Re^K \prod_{i=1}^{3} x_{\tau(i)}$, where $\tau(1) = 2, \tau(2) = 3$, $\tau(3) = 1$ and $\tau \in A_3$, we conclude that $\rho^*_{\Re K}(x_1) \otimes \rho^*_{\Re K}(x_2) = \rho^*_{\Re K}(x_2) \otimes \rho^*_{\Re K}(x_1)$ and hence $H/\rho^*_{\Re K}$ is abelian. $\qquad\square$

Theorem 6.26. *Let $(H, \circ)$ be a hypergroup. If the relation $\Re$ on $\mathcal{U}$ is defined as follows: for all $(n, m) \in \mathbb{N}^2$ and for all $(\prod_{i=1}^{n} x_i, \prod_{i=1}^{m} y_i) \in \mathcal{U}^2$,*

$$\prod_{i=1}^{n} x_i \Re \prod_{i=1}^{m} y_i \Leftrightarrow (|\prod_{i=1}^{n} x_i \setminus \prod_{i=1}^{m} y_i| < \infty$$

or

$$|\prod_{i=1}^{m} y_i \setminus \prod_{i=1}^{n} x_i| < \infty),$$

and

$$\prod_{i=1}^{n} x_i \cap \prod_{i=1}^{m} y_i \neq \emptyset,$$

where $|A|$ is the cardinal number of the set A, then $\rho^*_{\Re} = \beta^*$.

Proof. Since $\Re$ is symmetric, $\Re$ is regular and hence by Theorem 6.22(ii), $\rho^*_{\Re}$ is an equivalence relation. Suppose that $(x, y) \in \rho_{\Re}$ and hence there exists $(\prod_{i=1}^{n} x_i, \prod_{i=1}^{m} y_i) \in \Re$ such that $x \in \prod_{i=1}^{n} x_i$ and $y \in \prod_{i=1}^{m} y_i$. Without the lost of generality, suppose that $|\prod_{i=1}^{n} x_i \setminus \prod_{i=1}^{m} y_i| < \infty$, so $\prod_{i=1}^{n} x_i \setminus \prod_{i=1}^{m} y_i = \{b_1, b_2, ..., b_t\}$. Since H is a hypergroup, there exists $(c_1, d_1) \in H^2$ such that $y_m \in c_1 \circ b_1$, $b_1 \in a \circ d_1$, where $a \in \prod_{i=1}^{n} x_i \cap \prod_{i=1}^{m} y_i$, so we have:

$$\prod_{i=1}^{m} y_i \subseteq \prod_{i=1}^{m-1} y_i \circ c_1 \circ b_1$$
$$\subseteq \prod_{i=1}^{m-1} y_i \circ c_1 \circ a \circ d_1$$
$$\subseteq \prod_{i=1}^{m-1} y_i \circ c_1 \circ \prod_{i=1}^{n} x_i \circ d_1.$$

On the other hand,

$$b_1 \in a \circ d_1$$
$$\subseteq \prod_{i=1}^{m-1} y_i \circ y_m \circ d_1$$
$$\subseteq \prod_{i=1}^{m-1} y_i \circ c_1 \circ b_1 \circ d_1$$
$$\subseteq \prod_{i=1}^{m-1} y_i \circ c_1 \circ \prod_{i=1}^{n} x_i \circ d_1.$$

Denote $\prod_{i=1}^{k_1} z_{1,i} = \prod_{i=1}^{m-1} y_i \circ c_1 \circ \prod_{i=1}^{n} x_i \circ d_1$, thus $\{b_1\} \cup \prod_{i=1}^{m} y_i \subseteq \prod_{i=1}^{k_1} z_{1,i}$. Using again that H is a hypergroup, there exists $(c_2, d_2) \in H^2$ such that $z_{1,k_1} \in c_2 \circ b_2$ and $b_2 \in b_1 \circ d_2$. Denote $\prod_{i=1}^{k_2} z_{2,i} = \prod_{i=1}^{k_1-1} z_{1,i} \circ c_2 \circ \prod_{i=1}^{n} x_i \circ d_2$. Similarly as above, $\{b_2\} \cup \prod_{i=1}^{k_1} z_{1,i} \subseteq \prod_{i=1}^{k_2} z_{2,i}$ and hence $\{b_1, b_2\} \cup \prod_{i=1}^{m} y_i \subseteq \prod_{i=1}^{k_2} z_{2,i}$. After t steps, we obtain $\prod_{i=1}^{k_t} z_{t,i}$ such that $\{b_1, b_2, ..., b_t\} \cup \prod_{i=1}^{m} y_i \subseteq \prod_{i=1}^{k_t} z_{t,i}$, thus $\prod_{i=1}^{n} x_i \cup \prod_{i=1}^{m} y_i \subseteq \prod_{i=1}^{k_t} z_{t,i}$. Therefore $(x,y) \in \beta^*$ and hence $\rho_{\Re} \subseteq \beta^*$. This implies that $\rho_{\Re}^* \subseteq \beta^*$. Now let $(x,y) \in \beta^*$, so there exists $\prod_{i=1}^{n} x_i$ such that $x, y \in \prod_{i=1}^{n} x_i$. Since $\prod_{i=1}^{n} x_i \setminus \prod_{i=1}^{n} x_i = \emptyset$, $(x,y) \in \rho_{\Re}$ and hence $\beta^* \subseteq \rho_{\Re}^*$, thus $\rho_{\Re}^* = \beta^*$. $\square$

Remark 6.4. The relation $\overrightarrow{R}$ on $\mathcal{U}$ defined by

$$\prod_{i=1}^{n} x_i \;\overrightarrow{R}\; \prod_{i=1}^{m} y_i \Leftrightarrow \prod_{i=1}^{n} x_i \subseteq \prod_{i=1}^{m} y_i$$

is not symmetric and the induced strongly regular relation $\rho_{\overrightarrow{R}}^*$ coincides with the induced strongly regular relation $\rho_{\Re}^*$ of Theorem 6.26.

Theorem 6.27. *If $(H, \circ)$ is a (semi)hypergroup and $\Re$ is a strongly regular relation on $\mathcal{U}$, then a (semi)group structure can be defined on $H/\rho_{\Re}^*$ with respect to the operation:*

$$\rho_{\Re}^*(x) \otimes \rho_{\Re}^*(y) = \rho_{\Re}^*(z), \quad \text{where } z \in x \circ y.$$

Proof. We shall prove that the operation $\otimes$ is well defined. Set $\rho_{\Re}^*(x_0) = \rho_{\Re}^*(x_1)$ and $\rho_{\Re}^*(y_0) = \rho_{\Re}^*(y_1)$, it is necessary to verify that $\rho_{\Re}^*(x_0) \otimes \rho_{\Re}^*(y_0) = \rho_{\Re}^*(x_1) \otimes \rho_{\Re}^*(y_1)$. By hypothesis $(m, n) \in \mathbb{N}^2$, $(z_0, z_1, ..., z_m) \in H^{m+1}$ and

$(t_0, t_1, ..., t_n) \in H^{n+1}$ exist such that $z_0 = x_0$, $z_m = x_1$, $t_0 = y_0$, $t_n = y_1$, for all $1 \leqslant i \leqslant m$, $z_{i-1} \, \rho_{_\Re} \, z_i$ and for all $1 \leqslant j \leqslant n$, $t_{j-1} \, \rho_{_\Re} \, t_j$. Since $\Re$ is strongly regular, for all $u \in z_{s-1} \circ t_{s-1}$ and $v \in z_s \circ t_s$, where $1 \leqslant s \leqslant k$ and $k = min\{m, n\}$, we have $u \, \rho_{_\Re}^* \, v$. Therefore $\rho_{_\Re}^*(x_0) \otimes \rho_{_\Re}^*(y_0) = \rho_{_\Re}^*(z_1) \otimes \rho_{_\Re}^*(t_1) = ... = \rho_{_\Re}^*(z_k) \otimes \rho_{_\Re}^*(t_k) = \rho_{_\Re}^*(a_{k+i}) \otimes \rho_{_\Re}^*(b_{k+i})$, where $k + 1 \leqslant k + i \leqslant max\{m, n\}$ and:

$$(a_{k+i}, b_{k+i}) = \begin{cases} (x_1, t_{k+i}) & \text{if } k = m; \\ (z_{k+i}, y_1) & \text{if } k = n. \end{cases}$$

Hence $\otimes$ is well defined. $\square$

Definition 6.13. Let H be a hypergroup and let $p : H \to H/\rho_{_\Re}^*$ be the canonical projection. Denote by 1 the identity of the group $H/\rho_{_\Re}^*$. The set $p^{-1}(1)$ is called the $\Re$-*heart* of H and it is denoted by $\omega_{\Re,H}$.

Notice that if $\Re$ is the diagonal relation of $\mathcal{U}$ then the $\Re$-*heart* is just the heart of the hypergroup H.

Theorem 6.28. *If* $(H, \circ)$ *is a regular hypergroup and* $\Re$ *is a compatible relation with* $\circ$ *on* $\mathcal{U}$*, then* $\omega_{\Re,H}$ *is the smallest subhypergroup of* H*, which is also an* $\Re$*-part.*

Proof.   First, we check that $\omega_{\Re,H}$ is a subhypergroup of H. If $x, y \in \omega_{\Re,H}$ and $z \in x \circ y$, then $\rho_{_\Re}^*(z) = \rho_{_\Re}^*(x) \otimes \rho_{_\Re}^*(y) = 1$, the identity of the group $H/\rho_{_\Re}^*$. Hence $z \in \omega_{\Re,H}$. On the other hand, there exists $u \in H$, such that $x \in u \circ y$, whence $\rho_{_\Re}^*(x) = \rho_{_\Re}^*(u) \otimes \rho_{_\Re}^*(y)$, so $\rho_{_\Re}^*(u) = 1$. Therefore $u \in \omega_{\Re,H}$. This means that $\omega_{\Re,H} = \omega_{\Re,H} \circ y$ and similarly we obtain that $\omega_{\Re,H} = x \circ \omega_{\Re,H}$. It follows that $\omega_{\Re,H}$ is a subhypergroup of H.

By Theorems 6.23, 6.24 and Theorem 6.22, for all $x \in \omega_{\Re,H}$, $P_{\mathcal{L},\Re}(x) = \rho_{\mathcal{L},\Re}^*(x) = \rho_{_\Re}^*(x)$, which represents the identity of $H/\rho_{_\Re}^*$. On the other hand, $\rho_{_\Re}^*(x)$ represents the $\Re$-heart $\omega_{\Re,H}$, as a subset of H. Therefore, for all $x \in \omega_{\Re,H}$, $\omega_{\Re,H} = P_{\mathcal{L},\Re}(x)$, which is a $\mathcal{L}\Re_{\mathcal{U}}$-part of H, according to Theorems 6.23, 6.24. In fact, $P_{\mathcal{L},\Re}(x)$ is also an $\mathcal{R}\Re_{\mathcal{U}}$-part of H, by Theorem 6.22., hence it is an $\Re$-part of H. Moreover, $\omega_{\Re,H}$ is the smallest subhypergroup which is an $\Re$-part of H. Indeed, if K is a subhypergroup and an $\Re$-part of H, then for all $k \in K$ there is $e \in K$, such that $k \in e \circ k$, whence $\rho_{_\Re}^*(k) = \rho_{_\Re}^*(e) \otimes \rho_{_\Re}^*(k)$, so $e \in \omega_{\Re,H}$. Hence $P_{\mathcal{L},\Re}(e) = \omega_{\Re,H} \subseteq K$, since K is an $\Re$-part of H. $\square$

Theorem 6.29. *If* $(H, \circ)$ *is a regular hypergroup and* $\mathcal{RC}$ *is the set of all reflexive and compatible relations with* $\circ$ *on* $\mathcal{U}$*, then the heart of the hypergroup* H *is* $\omega_H = \cap_{\Re \in \mathcal{RC}} \, \omega_{\Re,H}$.

Proof. Notice that if $(x, y) \in \beta$ then $x, y \in \prod_{i=1}^{n} x_i$ for some elements x_i of H, where $i \in \{1, 2, ..., n\}$. Hence $\beta \subseteq \cap_{\Re \in \mathcal{RC}} \, \rho_{\Re}^*$. Conversely, $\cap_{\Re \in \mathcal{RC}} \, \rho_{\Re}^* \subseteq \beta$, since $\beta = \beta^*$ is ρ_{Id}^*, where Id is the diagonal relation on $\mathcal{U}$. Hence $\beta = \cap_{\Re \in \mathcal{RC}} \, \rho_{\Re}^*$.

From here it follows that $\omega_H = \cap_{\Re \in \mathcal{RC}} \, \omega_{\Re, H}$, since for all $x \in H$, $\beta(x) = 1$ if and only if $x \in \omega_H$, while for all $\Re \in \mathcal{RC}$, $\rho_{\Re}^*(x) = 1$ if and only if $x \in \omega_{\Re, H}$. $\square$

In the following theorem we introduce a strongly regular relation which can be a connection between general linear groups and hypergroups.

Theorem 6.30. *Let $(H, *)$ be a hypergroup and p be a prime number. If the relation $\Re_p$ on $\mathcal{U}$ is defined as follows:*

$$\Re_p = \left\{ (\prod_{i=1}^{n} x_i^s, \prod_{i=1}^{n} x_{\tau(i)}^t) \mid s, t \in \{1, p+1\}, n \in \mathbb{N}, \tau \in S_n \right\},$$

then $H/\rho_{\Re_p}^*$ is a p-elementary abelian group.

Proof. The proof follows from Theorem 6.22 and Theorem 6.27. $\square$

Corollary 6.2. *If $|H/\rho_{\Re_p}^*| = p^n$ then $Aut(H/\rho_{\Re_p}^*) \cong GL(n, p)$, where $GL(n, p)$ is the generalized linear group.*

Example 6.2. Let $p = 2$ and $H = S_3 \times S_3$, where S_3 is the permutation group of order 3, i.e.,

$$S_3 = \{(1), (12), (13), (23), (123), (132)\}.$$

Then $H/\rho_{\Re_p}^* \cong \mathbb{Z}_2 \times \mathbb{Z}_2$.

It is easy to see that for all $(x, y) \in S_3 \times S_3$, $\rho_{\Re_p}^*((x, y)) = \{(a, b) \in S_3 \times S_3 \mid xa, yb \text{ are even}\}$ therefore $H/\rho_{\Re_p}^* \cong \mathbb{Z}_2 \times \mathbb{Z}_2$.

Chapter 7

Join Spaces, Canonical Hypergroups and Lattices

Join spaces were introduced by W. Prenowitz [147; 148; 145; 146] to provide a common algebraic framework in which classical geometries could be axiomatized and studied. The underlying algebraic structure used was a hypergroup. Then this concept applied by him and J. Jantosciak both in Euclidian and in non-Euclidian geometry [149]. Using this notion, several branches of non-Euclidean geometry were rebuilt: descriptive geometry, projective geometry and spherical geometry. Then, several important examples of join spaces have been constructed in connection with binary relations, graphs, lattices, fuzzy sets and rough sets.

7.1 Join spaces and canonical hypergroups

In this section, we study the concept of join space. The main references for this section are [61; 92].

In order to define a join space, we recall the following notation: If a, b are elements of a hypergroupoid $(H, \circ)$, then we denote $a/b = \{x \in H \mid a \in x \circ b\}$. Moreover, by A/B we intend the set $\bigcup_{\substack{a \in A \\ b \in B}} a/b$.

Definition 7.1. A commutative hypergroup $(H, \circ)$ is called a *join space* if the following condition holds for all elements a, b, c, d of H:

$$a/b \cap c/d \neq \emptyset \ \Rightarrow \ a \circ d \cap b \circ c \neq \emptyset \quad \text{(transposition axiom)}.$$

Elements of H are called *points* and are denoted by $a, b, c, \ldots$. Sets of points are denoted by $A, B, C, \ldots$. The elementary algebra of join space theory for sets of points include the result

(1) $(A/B)/C = A/(B \circ C)$;

115

(2) $A \neq \emptyset$ implies $B \subseteq A/(A/B)$;

(3) $A \circ (B/C) \subseteq (A \circ B)/C$;

(4) $A/(B/C) \subseteq (A \circ C)/B$.

A set of points M is said to be *linear* if it is closed under join and extension $(a, b \in M$ imply $a \circ b \subseteq M$ and $a/b \subseteq M)$. If M is linear, then $M = M \circ M = M/M$. The linear sets are the closed subhypergroups of H. For a set of points A, the intersection of all linear sets containing A (the least linear set containing A) is denoted by $< A >$ and is called the linear space *spanned* by A. We use $< A, B >$ for $< A \cup B >$. Important results concerning linear sets are a formula for the linear span of two intersecting linear sets and a weak modularity property

(1) M and N linear and $M \cap N \neq \emptyset$ imply $< M, N > = M/N$;

(2) L, M and N linear, $L \cap M \neq \emptyset$ and $L \subseteq N$ imply $< L, M > \cap N = < L, M \cap N >$.

Both of these results depend on the transposition axiom.

A point e is said to be a *(scalar) identity* if $e \circ a = a$ for every point a. If H has an identity, it is unique. In a join space H with identity e the following hold:

(1) For each a there exists a unique a^{-1} such that $e \in a \circ a^{-1}$;

(2) $a^{-1} = e/a$;

(3) $a/b = a \circ b^{-1}$;

(4) $e \in M$ for any non-empty linear set M.

The transposition axiom is needed here to prove the uniqueness of the inverse point and that an extension reduce to a join. Join spaces with identity have been studied also in [126] by the name of canonical hypergroups.

Definition 7.2. A *spherical join space* is a join space with identity e that satisfies the axioms

$$a \circ a = a;$$
$$a/a = \{e, a, a^{-1}\}.$$

A *projective join space* is a join space with identity e that satisfies the axiom $a \circ a = a/a \subseteq \{e, a\}$.

Theorem 7.1. *If H is one of the classical join spaces, then the "line" spanned by points a and b, i.e., $< a, b >$ is given by*

(1) $a \cup b \cup a \circ b \cup a/b \cup b/a$, *for H descriptive;*

(2) $e \cup a \cup b \cup a^{-1} \cup b^{-1} \cup a \circ b \cup a \circ b^{-1} \cup a^{-1} \circ b \cup a^{-1} \circ b^{-1}$, *for H spherical;*

(3) $e \cup a \cup b \cup a \circ b$, *for H projective.*

Proof. It is straightforward. $\qquad\square$

Let M be a non-empty linear subset of the join space H. For any point a the *coset* of M in H containing a is denoted by $(a)_M$ and is given by $(a)_M = (a \circ M)/M$. The family $(H : M) = \{(a)_M \mid a \in H\}$ of all cosets of M in H is a partition of H. For the set of points A let

$$(A)_M = (A \circ M)/M = \cup\{(a)_M \mid a \in A\}.$$

Then, the join cosets $(a)_M$ and $(b)_M$ satisfy

$$(a)_M \circ (b)_M \subseteq (a \circ b)_M,$$

so that the equivalence relation on H corresponding to the partition of H into cosets of M is a regular equivalence relation.

Theorem 7.2. *The family* $(H : M)$ *of cosets M in H under the join operation*

$$(a)_M \star (b)_M = \{(x)_M \mid x \in (a)_M \circ (b)_M\}$$

is a join space with identity $M = (m)_M$, where $m \in M$.

Proof. Note that $(a)_M^{-1} = M/a$ and that $a \circ b \cap M \neq \emptyset$ if and only if $(a)_M^{-1} = (b)_M$. $\qquad\square$

$(H : M)$ under $\star$ is known as the *factor join space H modulo M.*

Analogues of the three classical isomorphism theorems of group theory hold for join spaces. Let $(H, \circ)$ and $(K, \cdot)$ be join spaces. A mapping $\varphi : H \to K$ is said to be a *good homomorphism* if $\varphi(a \circ b) = \varphi(a) \cdot \varphi(b)$. If φ is also one to one and onto, then φ is said to be an *isomorphism* and the notation $H \cong K$ is used. If φ is a good homomorphism and K has identity e, then $\{x \in H \mid \varphi(x) = e\}$ is called the *kernel* of φ and is denoted by $ker\varphi$.

Theorem 7.3 (First Isomorphism Theorem). *Let H and K be join spaces. Let K has identity. Let φ be a good homomorphism of H onto K. Then, $ker\varphi$ is a linear subset of H and $(H : ker\varphi) \cong K$.*

Theorem 7.4 (Second Isomorphism Theorem). *Let H be a join space. Let M and N be linear subsets such that $M \cap N \neq \emptyset$. Then,*

$$(< M, N >: M) \cong (N : M \cap N).$$

Theorem 7.5 (Third Isomorphism Theorem). *Let H be a join space. Let M and N be linear subsets such that $\emptyset \neq N \subseteq M$. Then, $(M : N)$ is a linear subset of $(H : N)$ and $(H : M) \cong ((H : N) : (M : N))$.*

A version of Jordan-Hölder theorem also holds for join spaces.

Theorem 7.6 (Jordan-Hölder Theorem). *Let H be a join space. Let M and N be linear subsets such that $\emptyset \neq N \subseteq M$. Suppose that*

$$N = A_0 \subseteq \cdots \subseteq A_m = M \text{ and } N = B_0 \subseteq \cdots \subseteq B_n = M$$

where each A_i and each B_j are linear and for $i = 1, \ldots, m$ and for $j = 1, \ldots, n$ that A_{i-1} and B_{j-1} are maximal proper linear subsets of A_i and B_j respectively. Then, $m = n$ and there exists a one to one corresponding between the families of factor spaces

$$\{(A_i : A_{i-1}) \mid i = 1, \cdots, m\} \text{ and } \{(B_j : B_{j-1}) \mid j = 1, \cdots, n\}$$

such that the correspondents are isomorphic.

Proof. The chains $A_0 \subseteq \cdots \subseteq A_m$ and $B_0 \subseteq \cdots \subseteq B_n$ are refined respectively by

$$A_{i,j} = \; < A_{i-1}, A_i \cap B_j > \text{ for } i = 1, \cdots, m \text{ and } j = 0, \cdots, n$$

and

$$B_{j,i} = \; < B_{j-1}, B_j \cap A_i > \text{ for } i = 0, \cdots, m \text{ and } j = 1, \cdots, n.$$

Thus,

$$A_0 = A_{1,0} \subseteq \cdots \subseteq A_{1,n} = A_1 = A_{2,0} \subseteq \cdots \subseteq A_{m,n} = A_m$$

and

$$B_0 = B_{1,0} \subseteq \cdots \subseteq B_{1,m} = B_1 = B_{2,0} \subseteq \cdots \subseteq B_{n,m} = B_n.$$

Next, by the second isomorphism theorem followed by the weak modularity property

$$
\begin{aligned}
(A_{i,j} : A_{i,j-1}) &= (< A_{i-1}, A_i \cap B_j >:< A_{i-1}, A_i \cap B_{j-1} >) \\
&= (<< A_{i-1}, A_i \cap B_{j-1} >, A_i \cap B_j >:< A_{i-1}, A_i \cap B_{j-1} >) \\
&\cong (A_i \cap B_j :< A_{i-1}, A_i \cap B_{j-1} > \cap (A_i \cap B_j)) \\
&= (A_i \cap B_j :< A_{i-1} \cap (A_i \cap B_j), A_i \cap B_{j-1} >) \\
&= (A_i \cap B_j :< A_{i-1} \cap B_j, A_i \cap B_{j-1} >).
\end{aligned}
$$

Similarly, we have

$$(B_{j,i} : B_{j,i-1}) \cong (B_j \cap A_i :< B_{j-1} \cap A_i, B_j \cap A_{i-1} >).$$

Hence,

$$(A_{i,j} : A_{i,j-1}) \cong (B_{j,i} : B_{j,i-1}) \text{ for } i = 1, \cdots, m \text{ and } j = 1, \cdots, n.$$

Therefore, given the maximality condition on A_{i-1} and B_{j-1} in A_i and B_j respectively, the conclusion of the theorem readily follows. $\qquad\square$

If N is a closed subhypergroup of a join space H and $\{x, y\} \subseteq H$, then we define the following binary relation: xJ_Ny if $x \circ N \cap y \circ N \neq \emptyset$.

Theorem 7.7. J_N *is an equivalence relation on H and the equivalence class of an element a is $J_N(a) = (a \circ N)/N$. In particular, $J_N(a) = N$ for all $a \in N$.*

Proof. Clearly, J_N is reflexive and symmetric. Now, suppose that $a \circ N \cap b \circ N \neq \emptyset$ and $b \circ N \cap c \circ N \neq \emptyset$. It follows that $b \in (a \circ N)/N \cap (c \circ N)/N$ and since $(H, \circ)$ is a join space, we obtain $a \circ N \cap c \circ N \neq \emptyset$, which means that aJ_Nc. Hence, J_N is also transitive, and so it is an equivalence relation on H. We check now that for all $a \in H$ we have $J_N(a) = (a \circ N)/N$. If $d \in J_N(a)$, then $d \circ N \cap a \circ N \neq \emptyset$. Hence, there exist $v \in a \circ N$ and $m \in N$ such that $v \in d \circ m$, whence it follows that $d \in v/m \subseteq (a \circ N)/N$. We obtain $J_N(a) \subseteq (a \circ N)/N$. Now, let $y \in (a \circ N)/N$. Then, there exist $u \in a \circ N$ and $m \in N$, such that $u \in y \circ m$, whence $y \circ N \cap a \circ N \neq \emptyset$, which means that $y \in J_N(a)$ and so, $(a \circ N)/N \subseteq J_N(a)$. Clearly, if $a \in N$, then $J_N(a) = N$, since N is closed. $\quad\square$

Canonical hypergroups are an important class of hypergroups, which are the additive structures of Krasner hyperrings and hyperfields. They were introduced and analyzed especially by Mittas [126; 127; 128; 129]. Several other classes of hyperstructures, connected with canonical hypergroups have been considered. Corsini [22; 25; 26] studied a particular class of canonical hypergroups, called sd-hypergroups, Roth [159; 160] used canonical hypergroups in order to prove theorems in the finite group character theory. Moreover, Prenowitz and Jantosciak [149] emphasized the role of canonical hypergroups in geometry, McMullen and Price [120; 121] used a generalization of them in harmonical analysis and particle physics. Serafimidis, Konstantinidou, Mittas [162; 163] and Corsini [23; 24] analyzed strongly canonical and i.p.s. hypergroups. Quasi-canonical hypergroups, also called polygroups, were investigated especially by Comer [18]–[20] in connections with graphs, relations, Boole and cylindric algebras. Feebly canonical hypergroups were studied by Corsini [25] and De Salvo [65]. Canonical hypergroups can be characterized using join spaces, which are very useful especially for their applications in geometry [149].

Definition 7.3. We say that a hypergroup $(H, \circ)$ is *canonical* if

    (1) it is commutative,

(2) it has a scalar identity (also called scalar unit), which means that

$$\exists e \in H, \ \forall x \in H, \ x \circ e = e \circ x = x,$$

(3) every element has a unique inverse, which means that for all $x \in H$, there exists a unique $x^{-1} \in H$, such that $e \in x \circ x^{-1} \cap x^{-1} \circ x$,

(4) it is reversible, which means that if $x \in y \circ z$, then there exist the inverses y^{-1} of y and z^{-1} of z, such that $z \in y^{-1} \circ x$ and $y \in x \circ z^{-1}$.

Clearly, the identity of a canonical hypergroup is unique. Indeed, if e is a scalar identity and e' is an identity of a canonical hypergroup $(H, \circ)$, then we have $e \in e \circ e' = \{e'\}$.

Some interesting examples of a canonical hypergroup is the following ones (see [22]).

Example 7.1. Let $C(n) = \{e_0, e_1, \cdots, e_{k(n)}\}$, where $k(n) = n/2$ if n is an even natural number and $k(n) = (n-1)/2$ if n is an odd natural number. For all e_s, e_t of $C(n)$, define $e_s \circ e_t = \{e_p, e_v\}$, where $p = \min\{s + t, n - (s + t)\}$, $v = |s - t|$. Then $(C(n), \circ)$ is a canonical hypergroup.

Example 7.2. Let (S, T) be a projective geometry, i.e., a system involving a set S of elements called *points* and a set T of sets of points called *lines*, which satisfies the following postulates:

- Any lines contains at least three points;
- Two distinct points a, b are contained in a unique line, that we shall denote by $L(a, b)$;
- If a, b, c, d are distinct points and $L(a, b) \cap L(c, d) \neq \emptyset$, then $L(a, c) \cap L(b, d) \neq \emptyset$.

Let e be an element which does not belong to S and let $S' = S \cup \{e\}$. We define the following hyperoperation on S':

- For all different points a, b of S, we consider $a \circ b = L(a, b) \backslash \{a, b\}$;
- If $a \in S$ and any line contains exactly three points, let $a \circ a = \{e\}$, otherwise $a \circ a = \{a, e\}$;
- For all $a \in S'$, we have $e \circ a = a \circ e = a$.

Then $(S', \circ)$ is a canonical hypergroup.

Theorem 7.8. *If $(H, \circ)$ is a canonical hypergroup, then the following implication holds for all x, y, z, t of H:*

$$x \circ y \cap z \circ t \neq \emptyset \ \Rightarrow \ x \circ z^{-1} \cap t \circ y^{-1} \neq \emptyset.$$

Proof. Let $u \in x \circ y \cap z \circ t$. Since H is reversible, we obtain $u^{-1} \in z^{-1} \circ t^{-1}$, whence $u \circ u^{-1} \subseteq x \circ y \circ z^{-1} \circ t^{-1}$. If e is an identity of H, we obtain $e \in (x \circ z^{-1}) \circ (t \circ y^{-1})^{-1}$. Hence there exists an element $v \in x \circ z^{-1} \cap t \circ y^{-1}$. $\square$

Theorem 7.9. *A commutative hypergroup is canonical if and only if it is a join space with a scalar identity.*

Proof. Suppose that $(H, \circ)$ is a canonical hypergroup. For all a, b of H we have $a/b = a \circ b^{-1}$. Then the implication $\Rightarrow$ follows by the above theorem.

Conversely, let us check that the inverse of an element is unique. Let e be the scalar identity. If $e \in a \circ b \cap a \circ c$, then $a \in e/b \cap e/c$, whence it follows that $e \circ c \cap e \circ b \neq \emptyset$, hence $b = c = a^{-1}$. Let us check now the reversibility of H. We have $a \in b \circ c$ if and only if $b \in a/c$. From $e \in b \circ b^{-1}$ we obtain $b \in e/b^{-1}$, hence $a \circ b^{-1} \cap e \circ c \neq \emptyset$, which means that $c \in a \circ b^{-1}$. Therefore, H is canonical. $\square$

Theorem 7.10. *If $(H, \circ)$ is a join space and N is a closed subhypergroup of H, then the quotient $(H/J_N, \otimes)$ is a canonical hypergroup, where for all $\overline{a}, \overline{b}$ of H/J_N, we have $\overline{a} \otimes \overline{b} = \{\overline{c} \mid c \in a \circ b\}$.*

Proof. First, we check that the hyperoperation $\otimes$ is well defined. In other words, we have to check that if $a_1 J_N a_2$ and $x \in H$, then for all $z \in a_1 \circ x$, there exists $w \in a_2 \circ x$, such that $z J_N w$. Indeed, from $a_1 J_N a_2$, it follows there exist m, n of N and v of H, such that $v \in a_1 \circ m \cap a_2 \circ n$. If $z \in a_1 \circ x$, then we have $a_1 \in z/x \cap v/m$, hence $z \circ N \cap v \circ x \neq \emptyset$, whence $z \circ N \cap N \circ a_2 \circ x \neq \emptyset$. It follows that there exists $w \in a_2 \circ x$, such that $z J_N w$. Therefore the hyperoperation $\otimes$ is well defined. Since $(H, \circ)$ is a join space, it follows that $(H/J_N, \otimes)$ is a join space, too. Moreover, notice that N is a scalar identity for $(H/J_N, \otimes)$, and according to the above theorem, we obtain that $(H/J_N, \otimes)$ is a canonical hypergroup. $\square$

7.2 Join spaces determined by lattices

Connections between hypergroups and lattices have been considered and analyzed by Nakano [136], Varlet [176], then by Mittas and Konstantinidou [130], Comer [17] and later by Kehagias [95], Kehagias and Konstantinidou [96], Konstantinidou and Serafimidis [98], Calugareanu and Leoreanu [13], Leoreanu and Radu [112], Tofan and Volf [175].

In this section, we consider and investigate two types of hyperoperations determined by lattices, see [111]. First, we show that we obtain a family of join spaces if the corresponding lattice is distributive. A second type of hypergroups associated with a lattice is analyzed in the next paragraph. Such hypergroups are join spaces if the corresponding lattice is modular.

In [175] Tofan and Volf introduced the following hypergroup, associated with a lattice, through the following map μ:

Let H be a non-empty set, $\mathbf{L} = (L, \vee, \wedge)$ a lattice, $\mu : H \to L$ a map, such that $\mu(H)$ is a sublattice of L. For all $a, b \in H$, set

$$a * b = \{c \mid \mu(a) \wedge \mu(b) \leq \mu(c) \leq \mu(a) \vee \mu(b)\}.$$

Since $\mu(H)$ is a sublattice of $\mathbf{L}$, we can as well assume that μ is an onto map, by taking $\mu(H)$ as L from the beginning. Further, if $a, a', b, b' \in H$ are such that $\mu(a) = \mu(a')$ and $\mu(b) = \mu(b')$, then $a * b = a' * b'$. From this and the surjectivity of μ it follows that to study the properties of $(H, *)$ we can replace it by $(L, *)$, where for all $c, d \in L$ we set $c * d$ to be the interval $[c \wedge d, c \vee d] = \{x \in L \mid c \wedge d \leq x \leq c \vee d\}$. This hyperstructure is frequently used in machine learning applications.

In [176], Varlet provided the following characterization of distributive lattices, in terms of join spaces:

Theorem 7.11. *Let* $\mathbf{L} = (L, \vee, \wedge)$ *be a lattice.* $\mathbf{L}$ *is a distributive lattice if and only if* $(L, *)$ *is a join space.*

Proof. Suppose that L is distributive. We check the associativity law for $*$. Let a, b, c be arbitrary in L. The least and the greatest elements of $a * (b * c)$ are $a \wedge b \wedge c$ and $a \vee b \vee c$. Let x be an element of $[a \wedge b \wedge c, a \vee b \vee c]$. If $y = (x \wedge (b \vee c)) \vee (b \wedge c)$ then $y \in b * c$. Moreover $c \in a * y$. Using distributy, we have

$$a \wedge ((x \wedge (b \vee c)) \vee (b \wedge c)) = (a \wedge x \wedge (b \vee c)) \vee (a \wedge b \wedge c) \leq x$$

and

$$a \vee ((x \wedge (b \vee c)) \vee (b \wedge c)) = (a \vee (b \wedge c) \vee x) \wedge (a \vee (b \wedge c) \vee (b \vee c)) \geq x.$$

Thus $x \in a * (b * c)$ and $a * (b * c) = [a \wedge b \wedge c, a \vee b \vee c]$. Similarly, we have $(a * b) * c = [a \wedge b \wedge c, a \vee b \vee c]$, so we obtain the associativity law.

Now, if $x \in a/b \cap c/d$, we check that there exists $y \in L$ such that $y \in a * d \cap b * c$, which means that $(a \wedge d) \vee (b \wedge c) \leq y \leq (a \vee d) \wedge (b \vee c)$. From $b \wedge x \leq a \leq b \vee x$ and $d \wedge x \leq c \leq d \vee x$, it follows that

$$a \wedge d \leq (b \vee x) \wedge d = (b \wedge d) \vee (x \wedge d) \leq (b \wedge d) \vee c \leq b \vee c.$$

Since $a \wedge d \leq b \vee c$, it follows that $(a \wedge d) \vee (b \wedge c) \leq b \vee c$. Similarly, $(b \wedge c) \vee (a \wedge edged) \leq a \vee d$. Therefore

$$(a \wedge d) \vee (b \wedge c) \leq (a \vee d) \wedge (b \vee c),$$

which means that $a * d \cap b * c \neq \emptyset$.

Conversely, notice that $a/b \cap b/d \neq \emptyset$ implies $a * d \cap b * b \neq \emptyset$ and since $b * b = \{b\}$, it follows that $b \in a * d$.

Hence $a/b \cap b/a \neq \emptyset$ implies that $b \in a * a = \{a\}$, whence $a = b$. Suppose that L is not distributive. Then it contains a five-element sublattice $\{a, b, c, d, e\}$ with $a \vee c = b \vee c = e$, $a \wedge c = b \wedge c = d$ and either $a > b$ or a, b, c are mutually non-comparable. In both cases, a/b contains a and c, but not d.

We have $c \in a/b \cap b/a$ and $a \neq b$, a contradiction.

Therefore L is a distributive lattice. $\qquad \square$

Remark 7.1. If **L** is a distributive lattice, then

$$(a * b) * c = a * (b * c) = [a \wedge b \wedge c, a \vee b \vee c].$$

Notice that for all $a \in L$, $a^{n+1} = a^n * a = \{a\}$ since

$$a * a = \{z \in L \mid a \wedge a \leq z \leq a \vee a\} = \{a\}.$$

First, we shall investigate when a join space is associated with a distributive lattice **L** as above.

Theorem 7.12. *Let* $\mathbf{L} = (L, \vee, \wedge)$ *be a lattice and* $P = \{A_i\}_{i \in L}$ *be a partition of a non-empty set* H. *We define the following hyperoperation on* H:

if $a \in A_i$ *and* $b \in A_j$, *then* $a \circ_P b = \bigcup_{i \wedge j \leq k \leq i \vee j} A_k$. *Then*

(1) $(H, \circ_P)$ *is a hypergroup;*

(2) **L** *is distributive if and only if* $(H, \circ_P)$ *is a join space.*

Proof. (1) It is sufficient to notice that

$$a \circ_P b \circ_P c = \bigcup_{i \wedge j \wedge k \leq s \leq i \vee j \vee k} A_s,$$

where $a \in A_i$, $b \in A_j$ and $c \in A_k$.

(2) $\Rightarrow$ Suppose that $a \in A_i$, $b \in A_j$, $c \in A_k$, $d \in A_t$, $u \in A_s$ are such that $a \in u \circ_P b$ and $c \in u \circ_P d$. We check that $a \circ_P d \cap b \circ_P c \neq \emptyset$. We have $s \wedge j \leq i \leq s \vee j$ and $s \wedge t \leq k \leq s \vee t$. Let $\alpha = (i \wedge t) \vee (k \wedge j)$. We have

$$k \wedge j \leq (s \vee t) \wedge j = (s \wedge j) \vee (t \wedge j) \leq (s \wedge j) \vee t \leq i \vee t,$$

hence $i \wedge t \leq \alpha \leq i \vee t$. Similarly, we get $k \wedge j \leq \alpha \leq k \vee j$.

Hence, for all $v \in A_\alpha$, we have $v \in a \circ_P d \cap b \circ_P c$, which means that $(H, \circ_P)$ is a join space.

$\Leftarrow$ As it is well known, **L** is not distributive only if it contains at least one of two 5-element sublattices. This can be expressed as the existence of pairwise distinct $i, j, k, t, s \in L$ such that $j \not< t, j \vee s = k = t \vee s, j \wedge s = i = t \wedge s$. Let $a \in A_i, b \in A_j, c \in A_k, d \in A_t, u \in A_s$. Then $a \in b \circ_P u, c \in d \circ_P u$, but $a \circ_P d$ and $b \circ_P c$ are disjoint. This means that $(H, \circ_P)$ is not a join space, which is a contradiction. Hence **L** is distributive. $\square$

Remark 7.2. If we consider the least partition of L: $P = \{\{a\} \mid a \in L\}$ then the hyperoperations "$*$" and "$\circ_P$" coincide.

Definition 7.4. We say that a hypergroup $(H, \circ)$ is an *L-hypergroup*, for a lattice L, if there is a partition P of H, such that "$\circ$" and "$\circ_P$" coincide.

We can conclude:

Corollary 7.1. *A join space $(L, \circ)$ is the join space associated with a distributive lattice **L** if and only if it is an L-hypergroup.*

Theorem 7.13. *Let **L** be a distributive lattice. If $n \in \mathbb{N}$, $n \geq 2$, and $(a_1, ..., a_n) \in L^n$, then $\prod\limits_{i=1}^{n} a_i$ is a subhypergroup of $(L, *)$.*

Proof.   Let us check that for any $a \in \prod\limits_{i=1}^{n} a_i$, we have $a * \prod\limits_{i=1}^{n} a_i = \prod\limits_{i=1}^{n} a_i$.

First, recall that $\prod\limits_{i=1}^{n} a_i = \{v \in L \mid a_1 \wedge ... \wedge a_n \leq v \leq a_1 \vee ... \vee a_n\}$. We have $a * \prod\limits_{i=1}^{n} a_i = \{v \in L \mid a \wedge a_1 \wedge ... \wedge a_n \leq v \leq a \vee a_1 \vee ... \vee a_n\} = \prod\limits_{i=1}^{n} a_i$, since $a \in \prod\limits_{i=1}^{n} a_i$. $\square$

We establish now some connections among the operations $\vee$ and $\wedge$ of the lattice **L** and the hyperoperation "$*$" of L.

For any $A, B \subset L$, denote

$$A \vee B = \{a \vee b \mid a \in A, \; b \in B\},$$
$$A \wedge B = \{a \wedge b \mid a \in A, \; b \in B\}.$$

Theorem 7.14. *Let **L** be a distributive lattice. For all a, b, c of L, we have $a * (b \vee c) = (a * b) \vee (a * c)$ and $a * (b \wedge c) = (a * b) \wedge (a * c)$.*

Proof. We check the first equality, since the second one follows by duality.

$\subseteq$: Let $u \in a * (b \vee c)$. Set $x = (a \vee b) \wedge u$ and $y = (a \vee c) \wedge u$. We have

$$x \vee y = u \wedge (a \vee b \vee c).$$

Since $u \in a * (b \vee c)$, it follows $x \vee y = u$. It is sufficient to notice that $x \in a * b$ and $y \in a * c$.

$\supseteq$: Let $v \in (a * b) \vee (a * c)$, that is, there are $x \in a * b$ and $y \in a * c$, such that $v = x \vee y$. Then

$$a \wedge (b \vee c) \leq x \vee y = v \leq a \vee b \vee c.$$

This completes the proof. $\square$

In what follows we introduce a new hyperoperation on a lattice $\mathbf{L} = (L, \vee, \wedge)$. Let $p, q \in L$ be arbitrary. For all $a, b \in L$ set

$$a *_{pq} b = [a \wedge b \wedge p, a \vee b \vee q].$$

We show that "$*$" is a special case of "$*_{pq}$". If $\mathbf{L}$ does not have a least element 0 or a greatest element 1, then we enlarge $\mathbf{L}$ to $\mathbf{L}'$, so that $\mathbf{L}'$ is bounded. Then the hyperoperation "$*_{pq}$" restricted to $\mathbf{L}$ is just "$*$".

Theorem 7.15.

(1) *If $p, p', q, q' \in L$ satisfy $p' \leq p$ and $q \leq q'$ then for all $a, b \in L$,*

$$a *_{pq} b \subseteq a *_{p'q'} b.$$

(2) *For all a, b, p of L, we have $a *_{pq} a \subseteq a *_{pq} b$.*

Proof. It is straightforward. $\square$

Theorem 7.16. *If $\mathbf{L}$ is distributive and $p, q \in L$, then $(L, *_{pq})$ is a join space.*

Proof. *Claim 1. Let $a, b, c \in L$ be arbitrary. Then*

(i) $(a *_{pq} b) *_{pq} c \subseteq [a \wedge b \wedge c \wedge p, a \vee b \vee c \vee q]$;

(ii) *If $\mathbf{L}$ is distributive then*

$$(a *_{pq} b) *_{pq} c = [a \wedge b \wedge c \wedge p, a \vee b \vee c \vee q] \qquad (7.1)$$

Proof of the Claim 1. (i) Let $u \in (a *_{pq} b) *_{pq} c$. Then there exists $v \in [a \wedge b \wedge p, a \vee b \vee q]$ such that $u \in [v \wedge c \wedge p, v \vee c \vee q]$. Clearly

$$a \wedge b \wedge c \wedge p \leq v \wedge c \wedge p \leq u \leq v \vee c \vee q \leq a \vee b \vee c \vee q.$$

(ii) Let $t \in [a \wedge b \wedge c \wedge p, a \vee b \vee c \vee q]$ be arbitrary. Set
$$s = (a \wedge b) \vee (a \wedge t) \vee (b \wedge t).$$
Clearly $s \leq a \vee a \vee b = a \vee b \leq a \vee b \vee q$. It is well known that in a distributive lattice
$$s = (a \vee b) \wedge (a \vee t) \wedge (b \vee t) \geq a \wedge a \wedge b \geq a \wedge b \wedge p \qquad (7.2)$$
Thus $s \in a *_{pq} b$. We prove that $t \in s *_{pq} c$. First, by distributivity and $a \wedge b \wedge c \wedge p \leq t$,
$$s \wedge c \wedge p = [(a \wedge b) \vee (a \wedge t) \vee (b \wedge t)] \wedge c \wedge p$$
$$= (a \wedge b \wedge c \wedge p) \vee (a \wedge t \wedge c \wedge p) \vee (b \wedge t \wedge c \wedge p) \leq t.$$
Similarly, using s from (7.2), distributivity and $a \vee b \vee c \vee q \geq t$ we obtain
$$s \vee c \vee q = [(a \vee b) \wedge (a \vee t) \wedge (b \vee t)] \vee c \vee q$$
$$= (a \vee b \vee c \vee q) \wedge (a \vee t \vee c \vee q) \wedge (b \vee t \vee c \vee q) \geq t.$$
Thus $t \in s *_{pq} c$. This proves $\supseteq$ in (7.1).

Claim 2. If **L** *is distributive, then for all* $a, b, c \in L$,
$$(a *_{pq} b) *_{pq} c = [a \wedge b \wedge c \wedge p, a \vee b \vee c \vee q] = a *_{pq} (b *_{pq} c).$$

Proof of the Claim 2. The first equality is Claim 1. Clearly, $(L, *_{pq})$ is commutative and so
$$a *_{pq} (b *_{pq} c) = (b *_{pq} c) *_{pq} a = [a \wedge b \wedge c \wedge p, a \vee b \vee c \vee q] = (a *_{pq} b) *_{pq} c.$$

We have shown that $*_{pq}$ is associative and commutative. For all $x, h \in L$, clearly $x, h \in x *_{pq} h$ and so $(L, *_{pq})$ is a commutative hypergroup.

To prove that $(L, *_{pq})$ is a join space, let $a, b, c, d, x \in L$ satisfy $a \in b *_{pq} x$ and $c \in d *_{pq} x$. Then
$$b \wedge x \wedge p \leq a \leq b \vee x \vee q, \qquad (7.3)$$
$$d \wedge x \wedge p \leq c \leq d \vee x \vee q. \qquad (7.4)$$
Set $z = (a \wedge d \wedge p) \vee (b \wedge c \wedge p)$. Clearly $a \wedge d \wedge p \leq z$. Moreover, from (7.3) and (7.4),
$$a \wedge d \wedge p \leq (a \wedge d \wedge p) \vee (b \wedge c \wedge p)) = z \leq ((b \vee x \vee q) \wedge d \wedge p) \vee (b \wedge (d \vee x \vee q) \wedge p)$$
$$= ((b \vee x \vee q) \wedge d \wedge p) \vee (b \wedge d \wedge p) \vee (b \wedge x \wedge p) \vee (b \wedge q \wedge p) \leq d \vee a \vee q$$
proving $z \in a *_{pq} d$. Similarly,
$$b \wedge c \wedge p \leq (a \wedge d \wedge p) \vee (b \wedge c \wedge p)) = z \leq ((b \vee x \vee q) \wedge d \wedge p) \vee (b \wedge (d \vee x \vee q) \wedge p)$$
$$= (b \wedge d \wedge p) \vee (x \wedge d \wedge p) \vee (q \wedge d \wedge p) \vee (b \wedge (d \vee x \vee q) \wedge p) \leq b \vee c \vee q$$
proving $z \in b *_{pq} c$. Thus $z \in (a *_{pq} d) \cap (b *_{pq} c)$. $\qquad \square$

Remark 7.3. Let **L** be distributive and $p, q \in L$. Then for all $a \in L$

$$a *_{pq} a *_{pq} a = [a \wedge a \wedge a \wedge p, a \vee a \vee a \vee q] = [a \wedge p, a \vee q] = a *_{pq} a.$$

Theorem 7.17. *Let* **L** *be distributive and* $p, q, a_1, ..., a_n \in L$. *Then the set* $A = a_1 *_{pq} ... *_{pq} a_n$ *is closed under* $*_{pq}$ *(in the sense that* $x *_{pq} y \subseteq A$ *for all* $x, y \in A$*).*

Proof. Let $x, y \in A = [a_1 \wedge ... \wedge a_n \wedge p, a_1 \vee ... \vee a_n \vee q]$. Then

$$x *_{pq} y = [x \wedge y \wedge p, x \vee y \vee q] \subseteq A.$$

This completes the proof. □

Remark 7.4. If $a, b, p, q \in L$ satisfy $a \wedge b \leq p$ and $a \vee b \geq q$ then $a *_{pq} b = a * b$.

Theorem 7.18. *Let* **L** *be distributive,* $a, b, p, q, r, s \in L$ *and* $u = p \vee r, v = q \wedge s$. *Then*

$$(a *_{pq} b) \cap (a *_{rs} b) = a *_{uv} b. \tag{7.5}$$

Proof. The left-hand side of (7.5) is the interval $[c, d]$ where

$$c = (a \wedge b \wedge p) \vee (a \wedge b \wedge r) = a \wedge b \wedge (p \vee r) = a \wedge b \wedge u$$

and similarly $d = a \vee b \vee v$. □

Theorem 7.19. *Let* **L** *be distributive,* $a, b, p, q \in L$. *Then*

$$(a *_{pq} a) \cap (b *_{pq} b) = (a \wedge b) *_{pq} (a \wedge b). \tag{7.6}$$

Proof. The left-hand side of (7.6) is

$$[a \wedge p, a \vee q] \cap [b \wedge p, b \vee q] = [(a \wedge p) \vee (b \wedge p), (a \vee q) \wedge edge(b \vee q)]$$
$$= [(a \vee b) \wedge p, (a \wedge b) \vee q] = (a \vee b) *_{pq} (a \wedge b).$$

This completes the proof. □

Theorem 7.20. *Let* **L** *be distributive,* $a, b, c, p, q \in L$. *Then*

(1) $a \wedge (b *_{pq} c) \subseteq (a \wedge b) *_{pq} (a \wedge c)$;

(2) $a \vee (b *_{pq} c) \subseteq (a \vee b) *_{pq} (a \vee c)$;

(3) $(a *_{pq} b) \vee (a *_{pq} c) = a *_{pq} (b \vee c)$;

(4) $(a *_{pq} b) \wedge (a *_{pq} c) = a *_{pq} (b \wedge c)$.

Proof.   (1) $a \vee (b *_{pq} c) = a \wedge [b \wedge c \wedge p, b \vee c \vee q] = [a \wedge b \wedge c \wedge p, a \wedge (b \vee c \vee q)]$
$= [a \wedge b \wedge c \wedge p, (a \wedge b) \vee (a \wedge c) \vee (a \wedge q)] \subseteq [(a \wedge b) \wedge (a \wedge c) \wedge p, (a \wedge b) \vee (a \wedge c) \vee q]$
$= (a \wedge b) *_{pq} (a \wedge c)$.

(2) It follows from (i) (for qp) by duality.

(3) $(a *_{pq} b) \vee (a *_{pq} c) = [a \wedge b \wedge p, a \vee b \vee q] \wedge [a \wedge c \wedge p, a \vee c \vee q]$
$= [(a \wedge b \wedge p) \vee (a \wedge c \wedge p), (a \vee b \vee q) \vee (a \vee c \vee q)] = [a \wedge (b \vee c) \wedge p, a \vee b \vee c \vee q]$
$= a *_{pq} (b \vee c)$.

(4) It follows from (3) (for qp) by duality. $\square$

Theorem 7.21. *Let* **L** *be distributive,* $a, b, c, p \in L$ *be such that* $a *_{pq} c = b *_{pq} c$ *and* $a \wedge c = b \wedge c$. *Then* $a \vee q = b \vee q$.

Proof.   By hypothesis, it follows that

$$a \vee c \vee q = b \vee c \vee q.$$

On the other hand,

$$(a \wedge c) \vee q = (b \wedge c) \vee q.$$

Then we get

$$(a \vee q) \vee (c \vee q) = (b \vee q) \vee (c \vee q)$$

and

$$(a \vee q) \wedge (c \vee q) = (b \vee q) \wedge (c \vee q)$$

and so $a \vee q = b \vee q$, by distributivity. $\square$

Similarly, we obtain

Theorem 7.22. *Let* **L** *be distributive,* $a, b, c, p \in L$ *be such that* $a *_{pq} c = b *_{pq} c$ *and* $a \vee c = b \vee c$. *Then* $a \wedge p = b \wedge p$.

Let **L** $= (L, \vee, \wedge)$ be a lattice. In [136] Nakano introduced the following hyperoperation on L: $a \circ b = \{c \in H \mid a \vee b = a \vee c = b \vee c\}$. Later, Comer [17] showed that:

Theorem 7.23. **L** *is a modular lattice if and only if* $(L, \circ)$ *is a join space.*

Proof. Set $S = \{y \in L | a \vee b \vee y = a \vee c \vee y = b \vee c \vee y = a \vee b \vee c\}$, for all $a, b, c \in L$. We have $(a \circ b) \circ c \subseteq S$. Indeed, if $y \in (a \circ b) \circ c$, then there exists $x \in L$ such that $a \vee x = b \vee x = a \vee b$ and $x \vee c = x \vee y = y \vee c$ so $(a \vee b) \vee y = (a \vee x) \vee y = a \vee (x \vee c) = a \vee (c \vee y) = (a \vee c) \vee y$ and $(a \vee b) \vee y = (b \vee x) \vee y = b \vee (x \vee y) = b \vee (c \vee y) = (b \vee c) \vee y$. Hence $y \in S$.

Now, if $y \in S$, then $b \vee c \leq b \vee y \vee a$, $b \vee c \leq c \vee y \vee a$ and $y \vee a \leq y \vee b \vee c$, $y \vee a \leq a \vee b \vee c$.

We have $S \subseteq a \circ (b \circ c)$. Indeed, if $y \in S$, then $z = (y \vee a) \wedge (b \vee c)$ satisfy the condition $z \in b \circ c$, $y \in a \circ z$. We have $b \vee z = b \vee [(y \vee a) \wedge (b \vee c)] = (b \vee y \vee a) \wedge (b \vee c) = b \vee c$. Similarly $c \vee z = b \vee c$.

In a similar way, $y \vee z = y \vee a$ and $a \vee z = y \vee a$. Hence $y \in a \circ z \subseteq a \circ (b \circ c)$. Now it follows the associativity law for $\circ$ since $(a \circ b) \circ c \subseteq S \subseteq a \circ (b \circ c)$. The set S is invariant to permutations of $\{a, b, c\}$, hence $(b \circ c) \circ a \subseteq S$, whence by commutativity we have $a \circ (b \circ c) \subseteq S$. Hence $a \circ (b \circ c) = S$. In a similar way, $(a \circ b) \circ c = S$. On the other hand, for all a, b element of L there is $x = a \vee b$ such that $a \in b \circ x$.

Conversely, if we suppose that L is not modular, then L contains a 5-elements sublattice isomorphic to $\{m, a, b, c, M\}$ where $m < b < a < M$, $m < c < M$ and a and c, respectively b and c are not comparable. In this case $c \in a \circ (b \circ c)$ but $c \notin (a \circ b) \circ c$, a contradiction.

Indeed, $c \in a \circ M \subseteq a \circ (b \circ c)$, since $M \in b \circ c$. Finally, $a \circ b \cap \{x \in L | x \leq b\} = \emptyset$ and $c \in (a \circ b) \circ c$. Hence if $x \in a \circ b$, then $x \leq c$ whence $x \leq m \leq b$ which is a contradiction with the above void intersection. $\square$

We define a new hyperoperation determined by a lattice.
For any $p \in L$ set

$$a \circ_p b = \{c \in L \mid a \vee b \vee p = a \vee c \vee p = b \vee c \vee p\}.$$

For $p \in L$ set $L_p = \{x \in L \mid x \geq p\}$ and denote by $\mathbf{L}_p$ the restriction of $\mathbf{L}$ to L_p.

Theorem 7.24. *Let $p \in L$. If $\mathbf{L}_p$ is modular, then $(L, \circ_p)$ is a join space.*

Proof. Define a self-map φ of L, by $x \mapsto x \vee p$. It is well known and immediate that φ is a join-semilattice homomorphism from $\mathbf{L}$ to $\mathbf{L}_p$. As both $(L, \circ)$ and $(L, \circ_p)$ are defined in terms of joins, clearly φ transfers the associativity, reproductive law and the join space condition from $(L, \circ)$ and $(L, \circ_p)$. $\square$

By an immediate check we obtain the following properties of the hyperoperation $\circ_p$. For all a, b, c, p, q of L:

(1) $(a \vee b) \circ_p c \subseteq (a \vee b) \circ_p (a \vee c)$;

(2) $(a \circ_p b) \vee (a \circ_p c) \subseteq a \circ_p (b \vee c)$;

(3) $(a \circ_p a) \cap (b \circ_p b) \subseteq (a \vee b) \circ_p (a \vee b)$;

(4) $(a \circ_p b) \cap (a \circ_q b) \subseteq a \circ_{p \vee q} b$.

Theorem 7.25. *Let* **L** *be a lattice and* a, b, c, p *be elements of* L. *Let* $a \circ_p c \cap b \circ_p c \neq \emptyset$ *and* $p \leq c$. *Then*

(1) *If* **L** *is modular and* $(a \vee b) \wedge c = b \wedge c$, *then* $a \vee b \vee p = b \vee p$;

(2) *If* **L** *is distributive and* $a \wedge c = b \wedge c$, *then* $a \vee p = b \vee p$.

Proof.   (1) Since $a \circ_p c \cap b \circ_p c \neq \emptyset$, it follows $a \vee p \vee c = b \vee p \vee c$, whence $a \vee b \vee p \vee c = b \vee p \vee c$. On the other hand, by modularity

$$((a \vee b) \vee p) \wedge c = ((a \vee b) \wedge c) \vee p = (b \wedge c) \vee p = (b \vee p) \wedge c.$$

Finally, from

$$a \vee b \vee p \vee c = b \vee p \vee c, ((a \vee b) \vee p) \wedge c = (b \vee p) \wedge c$$

and by modularity, we obtain $a \vee b \vee p = b \vee p$.

(2) Similarly as above, we obtain

$$a \vee p \vee c = b \vee p \vee c$$

and

$$(a \vee p) \wedge c = (a \wedge c) \vee p = (b \wedge c) \vee p = (b \vee p) \wedge c,$$

whence by distributivity $a \vee p = b \vee p$.   □

Remark 7.5. Let **L** be a modular lattice and $(L, \circ_p)$ be the associated join space. Let I be a non-empty subset of L and $p \in I$. Then $(I, \circ_p)$ is a subhypergroup of $(L, \circ_p)$ if and only if I is an lattice ideal of **L**.

Indeed, if $(I, \circ_p)$ is a subhypergroup, then for all $a, b \in I$, $a \vee b \in a \circ_p b$ and if $x \leq a$ then $x \in a \circ_p a$. Conversely, if I is an lattice ideal and if $a, b \in I$ then $a \circ_p b \subseteq I$, and $a \vee b \in I$ is such that $a \in b \circ_p (a \vee b)$.

7.3   The hypergroupoid $H_{\mathcal{L}, \mathbf{H}}$. A new proof of Varlet characterization

We analyze here a lattice-determined hypergroupoid/join space, as a generalization of a hypergroupoids associated with a fuzzy set, introduced in

[27] and studied then in [34; 95; 164; 175]. Then we give another proof to the Varlet characterization of distributive lattices, see [109].

Let $\mathcal{L} = \langle L; \cdot, + \rangle$ be a lattice with meet "·", join "+" and order relation $\leq$ and let a nonvoid subset H of L be fixed. For $x, y \in H$ set

$$x \circ y = [xy, x + y] \cap H \tag{7.7}$$

where, as usual, $[a, b]$ denotes the closed interval $\{x \in L \mid a \leq x \leq b\}$ of $\mathcal{L}$. It is immediate that for all $x, y \in H$

$$x, y \in x \circ y. \tag{7.8}$$

It follows that $(x, y) \mapsto x \circ y$ is a map from H^2 $(= H \times H)$ into the set $\mathcal{P}$ of non-empty subsets of H; expressed differently $\mathbf{H} = \mathbf{H}_{\mathcal{L}, \mathrm{H}} = (\mathbf{H}, \circ)$ is a hypergroupoid. Obviously $\mathbf{H}$ is commutative and idempotent; i.e., for all $x, y \in H$

$$x \circ y = y \circ x, \quad x \circ x = \{x\}.$$

Theorem 7.26. *For all $x, y \in H$ we have*

$$x \circ (x \circ y) = x \circ y.$$

Proof. Let $u \in x \circ (x \circ y)$. Then $u \in x \circ v$ for some $v \in x \circ y$. Thus

$$xv \leq u \leq x + v, \quad xy \leq v \leq x + y.$$

Now $xy \leq v$ implies $xy \leq xv \leq u$ and similarly $v \leq x + y$ implies $u \leq x + v \leq x + y$. Thus $xy \leq u \leq x + y$ and hence $u \in x \circ y$ proving $\subseteq$. The inclusion $\supseteq$ follows from (7.8). □

Extend $\circ$ from H^2 to $\mathcal{P}^2$ by setting for all $X, Y \in \mathcal{P}$

$$X \circ Y = \bigcup_{\substack{x \in X \\ y \in Y}} x \circ y$$

and for $x \in H$ abbreviate $\{x\} \circ X$ by $x \circ X$. A hypergroupoid $\mathbf{G} = (\mathbf{G}, \circ)$ satisfying the *reproductive law* $g \circ G = G = G \circ g$ for all $g \in G$ is a *quasihypergroup*. Notice that $\mathbf{H}$ is a quasihypergroup due to $y \in x \circ y$ for all $x, y \in H$. Next $\mathbf{G}$ is a *semihypergroup* if for all $x_1, x_2, y \in H$

$$x_1 \circ (y \circ x_2) = (x_1 \circ y) \circ x_2. \tag{7.9}$$

Finally $\mathbf{G}$ is a *hypergroup* if it is both a quasihypergroup and semihypergroup. We need the following easy and known result.

Theorem 7.27. *Let* $\mathbf{G} = (\mathbf{G}, \circ)$ *be a commutative hypergroupoid. Then*

(1) *If* $(x_1, x_2, y) \in G^3$ *satisfies (7.9) then* (x_2, x_1, y) *satisfies (7.9), and*

(2) *If for all* $(x_1, x_2, y) \in G^3$

$$x_1 \circ (y \circ x_2) \subseteq (x_1 \circ y) \circ x_2 \qquad (*)$$

then $\mathbf{G}$ *is a semihypergroup.*

Proof. (1) Let $(x_1, x_2, y) \in G^3$ satisfy (7.9). Then

$$x_2 \circ (y \circ x_1) = (x_1 \circ y) \circ x_2 = x_1 \circ (y \circ x_2) = (x_2 \circ y) \circ x_1$$

proving that (x_2, x_1, y) satisfies (7.9).

(2) Let $(*)$ hold and let $x_1, x_2, y \in G$. Applying $(*)$ to (x_1, x_2, y) and (x_2, x_1, y) we obtain

$$x_1 \circ (y \circ x_2) \subseteq (x_1 \circ y) \circ x_2 = x_2 \circ (y \circ x_1) \subseteq (x_2 \circ y) \circ x_1 = x_1 \circ (y \circ x_2).$$

This shows

$$x_1 \circ (y \circ x_2) = x_2 \circ (y \circ x_1) = (x_1 \circ y) \circ x_2$$

and (7.9). $\qquad\qquad\square$

The *dual* of an order $\mathcal{M} = (M, \leq)$ is $\mathcal{M}^{\partial} = (M, \geq)$. Thus the dual $\mathcal{L}^{\partial}$ of the lattice $\mathcal{L}$ is obtained by exchanging joins and meets. Directly from (7.7) we get:

Theorem 7.28. $\mathbf{H}_{\mathcal{L},\mathrm{H}} = \mathbf{H}_{\mathcal{L}^{\partial},\mathrm{H}}$; *i.e.,* $\mathcal{L}$ *and* $\mathcal{L}^{\partial}$ *yield the same hypergroupoid. In particular, a valid statement about* $\mathbf{H}_{\mathcal{L},\mathrm{H}}$ *remains true if* $\leq, <, \vee$ *and* $\wedge$ *are replaced throughout by* $\geq, >, \wedge$ *and* $\vee$.

In formulae involving $\circ$ and the set operations we adopt the convention that the operation $\circ$ takes precedence over the set operations; e.g., $(a \circ b \cup c) \setminus d \circ e$ stands for $((a \circ b) \cup c) \setminus (d \circ e)$.

Theorem 7.29. *Let* $x_1, x_2, y \in H$ *and*

$$U = x_1 \circ y \,\cup\, x_2 \circ y \,\cup\, x_1 \circ x_2.$$

Then, we have

$$x_1 \circ (y \circ x_2) \subseteq (x_1 \circ y) \circ x_2 \qquad (7.10)$$

if and only if for all

$$v \in x_2 \circ y \setminus x_1 \circ x_2, \quad w \in x_1 \circ v \setminus U \qquad (7.11)$$

there exists

$$u \in x_1 \circ y \setminus x_1 \circ x_2 \qquad (7.12)$$

such that

$$w \in x_2 \circ u. \qquad (7.13)$$

Proof. We start with the following claim. By definition, (7.10) means

$$\bigcup_{v \in x_2 \circ y} x_1 \circ v \subseteq \bigcup_{u \in x_1 \circ y} x_2 \circ u. \tag{7.14}$$

Claim 1. *The set U is a subset of both sides of* (7.14).

Denote the left-hand and right-hand sides of (7.14) by A and B.

(1) Let $w \in x_1 \circ y$. Then $w \in A$ since we can choose $v = y$ to get $y \in x_2 \circ y$ and $w \in x_1 \circ y$. Similarly, $w \in B$ because choosing $u = w$ we get $w \in x_1 \circ y$ and $w \in x_2 \circ w$.

(2) Let $w \in x_2 \circ y$. Then $w \in A$ as we can choose $v = w$ to get $w \in x_2 \circ y$ and $w \in x_1 \circ w$. Similarly, $w \in B$ since choosing $u = y$ we get $y \in x_1 \circ y$ and $w \in x_2 \circ y$.

(3) Let $w \in x_1 \circ x_2$. Then choosing $v = x_2$ we get $x_2 \in x_2 \circ y$ and $w \in x_1 \circ x_2$ proving $w \in A$.

Finally, $w \in B$ since choosing $u = x_1$ we get $x_1 \in x_1 \circ y$ and $w \in x_2 \circ x_1$. This proves the claim.

Let (7.10) hold. As mentioned above this is equivalent to (7.14). Let v and w satisfy (7.11). Then w belongs to the left-hand side of (7.14) and hence to the right-hand side of (7.14). Thus there exists $u \in x_1 \circ y$ such that $w \in x_2 \circ u$. Were $u \in x_1 \circ x_2$, then by (7.13) and Theorem 11.14.

$$w \in x_2 \circ u \subseteq x_2 \circ (x_1 \circ x_2) = x_1 \circ x_2 \subseteq U$$

contrary to (7.11). Thus (7.12) holds. This proves the necessity of (7.11)–(7.13).

Now, let (7.11)–(7.13) hold. Suppose w belongs to the left-hand side of (7.14). Then $w \in x_1 \circ v$ for some $v \in x_2 \circ y$. If $w \in U$, then by Claim 1 clearly w belongs to the right-hand side of (7.14) and we are done. Thus, let $w \in x_1 \circ v \setminus U$. Were $v \in x_1 \circ x_2$ then from Fact 1 we would get the contradiction

$$w \in x_1 \circ v \subseteq x_1 \circ (x_1 \circ x_2) = x_1 \circ x_2 \subseteq U.$$

Thus (7.11) holds. From (7.12) and (7.13) we get $w \in x_2 \circ u$ for some $u \in x_1 \circ y$ and so w belongs to the right-hand side of (7.10). This proves (7.14) and the proposition. $\square$

The conditions (7.11)–(7.13) are not transparent. We first look at the case when (7.11) is vacuous, i.e.,

$$v \in x_2 \circ y \setminus x_1 \circ x_2 \Rightarrow x_1 \circ v \subseteq U. \tag{7.15}$$

In [175], Theorem 2 (iii), it is stated (without proof):

Corollary 7.2. *The inclusion* (7.10) *holds whenever* x_1, y *and* x_2 *are not pairwise distinct.*

Proof.   Let $x_1, y, x_2 \in H$ be not pairwise distinct. If $y = x_2$ then $x_2 \circ x_2 = \{x_2\} \subseteq x_1 \circ x_2$ and (7.15) holds since its premiss is void. The same holds if $y = x_1$. Thus let $x_1 = x_2$. Then, by Theorem 7.26, for every $v \in x_2 \circ y$

$$x_1 \circ v = x_2 \circ v \subseteq x_2 \circ (x_2 \circ y) = x_2 \circ y \subseteq U$$

proving (7.15) and Corollary 5. $\square$

Corollary 7.3. *The inclusion* (7.14) *holds whenever* $\{x_1, y, x_2\}$ *is a chain in* $\mathcal{L}$.

Proof.   We show that $x_1 \circ v \subseteq U$ for all $v \in x_2 \circ y$. Set $a = x_1 y x_2$ and $b = x_1 + y + x_2$. Then $v \in [a, b]$ and $x_1 \circ v \subseteq [a, b] \cap H \subseteq U$. $\square$

As usual, in $\mathcal{L} = (L, \leq)$ we write $a \parallel b$ if $a, b \in L$ are incomparable; i.e., neither $a \geq b$ nor $a < b$ holds. For $N \subseteq H$ and $a \in H$ we write $a \parallel N$ provided $a \parallel n$ for all $n \in N$. We address the more complex situation of $x_1 < y > x_2$, $x_1 \parallel x_2$.

Corollary 7.4. (a) *Let* x_1, y, x_2 *satisfy* $x_1 < y > x_2$ *and* $x_1 \parallel x_2$ *and let* $s = x_1 + x_2$. *Then* x_1, y, x_2 *satisfy* (7.10) *if and only if for all* $v \in x_2 \circ y$ *and* $w \in x_1 \circ v$ *such that*

$$v \parallel s, \ w \parallel \{x_1, x_2, s\} \tag{7.16}$$

there exists u *satisfying*

$$u \in x_1 \circ y, \ u \parallel \{x_2, s\}. \tag{7.17}$$

(b) *Let* x_1, y, x_2 *satisfy* $x_1 > y < x_2$ *and* $x_1 \parallel x_2$ *and let* $s = x_1 x_2$. *Then* x_1, y, x_2 *satisfy* (7.10) *if and only if for all* $v \in x_2 \circ y$ *and* $w \in x_1 \circ v$ *satisfying* (7.16), *there exists* u *satisfying* (7.17).

Proof.   (a) Let $x_1 < y > x_2$ and $x_1 \parallel x_2$. Then $y \geq s = x_1 + x_2$. Let v, w and u satisfy (7.11)–(7.13). To prove (7.16) first suppose to the contrary that $v \leq s$. Then, due to $v \in x_2 \circ y$ we obtain $x_1 x_2 \leq x_2 = x_2 y \leq v \leq s = x_1 + x_2$ and thus we get the contradiction $v \in x_1 \circ x_2$.

Similarly, $v > s$ leads to $v > x_1$ and to the contradiction

$$w \in x_1 \circ v \subseteq x_1 \circ y \subseteq U.$$

Thus $v \not> s$ and $v \parallel s$.

Next suppose to the contrary that $w \geq x_i$ for some $i \in \{1, 2\}$. Then $w \in [x_i, y] \cap H = x_i \circ y \subseteq U$. This contradiction shows $w \not\geq x_i$ for $i = 1, 2$. Suppose to the contrary that $w \leq x_2$. We claim that then $v x_1 = x_1 x_2$. Indeed, from $v \geq x_2$ we get $v x_1 \geq x_1 x_2$ while $x_2 \geq w \geq v x_1$ shows $x_1 x_2 \geq$

vx_1. Now $x_1x_2 = vx_1 \leq w \leq x_2$ leads to the contradiction $w \in x_1 \circ x_2 \subseteq U$. Thus $w \not\leq x_2$. Suppose to the contrary that $w \leq x_1$. From $x_2 \leq v$ we get $x_1x_2 \leq vx_1 \leq w \leq x_1$ leading to the contradiction $w \in x_1 \circ x_2 \subseteq U$ and proving $w \not\leq x_1$.

Suppose to the contrary that $w \leq s$. From (7.11) and $v \geq x_2$ we see that $w \geq x_1v \geq x_1x_2$ leading to the contradiction $w \in x_1 \circ x_2 \subseteq U$ and proving $w \not\leq s$. Next suppose to the contrary that $w \geq s$. Then $x_1 \leq s \leq w \leq v + x_1 \leq x_1 + y \leq y$ leads to the contradiction $w \in x_1 \circ y \subseteq U$ and proving $w \not\geq s$. We have shown (7.16).

We prove (7.17). Suppose to the contrary that $u \geq x_2$. From (7.12) and (7.13) we see that $x_2 \leq w \leq u \leq y$ leading to the contradiction $w \in x_2 \circ y \subseteq U$. Thus $u \not\geq x_2$.

Next $u < x_2$ combined with (7.12) would lead to the contradiction $x_1 \leq u < x_2$. Thus $u \parallel x_2$.

We show that $u \parallel s$. Suppose to the contrary that $u \leq s$. From (7.12) we get $u \geq x_1$ and $x_2u \geq x_1x_2$ while $x_2 + u \leq s$. Thus we get the contradiction $w \in x_2 \circ u \subseteq x_1 \circ x_2 \subseteq U$ proving $u \not\leq s$. Next $u \geq s$ would imply $u \geq x_2$ contrary to $u \parallel x_2$ shown above. Thus $u \parallel s$ and (7.17) holds.

We have shown that in our situation the condition (7.11)–(7.13) imply the additional restrictions on v, w, u in (7.16) and (7.17).

It follows that the condition of the corollary is also sufficient.

(b) If $x_1 > y < x_2$ and $x_1 \parallel x_2$ the statement follows from (a) and Theorem 12.4. $\qquad\square$

The following corollary treats the case when the condition (7.16) of Corollary 7.4 is vacuous.

Corollary 7.5. *Let $x_1, y, x_2 \in H$ satisfy $x_1 < y > x_2$ ($x_1 > y < x_2$) and $x_1 \parallel x_2$ and let $s = x_1 + x_2$ ($s = x_1x_2$). Then (x_1, y, x_2) satisfies (7.14) provided for all $v \in x_2 \circ y$, $v \parallel s$ every $w \in x_1 \circ v$ is comparable with at least one of x_1, x_2, s.*

Now we address the remaining cases of two comparisons among x_1, y and x_2.

Corollary 7.6. *Let $x_1, y, x_2 \in H$ satisfy $x_1 < x_2 > y$ and $x_1 \parallel y$. Then (x_1, y, x_2) satisfy (7.10) if and only if for all $v, w \in H$ with*

$$y \leq v \leq x_2, \quad x_1 \not\leq v, \tag{7.18}$$

$$x_1v \leq w \leq x_1 + v, \quad x_1 \not\leq w, \quad y \not\leq w \not\leq x_1 + y \tag{7.19}$$

there exists $u \in H$ such that

$$x_1 y \leq u \leq x_1 + y, \quad x_1 \not\leq u \leq w. \tag{7.20}$$

Proof. Let $x_1, y, x_2 \in H$ be such that $x_1 < x_2 > y$, $x_1 \parallel y$. Let $v, w, u \in H$ satisfy (7.11)–(7.13). Then $v \in x_2 \circ y$ means the first part of (7.18) and $w \in x_1 \circ v$ is the first part of (7.19). Now, from $v \leq x_2 > x_1$ we get $w \leq x_1 + v \leq x_2$. Combining this with $w \notin x_1 \circ x_2$ we obtain $x_1 \not\leq v$ proving (7.18). Similarly, $w \notin x_2 \circ y$ turns into $y \not\leq w$. Next, $w \notin x_1 \circ y$ means that $x_1 y \leq w \leq x_1 + y$ does not hold. Here $x_1 y \leq x_1 v \leq w$ is true and so $w \not\leq x_1 + y$. Finally, the assumption $x_1 \leq w$ leads to $x_1 \leq w \leq x_1 + v \leq x_1 + x_2$ and the contradiction $w \in x_1 \circ x_2$. Thus (7.19) holds. The first part of (7.20) follows from (7.12). Were $x_1 \leq u$, then $x_1 \leq u \leq x_1 + y \leq x_2$ would lead to the contradiction $u \in x_1 \circ x_2$. Finally (7.13) and the above $u \leq x_2$ means $u \leq w \leq x_2$ proving the second part of (7.20). $\qquad \square$

The next corollary states the case when the conditions (7.18)–(7.19) are vacuous.

Corollary 7.7. *Let $x_1, y, x_2 \in H$ satisfy $x_1 < x_2 > y$ and $x_1 \parallel y$. Then (x_1, y, x_2) satisfies (7.10) whenever for all $v, w \in H$ with*

$$y \leq v \leq x_2, \quad x_1 v \leq w \leq x_1 + v$$

at least one of

$$x_1 \leq v, \quad x_1 \leq w, \quad y \leq w, \quad w \leq x_1 + y$$

holds.

The proof of the remaining case of two comparisons is similar to the proof of Corollary 7.9 and is omitted.

Corollary 7.8. *Let $x_1, y, x_2 \in H$ satisfy $x_2 < x_1 > y$ and $x_2 \parallel y$. Then (x_1, y, x_2) satisfies (7.10) if and only if for all $v, w \in H$ with*

$$x_2 y \leq v \leq x_2 + y, \quad v \leq w \leq x_1, \tag{7.21}$$

$$x_2 \not\leq w, \quad y \not\leq w \not\leq x_2 + y \tag{7.22}$$

there exists $u \in H$ such that

$$y \leq u \leq x_1, \quad x_2 \not\leq u. \tag{7.23}$$

Corollary 7.9. *Let* $x_1, y, x_2 \in H$ *satisfy* $x_2 < x_1 > y$ *and* $x_2 \parallel y$. *Then* (x_1, y, x_2) *satisfies (7.10) provided for all* $v, w \in H$ *satisfying (7.21) at least one of*

$$x_2 \leq w, \ y \leq w, \ w \leq x_2 + y \tag{7.24}$$

holds.

Corollary 7.10. *If* $x_1, y, x_2 \in H$ *are such that one of them is the join or meet of the others, then* (x_1, y, x_2) *satisfies (7.14).*

Proof. By Corollary 7.2, it suffices to assume that x_1, y, x_2 are pairwise distinct. By Theorem 12.4, it suffices to consider the case of joins. There are three cases:

(a) Let $x_1 = x_2 + y$. In Corollary 7.8 the condition (7.21) becomes $x_2 y \leq v \leq w \leq x_1 = x_2 + y$.

(b) Let $y = x_1 + x_2$. Then in Corollary 7.5 every $v \in H$, $x_2 \leq v \leq y$ is comparable to $s = y$.

(c) Let $x_2 = x_1 + y$. Then in Corollary 7.7 for all $v, w \in H$ with $y \leq v \leq x_2$ and $x_1 v \leq w \leq x_1 + v = x_2 = x_1 + y$ we have $w \leq x_1 + y$. $\qquad\square$

Remark 7.6. If $U = [x_1 y x_2, x_1 + y + x_2] \cap H$, then (x_1, y, x_2) satisfies (7.10).

Indeed, from $x_2 y \leq v \leq x_2 + y$, it follows that

$$x_1 y x_2 \leq x_1 v \leq x_1 + v \leq x_1 + x_2 + y,$$

hence $x_1 \circ v \subseteq U$, that is (7.15) holds.

Corollary 7.11. *If* $x_1, y, x_2 \in H$ *satisfy one of the following conditions:*

(1) $x_1 \parallel y \parallel x_2 \parallel x_1$ *and* $x_1 y = x_1 x_2 = x_2 y$, $x_1 + y = x_1 + x_2 = x_2 + y$;

(2) $x_1 \parallel \{y, x_2\}$, $y < x_2$ ($y > x_2$) *and* $x_1 y = x_1 x_2$, $x_1 + y = x_1 + x_2$;

then (x_1, y, x_2) *satisfies (7.10).*

Proof. This follows from $U = [x_1 y x_2, x_1 + y + x_2] \cap H$ and Remark 7.6. $\qquad\square$

7.4 The join space $\mathbf{H}_{\mathcal{L}, \mathbf{H}}$

From Corollary 7.3 we obtain:

Theorem 7.30. *If* H *is a chain in* $\mathcal{L}$, *then* $\mathbf{H}$ *is a hypergroup.*

Theorem 7.31. *If H is a chain in $\mathcal{L}$, then $\mathbf{H}$ is a join space.*

Proof. Let $b, d, x \in H$ be arbitrary. The above definition of a join space is symmetric in b and d and so we may assume that $b \leq d$. We have three cases with respect to the position of x.

(1) Let $x \leq b$. From $c \in d \circ x = [x, d]$ we get $x \leq c \leq d$ and from $a \in b \circ x$ similarly $x \leq a \leq b$. Now, $b \in (b \circ c) \cap (d \circ a)$.

(2) Let $b < x \leq d$. Then $x \leq c \leq d$, $b \leq a \leq x$ and $x \in (b \circ c) \cap (d \circ a)$.

(3) Let $d < x$. Then $b \leq c, d \leq c \leq x$, $b \leq a \leq x$ and $d \in (b \circ c) \cap (d \circ a)$.

Now, we are done. □

Theorem 7.32. *If H is a distributive sublattice of $\mathcal{L}$, then $\mathbf{H}$ is a hypergroup.*

Proof. According to Theorem 7.29, it is sufficient to check the conditions (7.12) and (7.13) for all $x_1, x_2, y \in H$ that satisfy the condition (7.11).

Let $v \in x_2 \circ y \setminus x_1 \circ x_2$ and $w \in x_1 \circ v \setminus U$. We have $x_2 y \leq v \leq x_2 + y$, $x_1 v \leq w \leq x_1 + v$. Consider $u = w(y + x_1) + y x_1$. We show $u \in x_1 \circ y$. Clearly, $x_1 y \leq u$. Next by distributivity and absorption

$$u = w(y + x_1) + y x_1 \leq (x_1 + v)(y + x_1) + y x_1 = x_1 + v y + y x_1$$
$$= x_1 + v y \leq x_1 + (x_2 + y)y = x_1 + y.$$

We show $w \in x_2 \circ u$. First,

$$x_2 u = x_2(w(y + x_1) + y x_1) = x_2 w(y + x_1) + x_1 x_2 y \leq w + x_1 v = w.$$

Now, we show $x_2 + u \geq w$. By the distributivity of "+" over ".",

$$x_2 + u = (x_2 + y x_1) + w(y + x_1) = (x_2 + y x_1 + w)(x_2 + y x_1 + (y + x_1)) \geq w.$$

Therefore, $w \in x_2 \circ u$, that is (7.13) holds.

Finally, notice that $u \notin x_1 \circ x_2$. Indeed, were $u \in x_1 \circ x_2$, then $w \in x_2 \circ u \subseteq x_1 \circ x_2 \subseteq U$, contrary to (7.11). Hence the condition (7.12) holds, too. □

The following result, obtained by Varlet [176], gives a characterization of distributive lattices, using join spaces. We here present a proof, based on Theorem 7.32.

Theorem 7.33. *If H is a sublattice of $\mathcal{L}$, then H is distributive if and only if $\mathbf{H}$ is a join space.*

Proof. According to Theorem 7.32, **H** is a hypergroup. Let $b, d, x \in H$ be such that $c \in d \circ x$ and let $a \in b \circ x$. We have $dx \le c \le d + x$ and $bx \le a \le b + x$ and by distributivity, we get

$$ad \le (b + x)d = bd + xd \le bd + c \le b + c,$$

hence $ad + bc \le b + c$. Clearly, $ad + bc \ge bc$ and so $ab + bc \in b \circ c$. Similarly $bc + ad \le a + d$ and $bc + ad \in a \circ d$.

Therefore, $ad + bc \le (a + d)(b + c)$, that is $a \circ d$ and $b \circ c$ intersect. Hence **H** is a join space.

Conversely, by the way of contraposition, suppose $(H, \vee, \wedge)$ is not distributive. Then H contains a five-element sublattice $\{a, b, c, d, e\}$, where $a + c = b + c = e$, $ac = bc = d$ and either $a > b$ or $a \parallel b \parallel c \parallel a$. We have $bc \le a \le b + c$, $ac \le b \le a + c$, that is $a \in b \circ c$ and $b \in a \circ c$, but $b \circ b = \{b\}$ and $c \circ c = \{c\}$ do not intersect, which contradicts the join space definition. Therefore H is distributive. $\square$

7.5 The Euler's totient function in canonical hypergroups

In Sections 7.5, 7.6 and 7.7 we introduce the Euler totient function in canonical and i.p.s. hypergroup theory. Also, we determine a way to construct finite i.p.s. hypergroups using the concept of an extension of an i.p.s. hypergrup by an i.p.s. hypergroup. The results of these sections are contained in [167].

Let $(G, \cdot)$ be a group and $a \in G$. The Euler's totient function it is represented as

$$\varphi(G) = |\{a \in G \mid o(a) = \exp(G)\}|, \qquad (7.25)$$

where $o(a)$ is the order of an element a of a finite group G and the exponent of a group G, denoted by $\exp(G)$, is defined as the least common multiple of the orders of all elements of the group. If there is no least common multiple, the exponent is zero. For example, if $G = (\mathbb{Z}_n, +)$, then $ord(\hat{a}) = \frac{n}{\gcd(a,n)}$. Let us enumerate some properties of Euler's totient function in group theory:

(1) φ is not an injective function, more exactly if G_1 and G_2 are two groups such that $\varphi(G_1) = \varphi(G_2)$, it does not imply that $G_1 = G_2$. For example, $\varphi(\mathbb{Z}_3) = \varphi(\mathbb{Z}_4) = 2$, but $\mathbb{Z}_3 \ne \mathbb{Z}_4$.

(2) If G is a cyclic group, then $\varphi(G) = \varphi(|G|)$.

(3) Let G be a finite group with $\exp(G) = m$, $m \in \mathbb{N}^*$. Then, we have $\varphi(G) = \varphi(m)\,k$, where k represents the number of cyclic subgroups of order m in G. The next lemma presents the multiplication property of Euler's totient function.

Lemma 7.1. *If $\{G_i\}_{i=\overline{1,k}}$ is a family of finite groups of coprime orders, then we have*

$$\varphi\left(\prod_{i=1}^{k} G_i\right) = \prod_{i=1}^{k} \varphi(G_i).$$

Let us give some examples involving remarkable groups: the dihedral group D_{2n}, the symmetric group S_n and the alternating group A_n, $n \geq 2$.

If $G = S_n$ respectively $G = A_n$, $n \geq 2$,

(a) $\varphi(S_n) = 0$, $n \geq 3$;

(b) $\varphi(A_n) = 0$, $n \geq 4$;

(c) $G = D_{2n}$, if n is odd, then $\varphi(D_{2n}) = 0$ and for n even, we have $\varphi(D_{2n}) = \varphi(n)$, for all $n \geq 3$.

Definition 7.5. We say that an element x of H is called *periodic*, if there is $k \in \mathbb{N}$ such that $x^k \subseteq \omega_H$.

If

$$p(x) = \min\{k \in \mathbb{N}\mid x^k \subseteq \omega_H\}, \tag{7.26}$$

then we say that k is the period of x. In a similar way as in group theory, the Euler's totient function has the next form

$$\varphi(H) = |\{x \in H\mid p(x) = \exp(H)\}|,$$

where $\exp(H)$ represents the least common multiple of the orders of all elements of the hypergroup H.

Let $(C(n), \circ)$ be the next canonical hypergroup, where

$$C(n) = \{e_0, e_1, ..., e_{k(n)}\}, \ n \in \mathbb{N}^*$$

and

$$k(n) = \begin{cases} \dfrac{n}{2} & \text{if } n \in 2\mathbb{N}; \\[2mm] \dfrac{n-1}{2} & \text{if } n \in 2\mathbb{N}+1. \end{cases}$$

For a pair (s,t) such that $s \leq k(n)$, $t \leq k(n)$, we consider

$$e_s \circ e_t = \{e_v, e_p\},$$

where $v = \min\{s+t, n-(s+t)\}$, $p = |s-t|$.

More results about canonical hypergroup $(C(n), \circ)$ are given by Leoreanu, [102]. She studied the subhypergroups of $(C(n), \circ)$ and established in which condition the theorem of Lagrange could be applied. We give here some results and we prove that the Euler's function associated to the canonical hypergroup $C(n)$ is equal to the heart of $C(n)$.

Theorem 7.34. *If $r \in \{0, 1, ..., k(n)\}$ such that $(r, n) = 1$, then*

$$C(n) = <e_r>.$$

Proof. Denote $S = <e_r>$. We check that for all $\{\alpha, \beta\} \subseteq \mathbb{N}$, such that $|\alpha r - \beta n| \in \{0, ..., k(n)\}$ we have $e_{|\alpha r - \beta n|} \in S$.

Let $\alpha_1 \in \mathbb{N}$ be such that $\alpha_1 r \leq k(n) < (\alpha_1 + 1)r$. Then, we have

$$\{e_r, e_{2r}, ..., e_{\alpha_1 r}\} \subseteq S \text{ and } e_{n-(\alpha_1+1)r} \in e_{\alpha_1 r} \circ e_r.$$

Let $\alpha_2 \in \mathbb{N}$ be such that $n - (\alpha_1 + \alpha_2)r \geq 0$ and $n - (\alpha_1 + \alpha_2 + 1)r < 0$. Since

$$e_{n-(\alpha_1+2)r} \in e_{n-(\alpha_1+1)r} \circ e_r,$$

$$\vdots$$

$$e_{n-(\alpha_1+\alpha_2)r} \in e_{n-(\alpha_1+\alpha_2-1)r} \circ e_r,$$

it follows that $\{e_{n-(\alpha_1+1)r}, e_{n-(\alpha_1+2)r}, ..., e_{n-(\alpha_1+\alpha_2)r}\} \subseteq S$.

Let $\alpha_3 \in \mathbb{N}$ be such that

$$(\alpha_1 + \alpha_2 + \alpha_3)r - n \leq k(n) \quad \text{and} \quad (\alpha_1 + \alpha_2 + \alpha_3 + 1)r - n > k(n).$$

Since

$$e_{(\alpha_1+\alpha_2+1)r-n} \in e_{n-(\alpha_1+\alpha_2)r} \circ e_r,$$
$$e_{(\alpha_1+\alpha_2+2)r-n} \in e_{(\alpha_1+\alpha_2+1)r-n} \circ e_r,$$

$$\vdots$$

$$e_{(\alpha_1+\alpha_2+\alpha_3)r-n} \in e_{(\alpha_1+\alpha_2+\alpha_3-1)r-n} \circ e_r,$$

it follows that

$$\{e_{(\alpha_1+\alpha_2+1)r-n}, ..., e_{(\alpha_1+\alpha_2+\alpha_3)r-n}\} \subseteq S.$$

Hence $e_{n-[(\alpha_1+\alpha_2+\alpha_3+1)r-n]} = e_{2n-(\alpha_1+\alpha_2+\alpha_3+1)r} \in e_{(\alpha_1+\alpha_2+\alpha_3)r-n} \circ e_r$.

Let $\alpha_4 \in \mathbb{N}$ be such that

$$2n - (\alpha_1 + \alpha_2 + \alpha_3 + \alpha_4)r \geq 0 \text{ and } 2n - (\alpha_1 + \alpha_2 + \alpha_3 + \alpha_4 + 1)r < 0.$$

Since

$$e_{(\alpha_1+\alpha_2+\alpha_3+1)r} \in e_{(\alpha_1+\alpha_2+\alpha_3)r-n} \circ e_r,$$
$$e_{2n-(\alpha_1+\alpha_2+\alpha_3+2)r} \in e_{2n-(\alpha_1+\alpha_2+\alpha_3+1)r} \circ e_r,$$
$$\vdots$$
$$e_{2n-(\alpha_1+\alpha_2+\alpha_3+\alpha_4)r} \in e_{2n-(\alpha_1+\alpha_2+\alpha_3+\alpha_4-1)r} \circ e_r,$$

it follows that

$$\{e_{2n-(\alpha_1+\alpha_2+\alpha_3+1)r}, \ ..., \ e_{2n-(\alpha_1+\alpha_2+\alpha_3+\alpha_4)r}\} \subseteq S.$$

Continuing in the same manner, we obtain all $(\alpha, \beta) \in \mathbb{N}^2$ for which $|\alpha r - \beta n| \in \{0, ..., k(n)\}$, that is we obtain all combinations of r and n of type $\alpha r - \beta n$ or $\beta n - \alpha r$ for which $e_{|\alpha r - \beta n|} \in S$.

On the other hand, from $(r, n) = 1$ it follows that there exists $(\alpha, \beta) \in \mathbb{N}^2$ such that $\alpha r - \beta n = 1$ or $\beta n - \alpha r = 1$, that means such that $|\alpha r - \beta n| = 1$.

Then $e_{|\alpha r - \beta n|} = e_1 \in S$ and since $< e_1 > = C(n)$, it follows that $S = C(n)$. $\qquad \square$

Corollary 7.12. *If n is a natural prime number, then $C(n)$ has only improper subhypergroups.*

Theorem 7.35. *If $n = 2p$, where p is a natural prime number, with $p \neq 2$, then hypergroups $C(n)$ satisfy the Lagrange's theorem.*

Remark 7.7. The only proper subhypergroups of the canonical hypergroup $C(n)$ has the next form, see [102].

$$S_1 = \{e_0, e_p\} \text{ and } S_2 = \left\{e_{2i} \mid i \in \{0, 1, 2, ..., \frac{p-1}{2}\}\right\}. \tag{7.27}$$

The Remark 7.7 is useful for us to calculate the Euler's totient function, when $n = 2p$, p is a natural prime number.

Theorem 7.36. *Let $(C(n), \circ)$ be a canonical hypergroup, when $n = 2p$, p is a natural prime number, then*

$$\varphi(C(n)) = |\omega_{C(n)}|, \text{ for all } n \in \mathbb{N}.$$

Proof. We use the definition of Euler's function in context of canonical hypergroups. More exactly, this function represents the number of the elements which have the periodicity equal with the exponent of hypergroup. So, we can write in the following way

$$\varphi(C(n)) = |\{e_j \in C(n) : p(e_j) = \exp(C(n))\}|,$$

where

$$p(e_j) = \min\{k \in \mathbb{N}^* : e_j^k \subseteq \omega_{C(n)}, j \in \{0, 1, ..., p\}\}.$$

It is known that $\omega_{C(n)}$ is a subhypergroup of the hypergroup $C(n)$ and according to Corollary 12.12, we want to prove that $\omega_{C(n)} = S_2$.

We notice that e_0 is the identity of the canonical hypergroup $C(n)$, because

$e_s \in e_s \circ e_0 \cap e_0 \circ e_s$, for all $s \in \{0, 1, ..., p\}$ equivalent with $e_s \in \{e_v, e_p\} \cap \{e_p, e_v\}$,

for any $s \in \{0, 1, ..., p\}$, where $v = \min\{s, n - s\}$, $p = |s - 0| = s$. We notice that $s \leq \frac{n}{2} = p$, which implies that $v = \min\{s, n - s\} = s$. So, $e_s \circ e_0 = e_0 \circ e_s = e_s$, for any $s \in \{0, 1, ..., p\}$. Using the definition of the heart of hypergrpup, we have $\omega_{C(n)} = \beta^*(e_0) = \{e_j \in C(n)| \ e_j \beta e_0\}$. We wish to check that $\omega_{C(n)} = S_2$, so we show that $e_{2i} \in \omega_{C(n)}$, $i \in \{0, 1, 2, ..., \frac{p-1}{2}\}$ and $e_p \notin \omega_{C(n)}$. So, we have

$$e_{2i} \in \omega_{C(n)} \Leftrightarrow e_{2i} \beta e_0 \Leftrightarrow \exists m \in \mathbb{N}^*, \exists \ e_{j_1}, e_{j_2}, ..., e_{j_m} \in C(n) : \{e_{2i}, e_0\} \subseteq \prod_{s=1}^{m} e_{j_s}.$$

We consider $e_{j_1} = e_{j_2} = e_i$, then we obtain

$$e_i \circ e_i = \{e_v, e_p\}, v = \min\{i + i, n - (i + i)\}, p = |i - i| = 0.$$
$$v = \min\{2i, n - 2i\}, \ p = 0.$$

We notice that $\min\{2i, n - 2i\} = 2i$ if and only if $2i \leq n - 2i$, which is equivalent with $i \leq \frac{p}{2}$, which is true, because $i \leq \frac{p-1}{2} \leq \frac{p}{2}$. Therefore,

$$e_i \circ e_i = \{e_{2i}, e_0\}, i \in \left\{0, 1, 2, ..., \frac{p-1}{2}\right\},$$

i.e., $e_{2i} \in \omega_{C(n)}$, $i \in \{0, 1, 2, ..., \frac{p-1}{2}\}$ whence $S_2 \subseteq \omega_{C(n)}$.

In what follows, we want to prove that the element e_p, where p is a prime number, $p > 2$ does not belong to the heart of the canonical hypergroup $C(n)$. So, we suppose that $e_p \in \omega_{C(n)}$ it follows that $m \in \mathbb{N}^*$, $e_{i_1}, e_{i_2}, ..., e_{i_m} \in C(n)$ such that

$$\{e_o, e_p\} \subseteq \prod_{j=1}^{m} e_{i_j}.$$

For the beginning, we consider $m = 2$. So, we get $e_0 \in e_{i_1} \circ e_{i_2}$ and $e_p \in e_{i_1} \circ e_{i_2}$, where $e_{i_1} \circ e_{i_2} = \{e_v, e_t\}$, $v = \min\{i_1 + i_2, n - (i_1 + i_2)\}$, $t = e_{|i_1 - i_2|}$. We notice that if i_1 and i_2 have the same parity, then v and t are even and if i_1 and i_2 have the different parities, then v and t are odd. The set $\{e_v, e_t\}$ contains elements with the same parity, which leads to e_0

and e_p does not belong to $e_{i_1} \circ e_{i_2}$, because p is an odd number. Now, if $m = 3$, we have $\{e_0, e_p\} \subseteq e_{i_1} \circ e_{i_2} \circ e_{i_3}$. We note that $e_{i_1} \circ e_{i_2} = A$,

$$e_{i_1} \circ e_{i_2} \circ e_{i_3} = A \circ e_{i_3} = \bigcup_{e_z \in A} (e_z \circ e_{i_3}).$$

So, $\{e_0, e_p\} \subseteq \in e_{i_1} \circ e_{i_2} \circ e_{i_3}$ which implies that exists $e_z, e_s \in A$ such that $e_0 \in e_z \circ e_{i_3}$ and $e_p \in e_s \circ e_{i_3}$. To obtain $e_0 \in e_z \circ e_{i_3}$ whence results that $z = i_3$ or $n = z + i_3$. If we have $z = i_3$ then $e_p \in e_s \circ e_z$. Then the elements of A have the same parity, so we can state that $e_s \circ e_z$ contains elements with even number index, so $e_p \notin e_s \circ e_z$. Now, we analyze the case when $n = z + i_3$ and we know that n is an even number, so z and i_3 are both even or odd which implies that s, z and i_3 have the same parity. Also, in this case, $e_p \notin e_s \circ e_{i_3}$. For $m > 3$, $\prod_{j=1}^{m} e_{i_j} = B \circ e_{i_m}$, where $B = e_{i_1} \circ ... \circ e_{i_{m-1}}$ and the process is the same as in case $m = 3$. In conclusion, for $n = 2p$, p prime number, $p > 3$, $e_p \notin \omega_{C(n)}$. Therefore,

$$\omega_{C(n)} = S_2 \Rightarrow |\omega_{C(n)}| = |S_2| = \frac{p+1}{2}.$$

Now we check that all elements with odd number index have the periodicity 2. Let $e_{2j+1} \in C(n)$, $j \in \{0, 1, 2, ..., \frac{p-1}{2}\}$ and

$$p(e_{2j+1}) = \min\left\{k \in \mathbb{N}^* : e_{2j+1}^k \subseteq \omega_{C(n)}\right\}.$$

We notice that $e_{2j+1} \circ e_{2j+1} = \{e_{\min\{4j+2, n-(4j+2)\}}, e_0\}$, but n is an even number, so we can say that $4j + 2$ and $n - (4j + 2)$ are even numbers, so

$$e_{2j+1} \circ e_{2j+1} \subseteq \omega_{C(n)}, \text{ for any } j \in \left\{0, 1, 2, ..., \frac{p-1}{2}\right\}.$$

Therefore, all the elements with even number index have the periodicity equal to 2, which leads to $exp(C(n)) = 2$ and

$$\varphi(C(n)) = \frac{p+1}{2} = |\omega_{C(n)}|.$$

This completes the proof. $\square$

Remark 7.8. We notice that p is not necessary a prime number, because we proved that the elements with odd number index do not belong to the heart of the canonical hypergroup $C(n)$ and the elements with even number index satisfies the condition: $\{e_{2i}, e_0\} \subseteq e_i \circ e_i$, which leads $e_{2i} \in \omega_{C(n)}, i \in \{0, 1, ..., [\frac{p}{2}]\}$, when $[x]$ represents the whole part of x. Also, the periodicity of the elements with odd number index is the same: $e_{2j+1} \circ e_{2j+1} \subseteq \omega_{C(n)}$, for any $j \in \{0, 1, 2, ..., [\frac{p-1}{2}]\}$.

So, we have the following results.

Theorem 7.37. *If $n = 2p$, p is an odd number then $\varphi\left(C(n)\right) = |\omega_{C(n)}|$.*

Theorem 7.38. *If $n = 2p$, p is an even number, then $\varphi\left(C(n)\right) = |C(n)| - |\omega_{C(n)}|$.*

Theorem 7.39. *If n is an odd number, then $\varphi\left(C(n)\right) = |C(n)| = |\omega_{C(n)}|$.*

Proof. Suppose that $n = 2p + 1$, p is a natural number, $p \geq 1$. Then $C(n) = \{e_0, e_1, ..., e_{k(n)}\}$, $k(n) = \frac{n-1}{2}$. Hence $C(n) = \{e_0, e_1, ..., e_p\}$. We consider the following sets:

$$E_0 = \left\{ e_{2i}, \ i \in \left\{ 0, 1, ..., \left[\frac{p}{2}\right] \right\} \right\}$$

$$E_1 = \left\{ e_{2j+1}, \ j \in \left\{ 0, 1, ..., \left[\frac{p-1}{2}\right] \right\} \right\}.$$

We notice that $C(n) = E_0 \cup E_1$ and we want to prove that all the elements belongs to $\omega_{C(n)}$. For the elements from E_0, we proceed as in Theorem 7.36. We have,

$$e_i \circ e_i = \{e_{2i}, e_0\} \Rightarrow e_{2i} \in \omega_{C(n)}, i \in \left\{ 0, 1, 2, ..., \left[\frac{p}{2}\right] \right\}.$$

We have $E_0 \subseteq \omega_{C(n)}$. For the elements which are in the set E_1, we show that the next relation hold on

$$\{e_{2j+1}, e_0\} = e_{p-j} \circ e_{p-j}, \ j \in \left\{ 0, 1, ..., \left[\frac{p-1}{2}\right] \right\}$$

$$e_{p-j} \circ e_{p-j} = \{e_{\min\{2p-2j, n-(2p-2j)\}}, e_{|p-j-p+j|}\}$$
$$= \{e_{\min\{2p-2j, 2p+1-(2p-2j)\}}, e_0\}.$$
$$= \{e_{\min\{2p-2j, 1+2j\}}, e_0\}.$$

We suppose that $\min\{2p - 2j, 1 + 2j\} = 2p - 2j$, which is equivalent to the inequality $2p - 2j \leq 1 + 2j$, i.e., $j \geq \frac{2p-1}{4}$. It is known that $j \leq \left[\frac{p-1}{2}\right] \leq \frac{p-1}{2}$, then

$$j \leq \frac{p-1}{2} < \frac{2p-1}{4}.$$

We obtain $\min\{2p - 2j, 1 + 2j\} = 2j + 1$. In conclusion,

$$\{e_{2j+1}, e_0\} = e_{p-j} \circ e_{p-j}, \ j \in \left\{ 0, 1, ..., \left[\frac{p-1}{2}\right] \right\} \Rightarrow E_1 \subseteq \omega_{C(n)}.$$

Therefore, $C(n) = E_0 \cup E_1 \subseteq \omega_{C(n)}$, but $\omega_{C(n)}$ is a subhypergroup of the canonical hypergroup $C(n)$. So, $C(n) = \omega_{C(n)}$. We can state that all the elements have the periodicity equal to one, which leads to $\varphi\left(C(n)\right) = |C(n)| = |\omega_{C(n)}|$. $\square$

Example 7.3. Let $n = 9$. So, $p = 4$ and $C(9) = \{e_0, e_1, e_2, e_3, e_4\}$.

$\circ$	e_0	e_1	e_2	e_3	e_4
e_0	e_0	e_1	e_2	e_3	e_4
e_1	e_1	$\{e_0, e_2\}$	$\{e_1, e_3\}$	$\{e_2, e_4\}$	$\{e_3, e_4\}$
e_2	e_2	$\{e_1, e_3\}$	$\{e_0, e_4\}$	$\{e_1, e_4\}$	$\{e_2, e_3\}$
e_3	e_3	$\{e_2, e_4\}$	$\{e_1, e_4\}$	$\{e_0, e_3\}$	$\{e_1, e_2\}$
e_4	e_4	$\{e_3, e_4\}$	$\{e_2, e_3\}$	$\{e_1, e_2\}$	$\{e_0, e_1\}$

We notice that

$$\{e_1, e_0\} = e_4 \circ e_4 \Rightarrow e_1 \in \omega_{C(9)};$$

$$\{e_3, e_0\} = e_3 \circ e_3 \Rightarrow e_3 \in \omega_{C(9)}.$$

For the elements from E_0, we have $\{e_{2i}, e_0\} = e_i \circ e_i$, $i \in \{0, 1, 2\}$ whence results $e_0, e_2, e_4 \in \omega_{C(9)}$. Therefore, $\omega_{C(9)} = C(9)$ and $\varphi(C(9)) = |C(9)| = |\omega_{C(9)}| = 5$.

7.6 The extension of canonical hypergroup by canonical hypergroup

In this section, we use the extension of polygroups by polygroups [19; 49] in canonical hypergroup theory. In [166] it is calculated the commutativity degree of this extension. We can see the extension as a new way to construct canonical and i.p.s. hypergroups.

Let $\mathcal{H}_1 = \ <H_1, \cdot, e,^{-1}>$, $\mathcal{H}_2 = \ <H_2, \cdot, e,^{-1}>$ be two canonical hypergroups of which elements have been renamed, so that $H_1 \cap H_2 = \{e\}$, where e is the identity of canonical hypergroups $\mathcal{H}_1$ and $\mathcal{H}_2$.

Definition 7.6. A system $\mathcal{H}_1[\mathcal{H}_2]$ is called the *extension of a canonical hypergroup* $\mathcal{H}_1$ by a canonical hypergroup $\mathcal{H}_2$ if:

$$\mathcal{H}_1[\mathcal{H}_2] = \ <M, *, e,^I>,$$

where

$$M = H_1 \cup H_2, \ e^I = e, \ x^I = x^{-1}, e * x = x * e = x, \text{ for any } x \in M.$$

For any $x, y \in M \backslash \{e\}$, we have

$$x * y = \begin{cases} x \cdot y & x, y \in H_1 \\ x & x \in H_2, y \in H_1 \\ y & x \in H_1, y \in H_2 \\ x \cdot y & x, y \in H_2, y \neq x^{-1} \\ x \cdot y \cup H_1 & x, y \in H_2, y = x^{-1} \end{cases} \quad (7.28)$$

If we consider $H_1 = \{e, a_1, a_2, ..., a_{n-1}\}$ and $H_2 = \{e, b_1, b_2, ..., b_{m-1}\}$, where $n, m \in \mathbb{N}^*$, then we obtain

$*$	e	a_1	...	a_{n-1}	b_1	...	b_i	...	b_{m-1}
e	e	a_1	...	a_{n-1}	b_1	...	b_i	...	b_{m-1}
a_1	a_1	a_1a_1	...	a_1a_{n-1}	b_1	...	b_i	...	b_{m-1}
$\vdots$	$\vdots$	$\vdots$	...	$\vdots$	$\vdots$	...	$\vdots$	...	$\vdots$
a_{n-1}	a_{n-1}	$a_{n-1}a_1$	...	$a_{n-1}a_{n-1}$	b_1	...	b_i	...	b_{m-1}
b_1	b_1	b_1	...	b_1	b_1b_1	...	$b_1b_i \cup H_1$	...	b_1b_{m-1}
$\vdots$	$\vdots$	$\vdots$	...	$\vdots$	$\vdots$	...	$\vdots$	...	$\vdots$
b_i	b_i	b_i	...	b_i	$b_ib_1 \cup H_1$	...	b_ib_i	...	b_ib_{m-1}
$\vdots$	$\vdots$	$\vdots$	...	$\vdots$	$\vdots$	...	$\vdots$	...	$\vdots$
b_{m-1}	b_{m-1}	b_{m-1}	...	b_{m-1}	$b_{m-1}b_1$	...	$b_{m-1}b_i$	...	$b_{m-1}b_{m-1}$

Without loss the generality, we suppose that for i fixed, $b_i = b_1^{-1}$ and for each element b_j, there is a unique element b_k such that $b_j = b_k^{-1}$ with i, j, $k \in \overline{1, m-1}$.

Theorem 7.40. *If $\mathcal{H}_1 = <H_1, \cdot, e, ^{-1}>$ and $\mathcal{H}_2 = <H_2, \cdot, e, ^{-1}>$ are two canonical hypergroups, then $\mathcal{H}_1[\mathcal{H}_2]$ is a canonical hypergroup.*

Proof. First of all we prove that $\mathcal{H}_1[\mathcal{H}_2]$ is a hypergroup, i.e., (1) $(a * b) * c = a * (b * c)$, for any a, b, $c \in \mathcal{H}_1[\mathcal{H}_2]$ and (2) $a * \mathcal{H}_1[\mathcal{H}_2] = \mathcal{H}_1[\mathcal{H}_2] * a$, for any $a \in \mathcal{H}_1[\mathcal{H}_2]$. We remark that the behavior of polygroups is similar to that of canonical hypergroups, so the proof of associativity is similar to that one for polygroups [49]. Therefore, we analyze the relation (2) and the commutativity.

For the commutativity of "$*$", we have $x * y = y * x$, for any $x, y \in M$. As $\mathcal{H}_1$ and $\mathcal{H}_2$ are canonical hypergroups, it follows that $x \cdot y = y \cdot x$, for any $x, y \in H_1$ and $x \cdot y = y \cdot x$, for any $x, y \in H_2$. If $x \in H_1$ and $y \in H_2$, then $x * y = y$ and $y * x = y$ so we equality holds. Now, if $x \in H_2$, $y \in H_1$ we obtain $x * y = y * x = x$. Also, if $x, y \in H_2$, where $x \neq y^{-1}$ and $y \neq x^{-1}$ the equality is obvious. It remain to show that if $y = x^{-1}$ it follows that $x = y^{-1}$. We have

$$y = x^{-1}, \text{ which implies } e \in y \cdot x, \text{ i.e., } x \in y^{-1} \cdot e, \text{ so } x = y^{-1}.$$

Therefore, for $x, y \in H_2$, where $x = y^{-1}$ it follows that $x * y = y * x$. In conclusion, $\mathcal{H}_1[\mathcal{H}_2]$ is commutative.

The second condition is the law of reproducibility:

$$a * \mathcal{H}_1[\mathcal{H}_2] = \mathcal{H}_1[\mathcal{H}_2] * a = \mathcal{H}_1[\mathcal{H}_2], \text{ for any } a \in \mathcal{H}_1[\mathcal{H}_2]. \tag{7.29}$$

So, we can write

$$a * M = M * a = M, \text{ for any } a \in M.$$

We have

$$a * M = \bigcup_{b \in M} (a * b) = \left(\bigcup_{b \in H_1} (a * b) \right) \cup \left(\bigcup_{b \in H_2} (a * b) \right).$$

If $a \in H_1$, then $\bigcup_{b \in H_1} (a * b) = H_1$, because $\mathcal{H}_1$ is a hypergroup and

$$\bigcup_{b \in H_2} (a * b) = \bigcup_{b \in H_2} (a \cdot b) = \bigcup_{b \in H_2} b = H_2.$$

So, in this case we obtain $a * M = H_1 \cup H_2 = M$. The condition $M * a = M$ follows from the commutativity relation " $*$ ". If $a \in H_2$, then $\bigcup_{b \in H_2} (a \cdot b) = H_2$, since $\mathcal{H}_2$ is a hypergroup, and

$$\bigcup_{b \in H_2} (a * b) = \bigcup_{b \neq a^{-1}} (a \cdot b) \cup \left(a * a^{-1} \right) = \left[H_2 \backslash (a \cdot a^{-1}) \right] \cup \left[(a \cdot a^{-1}) \cup H_1 \right]$$
$$= H_2 \cup H_1 = M.$$

Consequently, we have $a * M = M * a = M$, for any $a \in M$. This completes the proof. $\qquad \square$

In what follows, we establish a connection between the hearts of canonical hypergroups $\mathcal{H}_1$, $\mathcal{H}_2$ and $\mathcal{H}_1[\mathcal{H}_2]$.

Theorem 7.41. *If $\mathcal{H}_1 = \; < H_1, \cdot, e, ^{-1} >$ and $\mathcal{H}_2 = \; < H_2, \cdot, e, ^{-1} >$ are canonical hypergroups, then $\omega_{\mathcal{H}_1[\mathcal{H}_2]} = H_1 \cup \omega_{\mathcal{H}_2}$.*

Proof. First of all, we show that

$$H_1 \cup \omega_{\mathcal{H}_2} \subseteq \omega_{\mathcal{H}_1[\mathcal{H}_2]}. \tag{7.30}$$

Let $x \in H_1 \cup \omega_{\mathcal{H}_2}$, it follows that $x \in H_1$ or $x \in \omega_{\mathcal{H}_2}$. If $x \in H_1$, we notice that exists $z_1, z_2 \in H_2$ with $z_1 = z_2^{-1}$ such that

$$\{e, x\} \subseteq z_1 * z_2^{-1} = z_1 \cdot z_1^{-1} \cup H_1 \Rightarrow x \in \omega_{\mathcal{H}_1[\mathcal{H}_2]}.$$

If $x \in \omega_{\mathcal{H}_2}$. Using the definition of the heart of a hypergroup it follows that there are $b_1, b_2, ..., b_p \in H_2$, $p \geq 1$ as

$$\{e, x\} \subseteq b_1 b_2 ... b_p.$$

Because $e \in b_1 b_2 ... b_p$, means that exist $i, j \in \{1, 2, ..., p\}$ such that $b_i = b_j^{-1}$. We consider $z_k = b_k$, $k \in \{1, 2, ..., p\}$ and

$$S = z_1 * z_2 * ... * z_{i-1} * z_{i+1} * ... * z_{j-1} * z_{j+1} * ... * z_p$$

and

$$S * (z_i * z_i^{-1}) = (z_1 z_2 ... z_{i-1} z_i ... z_p) \cup (S * H_1)$$
$$= (z_1 z_2 ... z_{i-1} z_i ... z_p) \cup S$$
$$= z_1 z_2 ... z_{i-1} z_i ... z_p.$$

So, there is $z_k \in M$, $z_k = b_k$, $k \in \{1, 2, ..., p\}$ with the next property

$$\{e, x\} \subseteq z_1 * z_2 * ... * z_p \Rightarrow x \in \omega_{\mathcal{H}_1[\mathcal{H}_2]}.$$

Therefore, $H_1 \cup \omega_{\mathcal{H}_2} \subseteq \omega_{\mathcal{H}_1[\mathcal{H}_2]}$. In what follows, we prove the second inclusion:

$$\omega_{\mathcal{H}_1[\mathcal{H}_2]} \subseteq H_1 \cup \omega_{\mathcal{H}_2}. \tag{7.31}$$

Let $x \in \omega_{\mathcal{H}_1[\mathcal{H}_2]}$, which means that exists $z_1, z_2, ..., z_p \in M$ such that

$$\{e, x\} \subseteq z_1 * z_2 * ... * z_p. \tag{7.32}$$

We analyze several cases. If for any $i \in \{1, 2, ..., p\}$, we have $z_i \in H_1$, the relation (7.32) becomes

$$\{e, x\} \subseteq z_1 z_2 ... z_p \Rightarrow x \in H_1.$$

If there is $j \in \{1, 2, ..., s\}$, $s < p$, such that $z_j \in H_2$ and $z_i \in H_1$, $i \neq j$. The indexes could be renumbered.

$$z_1 * z_2 * ... * z_p = \underbrace{(z_1 * ... * z_j)}_{\in H_2} * \underbrace{(z_{j+1} \cdot z_{j+2} \cdot ... \cdot z_p)}_{\in H_1} = z_1 * ... * z_j,$$

because, if we consider $A = z_1 * ... * z_j$, $B = z_{j+1} \cdot z_{j+2} \cdot ... \cdot z_p$ and using the relation's " $*$ " definition, we have

$$A * B = \bigcup_{a \in A, b \in B} (a * b) = \bigcup_{a \in A} a = A.$$

The relation (7.32) becomes

$$\{e, x\} \subseteq z_1 * ... * z_j.$$

$e \in z_1 * ... * z_j$, it follows that there are $k, l \in \{1, 2..., j\}$ such that $z_k = z_l^{-1}$, so

$$z_1 * ... * z_j = (z_1 .. z_{k-1} \cdot z_{k+1} \cdot z_{l-1} \cdot z_{l+1} ... \cdot z_j) * (z_k * z_k^{-1}).$$

We proceed in a similar manner as in the previous case where $x \in \omega_{\mathcal{H}_2}$. Therefore we get that exists $z_j \in H_2$, $j \in \{1, 2, ..., s\}$, $s < p$ such that

$$\{e, x\} \subseteq z_1 \cdot z_2 ... \cdot z_j \Rightarrow x \in \omega_{\mathcal{H}_2}.$$

If $z_i \in H_2$, for any $i \in \{1, 2, ..., p\}$, then proceeding analogously we have $x \in \omega_{\mathcal{H}_2}$. So, $\omega_{\mathcal{H}_1[\mathcal{H}_2]} \subseteq H_1 \cup \omega_{\mathcal{H}_2}$. Consequently, we have

$$\omega_{\mathcal{H}_1[\mathcal{H}_2]} = H_1 \cup \omega_{\mathcal{H}_2}.$$

This completes the proof.   □

We wish to apply Euler's function for the extension. It is necessary to find a connection between the periodicity of elements from $\mathcal{H}_1[\mathcal{H}_2]$ and $\mathcal{H}_2$.

Theorem 7.42. *The periodicity of elements from canonical hypergroup $\mathcal{H}_1[\mathcal{H}_2]$ is the same as the periodicity of elements from canonical hypergroup $\mathcal{H}_2$*

$$p(x)_{\mathcal{H}_1[\mathcal{H}_2]} = p(x)_{\mathcal{H}_2}, \text{ for all } x \in \mathcal{H}_2.$$

Proof.   We recall the definition of periodicity as it follows

$$p(x) = \min\{k \in \mathbb{N}^* | x^k \subseteq \omega_H\}.$$

So, according to canonical hypergroup $\mathcal{H}_1[\mathcal{H}_2]$, the definition has the next form

$$p(x) = \min\{k \in \mathbb{N}^* | x^k \subseteq \omega_{\mathcal{H}_1[\mathcal{H}_2]}\} \qquad (7.33)$$
$$= \min\{k \in \mathbb{N}^* | x^k \subseteq H_1 \cup \omega_{\mathcal{H}_2}\}.$$

We denote $p(x)_{\mathcal{H}_1[\mathcal{H}_2]} = m$. For $x \in H_1$ we have $p(x) = 1$, but if $x \in H_2 \backslash H_1$ we obtain $x^m \subseteq H_1 \cup \omega_{\mathcal{H}_2}$. If $x^m \subseteq \omega_{\mathcal{H}_2}$, then $p(x)_{\mathcal{H}_2} = m$, which implies that $p(x)_{\mathcal{H}_1[\mathcal{H}_2]} = p(x)_{\mathcal{H}_2}$. If $x^m \subseteq H_1$ we use the definition of "$*$" and we have only the situation when

$$x * x^{-1} = x \cdot x^{-1} \cup H_1 = \{0\} \cup H_1, \text{ i.e., } p(x)_{\mathcal{H}_1[\mathcal{H}_2]} = 2.$$

But

$$x \cdot x^{-1} = \{0\} \subseteq \omega_{\mathcal{H}_2}, \text{ i.e., } p(x)_{\mathcal{H}_2} = 2.$$

So $p(x)_{\mathcal{H}_1[\mathcal{H}_2]} = p(x)_{\mathcal{H}_2}$, for all $x \in \mathcal{H}_2$.   □

In what follows, we give a connection between $\varphi(\mathcal{H}_1[\mathcal{H}_2])$ and $\varphi(\mathcal{H}_2)$.

Theorem 7.43. *Let $\mathcal{H}_1 = <H_1, \cdot, e, ^{-1}>$ and $\mathcal{H}_2 = <H_2, \cdot, e, ^{-1}>$ be canonical hypergroups, then $\varphi(\mathcal{H}_1[\mathcal{H}_2]) = \varphi(\mathcal{H}_2)$.*

Proof.   The proof is based on Theorems 7.41 and 7.42. From Theorem 7.42, we remark that the elements of H_1 have the periodicity one and in order to calculate the Euler's totient function in $\mathcal{H}_1[\mathcal{H}_2]$ we need only the elements of the canonical hypergroup $\mathcal{H}_2$. Therefore, $exp(\mathcal{H}_1[\mathcal{H}_2]) = \exp(\mathcal{H}_2)$ and we obtain

$$\varphi(\mathcal{H}_1[\mathcal{H}_2]) = |\{x \in M : p(x) = \exp(\mathcal{H}_1[\mathcal{H}_2])\}|$$
$$= |\{x \in H_2 : p(x) = \exp(\mathcal{H}_2)\}|$$
$$= \varphi(\mathcal{H}_2).$$

This completes the proof. □

Corollary 7.13. *If all the elements have the periodicity equal to one, then it follows that* $\omega_{\mathcal{H}_1[\mathcal{H}_2]} = H_1 \cup H_2$, *which implies that*

$$\varphi(\mathcal{H}_1[\mathcal{H}_2]) = |H_1| + |H_2| - 1.$$

7.7 The Euler's totient function in i.p.s. hypergroups

There is a strong connection between canonical hypergroups and i.p.s. hypergroup, so we aim to prove that the extension of i.p.s. hypergroup by i.p.s. hypergroup is an i.p.s. hypergroup.

Theorem 7.44. *If* $\mathcal{H}_1 = \; < H_1, \cdot, e, ^{-1} >$ *and* $\mathcal{H}_2 = \; < H_2, \cdot, e, ^{-1} >$ *are i.p.s. hypergroups, then* $\mathcal{H}_1[\mathcal{H}_2]$ *is an i.p.s. hypergroup.*

Proof. We know that $(H, \circ)$ is an i.p.s. hypergroup if $(H, \circ)$ is a canonical hypergroup that satisfies the condition $x \in a \circ x \Rightarrow x = a \circ x$, for any a, x in H. Using Theorem 7.40 we obtain that $\mathcal{H}_1[\mathcal{H}_2]$ is a canonical hypergroup. Therefore, we want to prove the following relation: if $x \in a * x$, then $x = a * x$, where x, a belongs to M and $M = H_1 \cup H_2$.

We analyze several cases. We consider a, x in H_1, then the condition $x \in a * x$ becomes $x \in a \cdot x$, but $\mathcal{H}_1$ is an i.p.s. hypergroup, which means that $x = a \cdot x = a * x$. Let a in H_1 and x in H_2; by using the definition "$*$", we have $a * x = x$. For $a \in H_2$ and $x \in H_1$, we have $a * x = a$, so $x \in a * x$ if and only if $x = a = e$. In this case, we obtain $x = a * x$. If we consider that a, x belongs to H_2 with $x \neq a^{-1}$, then the relation $x \in a * x$ becomes $x \in a \cdot x$. We know that $\mathcal{H}_2$ is an i.p.s. hypergroup, which means that $x = a \cdot x = a * x$. In the last case a, x are in H_2 with $x = a^{-1}$, then $x \in a * x$ could be write thus $a^{-1} \in a \cdot a^{-1} \cup H_1$. If $a \neq e$, it is obvious that $a^{-1} \notin H_1$, because a belongs to H_2. So, the only possibility is $a^{-1} \in a \cdot a^{-1}$, whence $a^{-1} = a \cdot a^{-1}$, due to i.p.s. hypergroup $\mathcal{H}_2$. We obtain that $a^{-1} = e$, which means that $a = x = e$. So, the relation $x \in a * x$ with $x = a^{-1}$ holds on if and only if $a = x = e$ whence $x = a * x$. Therefore, $\mathcal{H}_1[\mathcal{H}_2]$ is an i.p.s. hypergroup. □

Remark 7.9. Let $\mathcal{H}_1 = \left(H_1, \cdot, 0, ^{-1}\right)$, $\mathcal{H}_2 = \left(H_2, \cdot, 0, ^{-1}\right)$ be i.p.s. hypergroups, where $H_1 = \{0, 1, 2, 3\}$ and $H_2 = \{0, 4, 5, 6\}$. The tables for i.p.s. hypergroups $\mathcal{H}_1$ and $\mathcal{H}_2$ are the following

$\mathcal{H}_1$:

$\cdot$	0	1	2	3
0	0	1	2	3
1		2	0	3
2			1	3
3				$\{0, 1, 2\}$

$\mathcal{H}_2$:

$\cdot$	0	4	5	6
0	0	4	5	6
4		$\{0, 5\}$	4	6
5			0	6
6				$\{0, 4, 5\}$

So, using Definition 7.6 of "$*$", we obtain the next representation for $\mathcal{H}_1[\mathcal{H}_2]$:

$*$	0	1	2	3	4	5	6
0	0	1	2	3	4	5	6
1		2	0	3	4	5	6
2			1	3	4	5	6
3				$\{0, 1, 2\}$	4	5	6
4					$\{0, 5\} \cup H_1$	4	6
5						$\{0\} \cup H_1$	6
6							$\{0, 4, 5\} \cup H_1$

We prove that the i.p.s. hypergroup $\mathcal{H}_1[\mathcal{H}_2]$ is isomorph to the following i.p.s. hypergroup, which was determined by Corsini

$\cdot$	0	1	2	3	4	5	6
0	0	1	2	3	4	5	6
1		2	0	3	4	5	6
2			1	3	4	5	6
3				$\{0, 1, 2\}$	4	5	6
4					$\{0, 1, 2, 3\}$	5	6
5						$\{0, 1, 2, 3, 4\}$	6
6							H_6

where $H_6 = H \backslash \{6\}$ and $H = \{0, 1, 2, 3, 4, 5, 6\}$.

We consider the function $f : \mathcal{H}_1[\mathcal{H}_2] \to H$

$$f(x) = x, \ x \in H_1;$$
$$f(4) = 5, \ f(5) = 4, \ f(6) = 6.$$

We notice that for any x, $y \in H_1$ we have $f(x * y) = f(x) \cdot f(y)$. If $x \in H_1$ and $y \in H_2$, then $f(x * y) = f(y) = f(x) \cdot f(y)$. We analyze what happens when x, $y \in H_2$.

$$f(4 * 4) = f(\underbrace{\{0,1,2,3,5\}}_{A}) = \bigcup_{x \in A} f(x) = \{0,1,2,3,4\},$$
$$f(4) \cdot f(4) = 5 \cdot 5 = \{0,1,2,3,4\}.$$
$$f(4 * 5) = f(4) = 5; \ f(4) \cdot f(5) = 5 \cdot 4 = 5.$$
$$f(4 * 6) = f(6) = 6; \ f(4) \cdot f(6) = 5 \cdot 6 = 6;$$
$$f(5 * 5) = f(\underbrace{\{0,1,2,3\}}_{x \in B}) = \bigcup_{x \in B} f(x) = \{0,1,2,3\},$$
$$f(5) \cdot f(5) = 4 \cdot 4 = \{0,1,2,3\}.$$
$$f(5 * 6) = f(6) = 6, \ f(5) \cdot f(6) = 4 \cdot 6 = 6.$$
$$f(6 * 6) = f(H_6) = \bigcup_{x \in H_6} f(x) = \{0,1,2,3,4,5\};$$
$$f(6) \cdot f(6) = 6 \cdot 6 = H_6 = \{0,1,2,3,4,5\}.$$

Therefore we can state that $f(x*y) = f(x)*f(y)$, for any x, $y \in \mathcal{H}_1[\mathcal{H}_2]$ and $f(0) = 0$, where 0 is the identity for i.p.s. hypergroups $\mathcal{H}_1$ and $\mathcal{H}_2$, then we have a strongly homomorphism. Moreover, f is a bijective function. In consequence, the i.p.s. hypergroups $(\mathcal{H}_1[\mathcal{H}_2], *)$ and $(H, \cdot)$ are isomorphic.

In conclusion, the extension of i.p.s. hypergroup by an i.p.s. hypergroup represents an easy way to construct finite i.p.s. hypergroups, also we obtain similar results in the context of i.p.s. hypergroups as for canonical hypergroups.

Theorem 7.45. *The periodicity of elements from i.p.s. hypergroup $\mathcal{H}_1[\mathcal{H}_2]$ is the same as the periodicity of elements from i.p.s. hypergroup $\mathcal{H}_2$*

Theorem 7.46. *If $\mathcal{H}_1 = \ < H_1, \cdot, e,^{-1} >$ and $\mathcal{H}_2 = \ < H_2, \cdot, e,^{-1} >$ are i.p.s. hypergroups, then $\varphi(\mathcal{H}_1[\mathcal{H}_2]) = \varphi(\mathcal{H}_2)$.*

Chapter 8

Rosenberg Hypergroups

In this chapter, we associate a Rosenberg hypergroup with a complete hypergroup and we study it. Then the notion of weak mutually associative hypergroups is introduced and some properties of the hypergroups associated with binary relations are given. On the other hand, using some special type of Boolean matrices, the non-isomorphic Rosenberg hypergroups of order less than 7 are enumerated. Moreover, the regular and reversible Rosenberg hypergroups are identified. It is analyzed when different partial hypergroupoids associated with binary relations defined on a set H are reduced hypergroups. The cartesian product of two hypergroupoids associated with a binary relation is also analyzed.

8.1 On Rosenberg hypergroups

One of the most investigated hypergroups associated with binary relations is that introduced by Rosenberg [157] in 1998. It represents a theme of research of numerous papers, especially in finite case.

Rosenberg associated a partial hypergroupoid $\mathbb{H}_\rho = (H, \circ)$ with a binary relation ρ defined on a set H, where, for any $x, y \in H$,

$$x \circ x = L_x = \{z \in H \mid (x, z) \in \rho\} \quad \text{and} \quad x \circ y = L_x \cup L_y.$$

Definition 8.1. An element $x \in H$ is called *outer element* of ρ if there exists $h \in H$ such that $(h, x) \notin \rho^2$.

Theorem 8.1. $\mathbb{H}_\rho$ *is a hypergroup if and only if*

(1) ρ *has full domain;*
(2) ρ *has full range;*
(3) $\rho \subseteq \rho^2$;

(4) If $(a, x) \in \rho^2$ then $(a, x) \in \rho$, whenever x is an outer element of ρ.

Proof.   Notice that the associativity law can be written as follows:

$$(a, x) \in \rho \text{ or } (b, x) \in \rho^2 \text{ or } (c, x) \in \rho^2 \Leftrightarrow \atop (a, x) \in \rho^2 \text{ or } (b, x) \in \rho^2 \text{ or } (c, x) \in \rho. \tag{8.1}$$

If $\mathbb{H}_\rho$ is a semihypergroup and we suppose that $\rho \not\subset \rho^2$, then there exists $(b, x) \in \rho \setminus \rho^2$. Consider $a = x$ and $c = b$ in (8.1). Then the right-hand side in (8.1) is satisfied. On the left-hand side $(b, x) = (c, x) \notin \rho^2$, whence $(x, x) = (a, x) \in \rho$. From $(b, x) \in \rho$ and $(x, x) \in \rho$ we obtain the contradiction $(b, x) \in \rho^2$. Thus $\rho \subseteq \rho^2$. Suppose now that there exists an outer element $x \in H$ and $a \in H$ such that $(a, x) \in \rho^2 \setminus \rho$. On the other hand, there exists $b \in H$ such that $(b, x) \notin \rho^2$. Set $c = b$ in (8.1). Then the right-hand side of (8.1) holds, while the left-hand side does not hold since $(a, x) \notin \rho$ and $(b, x) \notin \rho^2$. Therefore the condition (4) is satisfied.

Conversely, let $\rho \subseteq \rho^2$ and $(a, x) \in \rho^2 \Rightarrow (a, x) \in \rho$ holds whenever x is an outer element of ρ. Let a, b, c, x be elements of H. If $(b, x) \in \rho^2$ then both sides of (8.1) are satisfied. Let $(b, x) \notin \rho^2$. Since $\rho \subseteq \rho^2$ we obtain $(a, x) \in \rho^2 \Leftrightarrow (a, x) \in \rho$. Similarly, $(c, x) \in \rho^2 \Leftrightarrow (c, x) \in \rho$, which together with $(b, x) \notin \rho^2$ proves (8.1). $\square$

Remark 8.1. If ρ is a quasiorder relation, then the hypergroupoid $\mathbb{H}_\rho$ associated with H is a hypergroup.

Theorem 8.2. *If $\mathbb{H}_\rho$ is a hypergroup, then the following statements hold:*

(1) ρ^2 *is a transitive relation;*

(2) *If ρ is symmetric, then ρ^2 is an equivalence relation on H;*

(3) *If ρ is symmetric and $|H/\rho^2| > 1$, then ρ is an equivalence relation on H.*

Proof.   (1) Suppose that there exist $(x, y), (y, z)$ elements of ρ^2, such that $(x, z) \notin \rho^2$. Then x is outer and so $(y, z) \in \rho$. Since $(x, z) \in \rho^2$, there exists $a \in H$ such that (x, a) and (a, y) are elements of ρ. Since $(a, z) \in \rho^2$ it follows that $(a, z) \in \rho$. Thus $(x, z) \in \rho^2$, which is a contradiction.

(2) Let ρ be symmetric and $x \in H$. We have $(x, y) \in \rho$ for some $y \in H$ and by symmetry, $(y, x) \in \rho$, whence $(x, x) \in \rho^2$. Moreover, ρ^2 is symmetric and transitive, so it is an equivalence relation on H.

(3) Let ρ be symmetric and $|H/\rho^2| > 1$. Then all elements of H are outer, so $\rho = \rho^2$. $\square$

Corollary 8.1. *Let ρ be a reflexive, symmetric and non-transitive relation on H. The following assertions are equivalent:*

(1) $\mathbb{H}_\rho$ *is a hypergroup;*
(2) *For any $x \in H$ we have $x \circ x \circ x = H$;*
(3) *There are no outer elements for ρ;*
(4) $\rho^2 = H \times H$.

Corollary 8.2. *Let ρ and σ be two binary relations on H with full domain and full range such that $\rho^2 = \rho$, $\sigma^2 = \sigma$ and $\rho\sigma = \sigma\rho$. Then $\mathbb{H}_{\rho\sigma}$ is a hypergroup.*

The results presented in the next sections are contained in [1; 40; 42; 106].

8.2 Complete hypergroups and Rosenberg hypergroups

We recall a representation theorem for complete hypergroups. In a complete hypergroup all hyperproducts are complete parts.

Theorem 8.3. *A hypergroup H is complete if and only if $H = \bigcup\limits_{g \in G} A_g$, where G and A_g satisfy the conditions:*

(1) $(G, \cdot)$ *is a group;*
(2) *for all $(g_1, g_2) \in G^2$, $g_1 \neq g_2$, we have $A_{g_1} \cap A_{g_2} = \varnothing$;*
(3) *if $(a, b) \in A_{g_1} \times A_{g_2}$, then $a \circ b = A_{g_1 g_2}$.*

Proof. If we set $G = H/\beta$ then for all $g \in G$ we have $A_g = \varphi_H^{-1}(g)$ and so we obtain 1) and 2). Moreover, if $(a, b) \in A_g \times A_h$ then $a \circ b = \mathcal{C}(a \circ b) = \varphi_H^{-1}(\varphi_H(a)\varphi_H(b)) = \varphi_H^{-1}(gh) = A_{gh}$. Conversely, it is immediate. $\square$

If G is a commutative group, then H is a complete commutative hypergroup, that is a join space.

Jantosciak [89] associated three equivalence relations with an arbitrary hypergroup $(H, \circ)$. These equivalence relations were analyzed in [39]. Let us recall them.

The operational relation, denoted by "$\sim_\circ$", is defined as follows:

$$x \sim_\circ y \Leftrightarrow a \circ x = a \circ y; \ x \circ a = y \circ a, \text{for all } a \in H.$$

The inseparability, denoted by "$\sim_i$", as follows:

$$x \sim_i y \Leftrightarrow \text{for } a, b \in H, \ x \in a \circ b \Leftrightarrow y \in a \circ b.$$

The essential indistinguishability, denoted by "$\sim_e$"

$$x \sim_e y \Leftrightarrow x \sim_o y \text{ and } x \sim_i y.$$

Denote by x_o, x_i, x_e the equivalence class of x with respect to "$\sim_o$", "$\sim_i$", "$\sim_e$", respectively.

We intend to establish a connection between these equivalence relations and the conjugacy relation in the context of complete hypergroups.

Let us recall the conjugacy relation in H :

$$a \sim_H b \iff \exists c \in H : \mathcal{C}(ca) = \mathcal{C}(bc).$$

Theorem 8.4. *The relation "$\sim_H$" is an equivalence relation.*

Proof. For $c = a$ we obtain the reflexivity. For symmetry, let $c \in H$ be such that $\mathcal{C}(ca) = \mathcal{C}(bc)$. We check that there is $d \in H$ such that $\mathcal{C}(db) = \mathcal{C}(ad)$. We have $cac'\omega_H = b\omega_H$, where $c' \in i(c)$. Hence, $c'cac'\omega_H = c'b\omega_H$, which means that $ac'\omega_H = c'b\omega_H$. Hence take $d = c'$ and we get the conclusion. Let us check now the transitivity. Let $d, e \in H$ be such that $da\omega_H = bd\omega_H$ and $eb\omega_H = ce\omega_H$. We find an element $f \in H$ such that $fa\omega_H = cf\omega_H$. Set $d' \in i(d)$ and $e' \in i(e)$. We have $dad'\omega_H = b\omega_H$ and $e'eb\omega_H = e'ce\omega_H$. Hence $dad'\omega_H = e'ce\omega_H$, whence $da\omega_H = e'ced\omega_H$, so $eda\omega_H = ced\omega_H$. There are $f, f_1 \in ed$ such that $f_1a\omega_H \cap cf\omega_H \neq \emptyset$. But $f\beta f_1$, so $f\omega_H = f_1\omega_H$. Hence $fa\omega_H \cap cf\omega_H \neq \emptyset$ and consequently we have $fa\omega_H = cf\omega_H$.

Thus "$\sim_H$" is an equivalence relation in H. $\square$

Denote by $[a]$ the conjugacy class of an element a of H. We have

$$[a] = \{b \in H \mid \exists c \in H : \mathcal{C}(ca) = \mathcal{C}(bc)\},$$

or equivalently

$$[a] = \{b \in H \mid \exists c \in H : b \in \mathcal{C}(cac'), \text{where } c' \in i(c)\},$$

whence

$$[a] = \bigcup_{\substack{c \in H \\ c' \in i(c)}} \mathcal{C}(cac').$$

Theorem 8.5. *If $(H, \circ)$ is a complete hypergroup and G is the associated group of H, then:*

$$x_o = x_i = x_e = A_g^x, \text{ where } g \in G \text{ and } x \in A_g.$$

Proof. Let H be a complete hypergroup and G be the associated group, such that $|G| = n$, $n \in \mathbb{N}^*$ and A_{g_i}, $i = \overline{1, n}$ are the associated non-empty subsets. Let $(x, y) \in H^2$ be such that $x \circ a = y \circ a$ and $a \circ x = a \circ y$, for all $a \in H$. Since H is a complete hypergroup, it follows that for $x \in H$ there is a unique $g_1 \in G$ such that $x \in A_{g_1}$ and for $y \in H$ there is a unique $g_2 \in G$ such that $y \in A_{g_2}$. Finally, for $a \in H$ there is a unique $g_a \in G$ such that $a \in A_{g_a}$ Hence

$$A_{g_1 g_a} = A_{g_2 g_a} \Rightarrow g_1 = g_2 \Rightarrow \{x, y\} \subseteq A_{g_1}.$$

Therefore, $x \circ y$ if x and y belong to the same set.

Clearly, for $x, y \in A_g$ it follows that $x \circ y$. Hence, $x_o = A_g^x$, where $g \in G$ and $x \in A_g$. In a similar way, we reason for relation "$\sim_i$". $\square$

In what follows, we associate the Rosenberg hypergroup to the complete hypergroup H, with respect to conjugacy relation "$\sim_H$", redenoted by ρ.

We recall the Rosenberg hyperoperation: Let $(\mathbb{H}_\rho, \circ_\rho)$ be the Rosenberg hypergroup, for which the hyperoperation "$\circ_\rho$" is defined as follows:

$$x \circ_\rho y = \{z \in H | \ (x, z) \in \rho \text{ or } (y, z) \in \rho\}. \tag{8.2}$$

We intend to analyze what relations "$\sim_o$", "$\sim_i$", "$\sim_e$" become in the context of Rosenberg hypergoup and to establish a connection with ρ, defined in a complete hypergroup H.

Let

$$P = \{x \in H | \ x \notin x \circ x\}; \tag{8.3}$$
$$K = \{e \in H | \ P \subset e \circ e\}.$$

Corsini [29] proved that:

Theorem 8.6. $\mathbb{H}_\rho$ *is regular if and only if* $K \neq \varnothing$.

Proof. First, suppose that e is an identity of the regular hypergroup $\mathbb{H}_\rho$. If $P = \emptyset$, then $P \subseteq e \circ e$. If $P \neq \emptyset$ then for all $x \in P$ we have $e \circ x = e \circ e \cup x \circ x$. Since $x \notin x \circ x$, it follows that $x \in e \circ e$ whence $P \subseteq e \circ e$ which means that $K \neq \emptyset$.

Conversely, if $P = \emptyset$ then for all $x \in H$ we have $x \in x \circ x$, whence for all x, y elements of H, we have $\{x, y\} \subseteq x \circ y$. Moreover, the set of inverses of an whichever element x is H. If $P \neq \emptyset$ and if $e \in H$ is such that $P \subseteq e \circ e$ then $x \in e \circ x$ for all $x \in P$ and $y \in e \circ y$, for all $y \in H - P$. Therefore all elements of K are identities. Moreover, if e is an identity, then for all $z \in H$, we have $e \in e \circ z$ which means that $\mathbb{H}_\rho$ is regular. $\square$

Remark 8.2. The hypergroup $(\mathbb{H}_\rho, \circ_\rho)$ is regular and reversible.

Therefore,

$$x \sim_{o_\rho} y \Leftrightarrow a \circ_\rho x = a \circ_\rho y; \; x \circ_\rho a = y \circ_\rho a, \text{for all } a \in H_\rho.$$

Theorem 8.7. *Let* $(\mathbb{H}_\rho, \circ_\rho)$ *be the hypergroup associated with the complete hypergroup* H. *We have*

$$x \sim_{o_\rho} y \text{ if and only if } (x, y) \in \rho.$$

Proof. Denoting the ρ-equivalence class of x by x_ρ, we have $x \circ_\rho y = x_\rho \cup y_\rho$. So, we have $x \sim_{o_\rho} y$ if and only if $a_\rho \cup x_\rho = a_\rho \cup y_\rho$, for every $a \in H$. This is clearly equivalent to $x_\rho = y_\rho$. In conclusion $(x, y) \in \rho$. $\square$

Theorem 8.8. *Let* $(\mathbb{H}_\rho, \circ_\rho)$ *be the hypergroup associated with a complete hypergroup* H. *Then*

$$x \sim_i y, \text{ if and only if } (x, y) \in \rho.$$

Proof. $x \sim_i y$ if and only if for $a, b \in H$, $x \in a \circ_\rho b \Leftrightarrow y \in a \circ_\rho b$. So, $x \in a_\rho \cup b_\rho \Leftrightarrow y \in a_\rho \cup b_\rho$ for all $a, b \in H$. Therefore $x \in a_\rho \Leftrightarrow y \in a_\rho$ for all $a \in H$, which holds if and only if $x \rho y$. $\square$

The results of the following two paragraphs belong to paper [40].

8.3 Weak mutually associativity for Rosenberg's hypergroups

Definition 8.2 (see [35]). Two partial hypergroupoids $(H, \circ_1)$ and $(H, \circ_2)$ are called *mutually associative*, or shortly m.a., if for all $(x, y, z) \in H^3$, we have:

$$(x \circ_1 y) \circ_2 z = x \circ_1 (y \circ_2 z) \text{ and } (x \circ_2 y) \circ_1 z = x \circ_2 (y \circ_1 z).$$

Definition 8.3. We say that two partial hypergroupoids $(H, \circ_1)$ and $(H, \circ_2)$ are *weak mutually associative* or shortly w.m.a., if for all $(x, y, z) \in H^3$, we have:

$$(x \circ_1 y) \circ_2 z \cup (x \circ_2 y) \circ_1 z = x \circ_1 (y \circ_2 z) \cup x \circ_2 (y \circ_1 z).$$

Example 8.1. Let ρ and ρ' be two binary relations on the set $H = \{x, y, z\}$ as follows:

$$\rho = \{(x, x), (x, y), (y, x), (z, y), (z, z)\} \text{ and } \rho' = \{(x, x), (y, x), (y, y), (z, x)\}.$$

The two Rosenberg's semihypergroups $\mathbb{H}_\rho$ and $\mathbb{H}_{\rho'}$ are weak mutually associative, but they are not mutually associative. Indeed, we have

$\mathbb{H}_\rho$	x	y	z
x	$\{x,y\}$	$\{x,y\}$	H
y	x,y	x	H
z	H	H	$\{y,z\}$

and

$\mathbb{H}_{\rho'}$	x	y	z
x	x	$\{x,y\}$	x
y	$\{x,y\}$	$\{x,y\}$	$\{x,y\}$
z	x	x,y	x

$\mathbb{H}_\rho$ is a hypergroup and $\mathbb{H}_{\rho'}$ is a semihypergroup and moreover

$$(y \circ_\rho y) \circ_{\rho'} x = x \neq \{x,y\} = y \circ_\rho (y \circ_{\rho'} x),$$

that is $\mathbb{H}_\rho$ and $\mathbb{H}_{\rho'}$ are not mutually associative.

Theorem 8.9. *Let two semihypergroups $(H, \circ_1)$ and $(H, \circ_2)$ be weak mutually associative. Then $(H, \circ)$ is a semihypergroup, where $x \circ y = x \circ_1 y \cup x \circ_2 y$ for all $(x, y) \in H^2$.*

Proof. Suppose that $(x, y, z) \in H^3$ and $a \in (x \circ y) \circ z$ are given. So $a \in v \circ z$ for an appropriate $v \in x \circ y$; thus we have:

(1) $a \in v \circ_1 z, v \in x \circ_1 y$ or
(2) $a \in v \circ_1 z, v \in x \circ_2 y$ or
(3) $a \in v \circ_2 z, v \in x \circ_1 y$ or
(4) $a \in v \circ_2 z, v \in x \circ_2 y$.

Consider the second case: $a \in v \circ_1 z, v \in x \circ_2 y$. Since $(H, \circ_1)$ and $(H, \circ_2)$ are w.m.a. we obtain that $a \in x \circ_2 (y \circ_1 z)$ or $a \in x \circ_1 (y \circ_2 z)$, hence $a \in x \circ (y \circ z)$.

The converse inclusion holds too and moreover, the other cases can be similarly proved. $\square$

Corollary 8.3. *Let ρ and ρ' be two relations on H such that the associated Rosenberg hypergroups $\mathbb{H}_\rho$ and $\mathbb{H}_{\rho'}$ are weak mutually associated hypergroups. Then $\mathbb{H}_{\rho \cup \rho'}$ is a hypergroup, too.*

Proof. Since $\mathbb{H}_\rho$ and $\mathbb{H}_{\rho'}$ are quasihypergroups, it follows that $\mathbb{H}_{\rho \cup \rho'}$ is a quasihypergroup. It remains to prove the associative law. Since, for any $x \in H$, $R^{\rho \cup \rho'}(x) = R^\rho(x) \cup R^{\rho'}(x)$, it follows that $x \circ_{\rho \cup \rho'} y = x \circ_\rho y \cup x \circ_{\rho'} y$ and then we conclude that the hyperoperation "$\circ_{\rho \cup \rho'}$" is also associative. Hence $\mathbb{H}_{\rho \cup \rho'}$ is a hypergroup. $\square$

Let ρ and ρ' be two relations on H and $X \subseteq H$; set $\rho(X) = \{y \mid (x,y) \in \rho$ for some $x \in X\}$ and $\rho\rho' = \{(x,y) \mid (x,u) \in \rho, (u,y) \in \rho'$, for some $u \in H\}$.

Theorem 8.10. *Let ρ and ρ' be two relations on the same set H. The associated Rosenberg's semihypergroups $\mathbb{H}_\rho$ and $\mathbb{H}_{\rho'}$ are w.m.a. if and only if, for all $(x,y,z) \in H^3$, we have*

$$\rho\rho'(x,y) \cup \rho'(z) \cup \rho'\rho(x,y) \cup \rho(z) = \rho'\rho(y,z) \cup \rho(x) \cup \rho\rho'(y,z) \cup \rho'(x). \quad (8.4)$$

Proof.   We have

(1) $(x \circ_\rho y) \circ_{\rho'} z = \{t \in H \mid (x,t) \in \rho\rho' \vee (y,t) \in \rho\rho' \vee (z,t) \in \rho'\} = \rho\rho'(x,y) \cup \rho'(z)$;

(2) $x \circ_\rho (y \circ_{\rho'} z) = \{t \in H \mid (y,t) \in \rho'\rho \vee (z,t) \in \rho'\rho \vee (x,t) \in \rho\} = \rho'\rho(y,z) \cup \rho(x)$;

(3) $(x \circ_{\rho'} y) \circ_\rho z = \{t \in H \mid (x,t) \in \rho'\rho \vee (y,t) \in \rho'\rho \vee (z,t) \in \rho\} = \rho'\rho(x,y) \cup \rho(z)$;

(4) $x \circ_{\rho'} (y \circ_\rho z) = \{t \in H \mid (y,t) \in \rho\rho' \vee (z,t) \in \rho\rho' \vee (x,t) \in \rho'\} = \rho\rho'(y,z) \cup \rho'(x)$.

Hence the assertion holds.   □

Theorem 8.11. *Let ρ and ρ' be two relations on H such that $\rho' \subseteq \rho \subseteq \rho\rho'$ and $\rho'\rho \cap \{(x,x) \mid x \in H\} = \emptyset$. If $\mathbb{H}_\rho$ is a hypergroup and $\mathbb{H}_\rho$, $\mathbb{H}_{\rho'}$ are weak mutually associative, then also $\mathbb{H}_{\rho\rho'}$ is a hypergroup.*

Proof.   Since $\rho \subseteq \rho\rho'$ we have $\rho^2 \subseteq \rho\rho'\rho$ and so $\rho\rho' \subseteq \rho^2\rho' \subseteq (\rho\rho')^2$. Now let us consider x an outer element for $\rho\rho'$ so x is an outer element for ρ. If $(a,x) \in (\rho\rho')^2$ then there exists $b \in H$ such that $(a,b) \in \rho\rho'$ and $(b,x) \in \rho\rho'$. Then $b \in (a \circ_\rho b) \circ_{\rho'} b$. Since H_ρ, $H_{\rho'}$ are weak mutually associative we have $b \in a \circ_\rho (b \circ_{\rho'} b)$ or $b \in a \circ_{\rho'} (b \circ_\rho b)$. If $b \in a \circ_\rho (b \circ_{\rho'} b)$ then $b \in a \circ_\rho c$ for an appropriate $c \in b \circ_{\rho'} b$ therefore $(a,b) \in \rho$ or $(c,b) \in \rho$ and $(b,c) \in \rho'$. If $(c,b) \in \rho$ and $(b,c) \in \rho'$ then $(b,b) \in \rho'\rho$ which is a contradiction so $(a,b) \in \rho$ follows. If $b \in a \circ_{\rho'} (b \circ_\rho b)$ as above we can conclude that $(a,b) \in \rho'$ or $(c,b) \in \rho'$ and $(b,c) \in \rho$ if $(c,b) \in \rho'$ and $(b,c) \in \rho$ we have $(c,c) \in \rho'\rho$ which is also a contradiction. So $(a,b) \in \rho' \subseteq \rho$. Thus $(a,x) \in \rho^2$ and, since x is an outer element for ρ, it results $(a,x) \in \rho \subseteq \rho\rho'$, then $\mathbb{H}_{\rho\rho'}$ is a hypergroup.   □

8.4   Binary relations and reduced hypergroups

We say that a hypergroup $(H, \circ)$ is *reduced* if and only if, for all $x \in H$, $\widehat{x}_e = \{x\}$.

Theorem 8.12 (Proposition 3, [90]). *For any hypergroup $(H, \circ)$, the quotient hypergroup $H/\sim_e, \star)$ is a reduced hypergroup, where the hyperoperation $\star$ on $H/\sim_e$ is defined by*

$$\widehat{x}_e \star \widehat{y}_e = \{\widehat{z}_e \mid z \in x \circ y\}.$$

The quotient hypergroup $(H/\sim_e, \star)$ is called the reduced form of the hypergroup $(H, \circ)$.

The results presented in this paragraph belong to paper [42].

Necessary and sufficient conditions are determined such that the hypergroup $\mathbb{H}_\rho$, associated with a binary relation ρ, is reduced. Moreover, given two binary relations ρ and σ defined on H, we investigate when the hypergroups $\mathbb{H}_{\rho \cap \sigma}$, $\mathbb{H}_{\rho \cup \sigma}$, $\mathbb{H}_{\rho\sigma}$ are reduced.

Let ρ be a binary relation defined on a non-empty set H.

For all $x \in H$, we denote

$$L_x^\rho = \{z \in H \mid (x, z) \in \rho\} \quad \text{and} \quad R_x^\rho = \{z \in H \mid (z, x) \in \rho\}.$$

When it is clear what is the relation we talk about, then we use the notations L_x and R_x instead of L_x^ρ and R_x^ρ.

If ρ is a relation such that the associated hypergroupoid $\mathbb{H}_\rho$ is a hypergroup, then for all $x \in H$ we have $L_x \neq \emptyset$ and $R_x \neq \emptyset$.

It is easy to see that

(1) ρ is reflexive if and only if, for any $x \in H, x \in L_x$;

(2) ρ is symmetric if and only if for all $x \in H, L_x = R_x$;

(3) ρ is transitive if and only if for all $x, y \in H$ with $L_x \cap R_y \neq \emptyset$ it follows that $y \in L_x$.

Let ρ and σ two distinct binary relations defined on H. We have:

(1) $L_x^{\rho \cap \sigma} = \{z \in H \mid (x, z) \in \rho \cap \sigma\} = L_x^\rho \cap L_x^\sigma$,

$\quad R_x^{\rho \cap \sigma} = \{z \in H \mid (z, x) \in \rho \cap \sigma\} = R_x^\rho \cap R_x^\sigma$,

(2) $L_x^{\rho \cup \sigma} = \{z \in H \mid (x, z) \in \rho \cup \sigma\} = L_x^\rho \cup L_x^\sigma$,

$\quad R_x^{\rho \cup \sigma} = \{z \in H \mid (z, x) \in \rho \cup \sigma\} = R_x^\rho \cup R_x^\sigma$,

(3)

$$L_x^{\rho\sigma} = \{z \in H \mid (x, z) \in \rho\sigma\}$$
$$= \{z \in H \mid \exists t \in H : (x, t) \in \rho, (t, z) \in \sigma\}$$
$$= \{z \in L_t^\sigma \mid t \in L_x^\rho\},$$

$$R_x^{\rho\sigma} = \{z \in H \mid (z, x) \in \rho\sigma\}$$
$$= \{z \in H \mid \exists t \in H : (z, t) \in \rho, (t, x) \in \sigma\}$$
$$= \{z \in R_t^\rho \mid t \in R_x^\sigma\},$$

(4) If for all $x \in H$, $L_x^\rho = L_x^\sigma$, then $\rho = \sigma$.

Theorem 8.13. *Let $\mathbb{H}_\rho$ be the hypergroup associated with the binary relation ρ defined on H. For all $x, y \in H$, the following implications hold:*

(1) $x \sim_o y \Leftrightarrow L_x = L_y$;
(2) $x \sim_i y \Leftrightarrow R_x = R_y$.

Proof. (1) By the definition of the relation "$\sim_o$", we have $x \sim_o y$ is equivalent with $x \circ a = y \circ a$, for all $a \in H$, which means that $L_x \cup L_a = L_y \cup L_a$. If $L_x = L_y$, it is clear that $x \sim_o y$.

Now, we suppose $x \sim_o y$, hence, for any $a \in H$, $L_x \cup L_a = L_y \cup L_a$.

- For $a = x$ it follows that $L_x = L_x \cup L_y$, so $L_y \subseteq L_x$.
- For $a = y$ it follows that $L_x \cup L_y = L_y$, so $L_x \subseteq L_y$.

Hence $L_x = L_y$.

(2) Consider $x, y \in H$, $x \sim_i y$. This means that $x \in a \circ b \Leftrightarrow y \in a \circ b$, for $a, b \in H$, that is $x \in L_a \cup L_b \Leftrightarrow y \in L_a \cup L_b$. For all $x \in H$, $R_x \neq \emptyset$, therefore there exists $a \in H$ such that $a \in R_x$, that is $x \in L_a$; it follows that $x \in L_a = a \circ a$ and since $x \sim_i y$, we obtain $y \in L_a$, that is $a \in R_y$. Similarly we obtain $R_y \subseteq R_x$ and then $R_x = R_y$.

Now, if $R_x = R_y$ we have $x \in L_z \Leftrightarrow y \in L_z$, for $z \in H$, therefore $x \in z \circ t \Leftrightarrow y \in z \circ t$, for $z, t \in H$, which means that $x \sim_i y$.

Now, we investigate when two different elements $x, y \in H$ are in the relation $x \sim_e y$ in the hypergroups $\mathbb{H}_{\rho\cap\sigma}$ and $\mathbb{H}_{\rho\sigma}$. $\square$

Theorem 8.14. *Let ρ and σ be two quasiorder relations on a non empty set H. For any $x, y \in H$, $x \sim_e y$ in $\mathbb{H}_{\rho\cap\sigma}$ if and only if $x \sim_e y$ in $\mathbb{H}_\rho$ and $x \sim_e y$ in $\mathbb{H}_\sigma$.*

Proof. Since ρ and σ are two quasiorder relations, it follows that the hypergroupoids associated with ρ, σ and $\rho \cap \sigma$ are hypergroups.

First, we suppose $x \sim_e y$ in $\mathbb{H}_\rho$ and $x \sim_e y$ in $\mathbb{H}_\sigma$. We have $L_x^\rho = L_y^\rho$, $R_x^\rho = R_y^\rho$, $L_x^\sigma = L_y^\sigma$ and $R_x^\sigma = R_y^\sigma$, so $L_x^{\rho \cap \sigma} = L_y^{\rho \cap \sigma}$ and $R_x^{\rho \cap \sigma} = R_y^{\rho \cap \sigma}$, that is $x \sim_e y$ in $\mathbb{H}_{\rho \cap \sigma}$.

Conversely, suppose that $x \sim_e y$ in $\mathbb{H}_{\rho \cap \sigma}$, that is $x \sim_o y$ and $x \sim_i y$ in $\mathbb{H}_{\rho \cap \sigma}$. It is enough to show the implications:

(1) $L_x^\rho \cap L_x^\sigma = L_y^\rho \cap L_y^\sigma \Longrightarrow L_x^\rho = L_y^\rho$ and $L_x^\sigma = L_y^\sigma$;

(2) $R_x^\rho \cap R_x^\sigma = R_y^\rho \cap R_y^\sigma \Longrightarrow R_x^\rho = R_y^\rho$ and $R_x^\sigma = R_y^\sigma$.

Since ρ and σ are reflexive relations, we write $x \in L_x^\rho \cap L_x^\sigma$, so $x \in L_y^\rho \cap L_y^\sigma$, that is $(y, x) \in \rho \cap \sigma$ and similarly, $(x, y) \in \rho \cap \sigma$.

Let us consider $z \in L_x^\rho$, that is $(x, z) \in \rho$ and since $(y, x) \in \rho$, by the transitivity of ρ, it results $(y, z) \in \rho$, $z \in L_y^\rho$. We have $L_x^\rho \subseteq L_y^\rho$ and similarly $L_y^\rho \subseteq L_x^\rho$. We obtain $L_x^\rho = L_y^\rho$ and, in the same way, $L_x^\sigma = L_y^\sigma$. Similarly, we obtain (2). $\square$

Theorem 8.15. *Let ρ and σ be two binary relations on H with full domain and full range such that $\rho^2 = \rho$, $\sigma^2 = \sigma$ and $\rho\sigma = \sigma\rho$. If, for $x, y \in H$, $x \sim_o y$ in $\mathbb{H}_\rho$ and $x \sim_i y$ in $\mathbb{H}_\sigma$, then $x \sim_e y$ in $\mathbb{H}_{\rho\sigma}$.*

Moreover, $x \sim_e y$ in $\mathbb{H}_\rho$ and $x \sim_e y$ in $\mathbb{H}_\sigma$ imply that $x \sim_e y$ in $\mathbb{H}_{\rho\sigma}$.

Proof. The hypergroupoids $\mathbb{H}_\rho$, $\mathbb{H}_\sigma$ and $\mathbb{H}_{\rho\sigma}$ are hypergroups.

Let us consider $x, y \in H$ such that $x \sim_o y$ in $\mathbb{H}_\rho$ and $x \sim_i y$ in $\mathbb{H}_\sigma$, so we have $L_x^\rho = L_y^\rho$ and $R_x^\sigma = R_y^\sigma$. It is enough to check the implications

(1) $L_x^\rho = L_y^\rho \Rightarrow L_x^{\rho\sigma} = L_y^{\rho\sigma}$;

(2) $R_x^\sigma = R_y^\sigma \Rightarrow R_x^{\rho\sigma} = R_y^{\rho\sigma}$.

Let $z \in L_x^{\rho\sigma}$. There exists $t \in L_x^\rho$ such that $z \in L_t^\sigma$, so there exists $t \in L_y^\rho$ such that $z \in L_t^\sigma$; therefore $z \in L_y^{\rho\sigma}$. Similarly $L_y^{\rho\sigma} \subseteq L_x^{\rho\sigma}$. In the same way we can show the second implication.

Thus, if $x \sim_o y$ in $\mathbb{H}_\sigma$ and $x \sim_i y$ in $\mathbb{H}_\rho$, it follows that $x \sim_e y$ in $\mathbb{H}_{\sigma\rho}$ and since $\rho\sigma = \sigma\rho$ we obtain the last assertion. $\square$

8.5 Reduced hypergroups associated with binary relations

A necessary and sufficient condition for the hypergroup $\mathbb{H}_\rho$ in order to be reduced is presented here. Secondly, it is proved that the hypergroupoid $\mathcal{H}_\rho$

associated with a binary relation defined by P. Corsini [29] is not a reduced hypergroup.

All results of this and the next two paragraphs are contained in [42].

Theorem 8.16. *The hypergroup* $\mathbb{H}_\rho$ *is reduced if and only if, for any* $x, y \in H$, x *different from* y, *either* $L_x \neq L_y$ *or* $R_x \neq R_y$.

Proof. The hypergroup $\mathbb{H}_\rho$ is reduced if and only if, for all $x \neq y$, it is true $x \nsim_o y$ or $x \nsim_i y$ which means that $L_x \neq L_y$ or $R_x \neq R_y$. $\square$

Theorem 8.17. *If* ρ *is an equivalence on* H, *then the hypergroupoid* $\mathbb{H}_\rho$ *is a reduced hypergroup if and only if* $\rho = \Delta_H = \{(x, x) \mid x \in H\}$.

Proof. If ρ is an equivalence on H, then $(\mathbb{H}_\rho, \circ)$ is a hypergroup. Since ρ is symmetric, we have $L_x = R_x$ for all $x \in H$ and then $\mathbb{H}_\rho$ is reduced if and only if for all $x \neq y$, $L_x \neq L_y$. We show that this condition is equivalent with the following one: for all $x \in H$, $L_x = \{x\}$ and we have $\rho = \Delta_H$.
If, for all $x \in H$, $L_x = \{x\}$, it follows that for all $x \neq y$ that $L_x \neq L_y$.
Conversely, let $y \neq x, y \in L_x$; we obtain $\{x, y\} \subseteq L_y$. For all $z \in L_y \setminus \{x, y\}$ we have $(y, z) \in \rho, (x, y) \in \rho$ and by transitivity it follows that $(x, z) \in \rho$, so $z \in L_x$. Similarly, it results $L_x \subseteq L_y$, thus $L_x = L_y$, a contradiction. $\square$

Theorem 8.18. *If* ρ *is a non-symmetric quasiorder on* H, *then the hypergroup* $(\mathbb{H}_\rho, \circ)$ *is reduced if and only if, for any* $x \neq y$, $L_x \neq L_y$.

Proof. If ρ is a quasiorder on H then, for all $x \neq y \in H$, we have the implication $x \sim_o y \Rightarrow x \sim_i y$.
Indeed, if we suppose $L_x = L_y$ and $R_x \neq R_y$, there exists $z \in R_x, z \notin R_y$; then $(z, x) \in \rho$ and $(z, y) \notin \rho$. But ρ is reflexive and then $y \in L_y = L_x$; thus $(x, y) \in \rho$ and by transitivity we obtain $(z, y) \in \rho$, which is false.

Hence, for all $x \neq y$, the condition "$L_x \neq L_y$ or $R_x \neq R_y$" is equivalent with "$L_x \neq L_y$". $\square$

Theorem 8.19. *If* ρ *is a reflexive symmetric non-transitive relation on* H, *such that* $\rho^2 = H \times H$, *then the hypergroup* $(\mathbb{H}_\rho, \circ)$ *is reduced if and only if* $L_x \neq L_y$, *for all* $x, y \in H$, x *different from* y.

Proof. It is enough to prove that for all $x \neq y$, $x \sim_o y \Rightarrow x \sim_i y$.

If we suppose there exists $a \in H$ such that $x \in L_a$ and $y \notin L_a$, then, by the symmetry, we have $a \in L_x = L_y$ and thus $a \in L_y$, so $y \in L_a$, a contradiction.

Given a binary relation ρ on H, Corsini [29] defined another hyperoperation: for all $x, y \in H$,

$$x \otimes_\rho y = L_x \cap R_y$$

and he proved that $\mathcal{H}_\rho = (H, \otimes_\rho)$ is a hypergroupoid if and only if $\rho^2 = H \times H$.

In case that the Corsini's hyperoperation $\otimes_\rho$ is left or right reproductive, then $\mathcal{H}_\rho$ is the total hypergroup. Hence, the unique hypergroup obtained in this manner is the total hypergroup, which clearly is not reduced. $\qquad \square$

8.6 The hypergroups $\mathbb{H}_{\rho \cap \sigma}$, $\mathbb{H}_{\rho \cup \sigma}$, $\mathbb{H}_{\rho \sigma}$ as reduced hypergroups

Let ρ and σ be two binary relations defined on a non-empty set H. The hypergroups $\mathbb{H}_{\rho \cap \sigma}$, $\mathbb{H}_{\rho \sigma}$ and $\mathbb{H}_{\rho \cup \sigma}$ are reduced independently if $\mathbb{H}_\rho$ and $\mathbb{H}_\sigma$ are or are not reduced hypergroups, as we can see below.

Theorem 8.20. *Let ρ and σ be two quasiorder relations on H. If the hypergroups $\mathbb{H}_\rho$ and $\mathbb{H}_\sigma$ are reduced, then the hypergroup $\mathbb{H}_{\rho \cap \sigma}$ is reduced too.*

Proof. If we suppose that the hypergroup $\mathbb{H}_{\rho \cap \sigma}$ is not reduced, then it results there exist $x \neq y$ in H such that $x \sim_e y$ in $\mathbb{H}_{\rho \cap \sigma}$ and therefore $x \sim_e y$ in $\mathbb{H}_\rho$, $x \sim_e y$ in $\mathbb{H}_\sigma$, which is impossible because the hypergroups $\mathbb{H}_\rho$ and $\mathbb{H}_\sigma$ are reduced. $\qquad \square$

Remark 8.3. If the hypergroup $\mathbb{H}_{\rho \cap \sigma}$ is reduced, then the hypergroups $\mathbb{H}_\rho$ and $\mathbb{H}_\sigma$ can be reduced or not, as one sees from the following examples.

Example 8.2. Let $H = \{1, 2, 3, 4\}$.

(1) If $\rho \cap \sigma = \Delta_H = \{(1,1), (2,2), (3,3), (4,4)\}$ and ρ, σ are equivalences on H different from the diagonal relation Δ_H, then the hypergroup $\mathbb{H}_{\rho \cap \sigma}$ is reduced, but neither $\mathbb{H}_\rho$ nor $\mathbb{H}_\sigma$ is a reduced hypergroup (see Proposition 3.2);

(2) Set $\rho = \Delta_H \cup \{(1,2)\}$ and $\sigma = \Delta_H \cup \{(1,3)\}$. Then $\rho \cap \sigma = \Delta_H$, so $\mathbb{H}_{\rho \cap \sigma}$ is a reduced hypergroup and also $\mathbb{H}_\rho$ and $\mathbb{H}_\sigma$;

(3) Set $\rho = \Delta_H \cup \{(1,2), (2,1), (1,3), (2,3)\}$, $\sigma = \Delta_H \cup \{(1,2), (3,4)\}$, so $\rho \cap \sigma = \Delta_H \cup \{(1,2)\}$. It results the hypergroups $\mathbb{H}_{\rho \cap \sigma}$ and $\mathbb{H}_\sigma$ are reduced, but the hypergroup $\mathbb{H}_\rho$ is not ($L_1^\rho = L_2^\rho$, $R_1^\rho = R_2^\rho$).

Theorem 8.21. *Let ρ and σ be two binary relations on H with full domain and full range such that $\rho^2 = \rho$, $\sigma^2 = \sigma$ and $\rho\sigma = \sigma\rho$. If the hypergroup $\mathbb{H}_{\rho\sigma}$ is reduced, then both hypergroups $\mathbb{H}_\rho$ and $\mathbb{H}_\sigma$ are reduced.*

Proof.   We have the implications:

(1) $L_x^{\rho\sigma} \neq L_y^{\rho\sigma} \Rightarrow L_x^\rho \neq L_y^\rho$ and $L_x^\sigma \neq L_y^\sigma$;

(2) $R_x^{\rho\sigma} \neq R_y^{\rho\sigma} \Rightarrow R_x^\rho \neq R_y^\rho$ and $R_x^\sigma \neq R_y^\sigma$.

If $\mathbb{H}_{\rho\sigma}$ is a reduced hypergroup then, for all $x \neq y$, we have $x \nsim_e y$, so, for any $x \neq y$, $L_x^{\rho\sigma} \neq L_y^{\rho\sigma}$ or $R_x^{\rho\sigma} \neq R_y^{\rho\sigma}$. It follows $(L_x^\rho \neq L_y^\rho$ and $L_x^\sigma \neq L_y^\sigma)$ or $(R_x^\rho \neq R_y^\rho$ and $R_x^\sigma \neq R_y^\sigma)$ and therefore the hypergroups $\mathbb{H}_\rho$ and $\mathbb{H}_\sigma$ are reduced. $\square$

Remark 8.4. In the same hypothesis as above if $\mathbb{H}_\rho$ and $\mathbb{H}_\sigma$ are reduced hypergroups, then the hypergroup $\mathbb{H}_{\rho\sigma}$ is reduced or not, as the following examples show.

Example 8.3. We consider the following two situations.

(1) Set $H = \{1, 2, 3, 4\}$, $\rho = \Delta_H \cup \{(1, 2)\} = \rho^2$ and $\sigma = \Delta_H \cup \{(1, 3)\} = \sigma^2$.

Clearly, $\mathbb{H}_\rho$ and $\mathbb{H}_\sigma$ are reduced hypergroups. Since $\rho\sigma = \Delta_H \cup \{(1, 2), (1, 3)\} = \sigma\rho$, it follows that that the hypergroup $\mathbb{H}_{\rho\sigma}$ is reduced;

(2) Set $H = \{1, 2, 3\}$, $\rho = \Delta_H \cup \{(2, 1), (2, 3)\} = \rho^2$ and $\sigma = \Delta_H \cup \{(1, 3), (1, 2)\} = \sigma^2$.

Again it results that $\mathbb{H}_\rho$ and $\mathbb{H}_\sigma$ are reduced hypergroups; we obtain $\rho\sigma = \Delta_H \cup \{(1, 2), (1, 3), (2, 1), (2, 3)\} = \sigma\rho$, and then $L_1^{\rho\sigma} = H = L_2^{\rho\sigma}$, $R_1^{\rho\sigma} = \{1, 2\} = R_2^{\rho\sigma}$, therefore the hypergroup $\mathbb{H}_{\rho\sigma}$ is not reduced.

Remark 8.5. Let ρ and σ be two binary relations defined on H such that the hypergroupoids $\mathbb{H}_\rho$, $\mathbb{H}_\sigma$ and $\mathbb{H}_{\rho\cup\sigma}$ are hypergroups. If $\mathbb{H}_\rho$ and $\mathbb{H}_\sigma$ are reduced hypergroups, then the hypergroup $\mathbb{H}_{\rho\cup\sigma}$ can be reduced or not and conversely, if $\mathbb{H}_{\rho\cup\sigma}$ is a reduced hypergroup, it doesn't result that the hypergroups $\mathbb{H}_\rho$ and $\mathbb{H}_\sigma$ are reduced, too, as it follows from the following examples.

Example 8.4. We present the following situations.

(1) Set $H = \{1, 2, 3\}$, $\rho = \Delta_H \cup \{(1, 2)\} = \rho^2$ and $\sigma = \Delta_H \cup \{(2, 1)\} = \sigma^2$; we find $\rho \cup \sigma = \Delta_H \cup \{(1, 2), (2, 1)\} = (\rho \cup \sigma)^2$. It is clear that

$\mathbb{H}_\rho$ and $\mathbb{H}_\sigma$ are reduced hypergroups, but the hypergroup $\mathbb{H}_{\rho\cup\sigma}$ is not reduced, since $L_1^{\rho\cup\sigma} = \{1,2\} = L_2^{\rho\cup\sigma}$, $R_1^{\rho\cup\sigma} = \{1,2\} = R_2^{\rho\cup\sigma}$ (see the Proposition 3.3);

(2) Set $H = \{1,2,3\}$, $\rho = \Delta_H \cup \{(1,2)\} = \rho^2$ and $\sigma = \Delta_H \cup \{(1,3)\} = \sigma^2$; we obtain $\rho \cup \sigma = \Delta_H \cup \{(1,2),(1,3)\} = (\rho \cup \sigma)^2$. It follows that all the three hypergroups $\mathbb{H}_\rho$, $\mathbb{H}_\sigma$ and $\mathbb{H}_{\rho\cup\sigma}$ are reduced;

(3) Set again $H = \{1,2,3\}$ and the relations $\rho = \Delta_H \cup \{(1,2),(2,1)\} = \rho^2$, $\sigma = \Delta_H \cup \{(1,3)\} = \sigma^2$, therefore $\rho \cup \sigma = \Delta_H \cup \{(1,2),(2,1),(1,3)\}$ which is different from $(\rho \cup \sigma)^2 = \Delta_H \cup \{(1,2),(2,1),(1,3),(2,3)\}$. In this case the hypergroup $\mathbb{H}_\rho$ is not reduced, the hypergroup $\mathbb{H}_\sigma$ is reduced and the hypergroup $\mathbb{H}_{\rho\cup\sigma}$ is reduced, too. The hypergroupoid $\mathbb{H}_{\rho\cup\sigma}$ is a hypergroup because $\rho \cup \sigma \subset (\rho \cup \sigma)^2$ and for the outer elements 1 and 2 of $\rho \cup \sigma$, the condition (iv) of the Theorem 1.3 holds.

8.7 The cartesian product of the reduced hypergroups

Let $(H_1, \circ_1)$, $(H_2, \circ_2)$ be two hypergroups. On the cartesian product $H_1 \times H_2$ we define the hyperoperation

$$(x_1, x_2) \otimes (y_1, y_2) = (x_1 \circ_1 y_1, x_2 \circ_2 y_2)$$

and we obtain the hypergroup $(H_1 \times H_2, \otimes)$.

Theorem 8.22. *In the hypergroup $(H_1 \times H_2, \otimes)$, the following implications hold:*

(1) $(x_1, x_2) \sim_o (y_1, y_2) \Leftrightarrow x_1 \sim_o y_1$ *in* H_1 *and* $x_2 \sim_o y_2$ *in* H_2;

(2) $(x_1, x_2) \sim_i (y_1, y_2) \Leftrightarrow x_1 \sim_i y_1$ *in* H_1 *and* $x_2 \sim_i y_2$ *in* H_2.

Proof. (1) By the definition of the relation $\sim_o$ we have $(x_1, x_2) \sim_o (y_1, y_2)$ if and only if, for any $(a_1, a_2) \in H_1 \times H_2$, it is true: $(x_1, x_2) \otimes (a_1, a_2) = (y_1, y_2) \otimes (a_1, a_2)$ and $(a_1, a_2) \otimes (x_1, x_2) = (a_1, a_2) \otimes (y_1, y_2)$, which is equivalent with $x_1 \circ_1 a_1 = y_1 \circ_1 a_1$, $x_2 \circ_2 a_2 = y_2 \circ_2 a_2$ and $a_1 \circ_1 x_1 = a_1 \circ_1 y_1$, $a_2 \circ_2 x_2 = a_2 \circ_2 y_2$, that is, $x_1 \sim_o y_1$ and $x_2 \sim_o y_2$.

(2) By the definition of the relation $\sim_i$ we get $(x_1, x_2) \sim_i (y_1, y_2)$ if and only if, for $(a_1, a_2), (b_1, b_2) \in H_1 \times H_2$, we have $(x_1, x_2) \in (a_1, a_2) \otimes (b_1, b_2)$ equivalently $(y_1, y_2) \in (a_1, a_2) \otimes (b_1, b_2)$, therefore $x_1 \in a_1 \circ_1 b_1$ and $x_2 \in a_2 \circ_2 b_2$ if and only if $y_1 \in a_1 \circ_1 b_1$ and $y_2 \in a_2 \circ_2 b_2$, that is, $x_1 \sim_i y_1$ and $x_2 \sim_i y_2$. $\square$

Theorem 8.23. *The hypergroup* $(H_1 \times H_2, \otimes)$ *is reduced if and only if the hypergroups* $(H_1, \circ_1)$ *and* $(H_2, \circ_2)$ *are reduced.*

Proof. First, we suppose that $(H_1 \times H_2, \otimes)$ is a reduced hypergroup and that H_1 is not reduced. Then there exists $x_1 \neq y_1$ in H_1 such that $x_1 \sim_e y_1$, that is, $x_1 \sim_o y_1$ and $x_1 \sim_i y_1$. It follows that, for any $x_2 \in H_2$, we have $(x_1, x_2) \sim_o (y_1, x_2)$ and $(x_1, x_2) \sim_i (y_1, x_2)$ that is $(x_1, x_2) \sim_e (y_1, y_2)$ with $(x_1, x_2) \neq (y_1, x_2)$; this means that $(H_1 \times H_2, \otimes)$ is not reduced, which is in contradiction with the hypothesis.

Conversely, we suppose that $(H_1, \circ_1)$ and $(H_2, \circ_2)$ are reduced hypergroups, but $(H_1 \times H_2, \otimes)$ is not. Then there exist $(x_1, x_2) \neq (y_1, y_2) \in H_1 \times H_2$ such that $(x_1, x_2) \sim_e (y_1, y_2)$. We find $x_1 \sim_e y_1$ and $x_2 \sim_e y_2$. Since $(H_1, \circ_1)$ and $(H_2, \circ_2)$ are reduced, it follows that $x_1 = y_1, x_2 = y_2$, thus $(x_1, x_2) = (y_1, y_2)$ which is false. $\qquad \square$

Theorem 8.24. *Let* ρ_1, ρ_2 *be two binary relations defined on the non-empty sets* H_1, H_2 *such that the associated hypergroupoids* $(\mathbb{H}_1)_{\rho_1}$ *and* $(\mathbb{H}_2)_{\rho_2}$ *are hypergroups.*

(1) *If* $(\mathbb{H}_1)_{\rho_1}$ *and* $(\mathbb{H}_2)_{\rho_2}$ *are reduced hypergroups and, for* $j \in \{1, 2\}$, *the implication* $\rho_j^2 \neq H_j^2 \implies \rho_{3-j} = \rho_{3-j}^2$ *holds, then* $(H_1 \times H_2)_{\rho_1 \times \rho_2}$ *is a reduced hypergroup.*

(2) *If* $(H_1 \times H_2)_{\rho_1 \times \rho_2}$ *is a reduced hypergroup, then at least one of the hypergroups* $(\mathbb{H}_1)_{\rho_1}$ *and* $(\mathbb{H}_2)_{\rho_2}$ *is reduced.*

Proof. (1) If we suppose that $(H_1 \times H_2)_{\rho_1 \times \rho_2}$ is not reduced, then there exist $(x_1, x_2) \neq (y_1, y_2) \in H_1 \times H_2$ such that $L_{(x_1, x_2)} = L_{(y_1, y_2)}$ and $R_{(x_1, x_2)} = R_{(y_1, y_2)}$, that is $L_{x_1} = L_{y_1}$, $L_{x_2} = L_{y_2}$, $R_{x_1} = R_{y_1}$, $R_{x_2} = R_{y_2}$. This implies that $x_1 \sim_e y_1$ in $(\mathbb{H}_1)_{\rho_1}$ and $x_2 \sim_e y_2$ in $(\mathbb{H}_2)_{\rho_2}$, but since $(\mathbb{H}_1)_{\rho_1}$ and $(\mathbb{H}_2)_{\rho_2}$ are reduced, it follows that $x_1 = y_1$ and $x_2 = y_2$, therefore $(x_1, x_2) = (y_1, y_2)$, which is false.

(2) Now, if $(H_1 \times H_2)_{\rho_1 \times \rho_2}$ is a reduced hypergroup and if we suppose that both hypergroups $(\mathbb{H}_1)_{\rho_1}$ and $(\mathbb{H}_2)_{\rho_2}$ are not reduced, it follows that there exist $x_1 \neq y_1 \in H_1$ and $x_2 \neq y_2 \in H_2$ such that $x_1 \sim_e y_1$ in $(\mathbb{H}_1)_{\rho_1}$ and $x_2 \sim_e y_2$ in $(\mathbb{H}_2)_{\rho_2}$. We obtain $L_{x_1} = L_{y_1}$, $R_{x_1} = R_{y_1}$ and $L_{x_2} = L_{y_2}$, $R_{x_2} = R_{y_2}$, which lead to the relations $L_{(x_1, x_2)} = L_{(y_1, y_2)}$ and $R_{(x_1, x_2)} = R_{(y_1, y_2)}$, a contradiction with the hypothesis $(H_1 \times H_2)_{\rho_1 \times \rho_2}$ is reduced. $\qquad \square$

8.8   Enumeration of finite Rosenberg hypergroups

In [122] Migliorato determined the structure of all non-isomorphic hyper-groups of order 3 and of total regular abelian hypergroups. Later on Bayon and Lygeros [9; 10; 11] computed the number of finite abelian hypergroups and H_v-groups.

Turning back to Rosenberg hypergroups, several algorithms were presented by Spartalis and Mamaloukas [170], Massouros and Tsitouras [118; 117], Cristea et al. [40] and Jafarpour et al. [86], where particular types of such hypergroups were computed. More exactly, in [40] the authors determined the number of Rosenberg hypergroups that are (general) mutually associative or complementary hypergroups. All regular reversible Rosenberg hypergroups of order less than 5 were computed, up to isomorphism [86], but not those of bigger order. We present here another method, consisting in representing every Boolean matrix by a non-negative integer.It is computed the number of idempotent Rosenberg hypergroups, regular Rosenberg hypergroups and reversible Rosenberg hypergroups of order 6, up to isomorphism.

All results from this paragraph and the next one belong to paper [1].

The rest of the section gathers properties of regular reversible Rosenberg hypergroups based on the representation of binary relations by Boolean matrices, having all elements 0 or 1 and satisfying the rules in a Boolean algebra: $0 + 1 = 1 + 0 = 1 + 1 = 1$, while $0 + 0 = 0$, and $0 \cdot 0 = 0 \cdot 1 = 1 \cdot 0 = 0$, $1 \cdot 1 = 1$ [170]. Let ρ be a binary relation defined on a finite set $H = \{a_1, ..., a_n\}$. The associated Boolean matrix $M(\rho) = (a_{ij})$, $i, j \in \{1, 2, \ldots, n\}$, is obtained as follows: $a_{ij} = 1$, if $(a_i, a_j) \in \rho$ and $a_{ij} = 0$, if $(a_i, a_j) \notin \rho$. It is clear that $M(\rho^2) = M^2(\rho)$ and in this way, 2^{n^2} Rosenberg partial hypergroupoids can be defined on every set with n elements.

We recall now some definitions and results from [40; 86].

Definition 8.4. The matrix $M(\rho)$ is called *very good* if the Rosenberg hypergroupoid H_ρ is a hypergroup.

Definition 8.5. A Rosenberg hypergroup H_ρ is called *i-Rosenberg hypergroup* if $M(\rho)$ is an idempotent very good matrix (i.e., $M(\rho)^2 = M(\rho)$).

Definition 8.6.

(1) A very good matrix $M(\rho)$ is called *i-very good* if and only if H_ρ is an i-Rosenberg hypergroup.

(2) A very good matrix $M(\rho)$ is called *regular* if and only if H_ρ is a regular hypergroup.

(3) A regular matrix $M(\rho)$ is called *reversible* if and only if H_ρ is a regular reversible hypergroup.

An $n \times 1$ matrix (one column and n rows) is called a column vector and for a given matrix $M = (a_{ij})$, $i,j \in \{1, 2, \ldots, n\}$, M_j is the column vector (a_{nj}) and M_j^2 is the j-column vector of the matrix $M(\rho^2)$. In particular, (0) is the column vector with all elements equal to 0, and (1) is the column vector with all elements equal to 1. The transpose of a matrix M is the matrix M^T, formed by turning rows into columns and vice versa.

Theorem 8.25. *A matrix $M = M(\rho)$ is a very good matrix if and only if, for any j, with $1 \le j \le n$, the following assertions hold:*

(1) $M_j \ne (0)$;

(2) $M_j^T \ne (0)$;

(3) *if $M_j^2 \ne (1)$, then $M_j = M_j^2$.*

Proof. Suppose that (3) holds and $(a_i, a_j) \in \rho$; if $(a_i, a_j) \notin \rho^2$ we have $M_j^2 \ne (1)$. By using (3) we conclude that $M_j^2 = M_j$ and so $(a_i, a_j) \in \rho^2$ follows, which is a contradiction and therefore $\rho \subseteq \rho^2$. Now, suppose that a_j is an outer element of ρ, that is there exists $a_i \in H$ such that $(a_i, a_j) \notin \rho^2$ which means $M_j^2 \ne (1)$. As above, from $(a_k, a_j) \in \rho^2$ we can conclude that $(a_k, a_j) \in \rho$. Similarly, the converse assertion holds, too. $\square$

Remark 8.6. The matrix $M = (a_{ij})$, with $a_{ij} = 1$, for any $i, j \in \{1, 2, \ldots, n\}$, is a very good matrix and the corresponding hypergroup is the total hypergroup which is a Rosenberg hypergroup.

In what follows, we give in terms of matrices a necessary and sufficient condition such that two Rosenberg's hypergroups associated with two binary relations on the same set H are isomorphic.

Theorem 8.26. *Let $H = \{a_1, \ldots, a_n\}$ be a finite set, ρ and ρ' be two binary relations on H and $M(\rho) = (t_{ij})$, $M(\rho') = (t'_{ij})$ be their associated matrices. The hypergroups $\mathbb{H}_\rho$ and $\mathbb{H}_{\rho'}$ are isomorphic if and only if $t_{ij} = t'_{\sigma(i)\sigma(j)}$, for σ a permutation of the set $\{1, 2, \ldots, n\}$.*

Proof. Let $\theta : \mathbb{H}_\rho \to \mathbb{H}_{\rho'}$ be an isomorphism, then $\theta(a_i \circ_\rho a_j) = \theta(a_i) \circ_{\rho'} \theta(a_j)$ and so $\{\theta(a_j) | (a_i, a_j) \in \rho\} = \{a_t | (\theta(a_i), a_t) \in \rho'\}$; thus we have $(a_i, a_j) \in \rho$ if and only if $(\theta(a_i), \theta(a_j)) \in \rho'$. Consequently $t_{ij} = t'_{\sigma(i)\sigma(j)}$, where $\theta(a_j) = a_{\sigma(j)}$, for σ a permutation of the set $\{1, 2, \ldots, n\}$.

Conversely, for σ a permutation of the set $\{1, 2, \ldots, n\}$, we have

$$(a_i, a_j) \in \rho \Leftrightarrow (a_{\sigma(i)}, a_{\sigma(j)}) \in \rho'.$$

Consider the map $\varphi : \mathbb{H}_\rho \to \mathbb{H}_{\rho'}$ with $\varphi(a_i) = a_{\sigma(i)}$. Clearly φ is a bijection and $\varphi(a_i \circ_\rho a_i) = \{\varphi(a_j) \mid (a_i, a_j) \in \rho\} = \{a_{\sigma(j)} \mid (a_i, a_j) \in \rho\} = \{a_{\sigma(j)} \mid (a_{\sigma(i)}, a_{\sigma(j)}) \in \rho'\} = a_{\sigma(i)} \circ_{\rho'} a_{\sigma(i)} = \varphi(a_i) \circ_{\rho'} \varphi(a_i)$. Moreover, for all $(a_i, a_j) \in H^2$ we have $\varphi(a_i \circ_\rho a_j) = \varphi(a_i \circ_\rho a_i) \cup \varphi(a_j \circ_\rho a_j) = \varphi(a_i) \circ_{\rho'} \varphi(a_i) \cup \varphi(a_j) \circ_{\rho'} \varphi(a_j) = \varphi(a_i) \circ_{\rho'} \varphi(a_j)$. $\qquad\square$

Now, we check when a very good matrix is regular.

Theorem 8.27 [86]. *Let $M(\rho) = (t_{ij})_{n \times n}$ be a very good matrix. Then $M(\rho)$ is regular if and only if there exists k, $1 \leq k \leq n$, such that, for all i, $1 \leq i \leq n$, we have $t_{ii}^c \leq t_{ki}$, where $t_{ij}^c = 0$ if and only if $t_{ij} = 1$.*

Finally, we present how to combine two idempotent very good matrices in order to obtain another idempotent very good matrix.

Theorem 8.28 [86]. *Let $M = (t_{ij})_{n \times n}$, $M' = (t'_{ij})_{m \times m}$ be two idempotent very good matrices. Then $M \oplus M' = (m_{ij})_{k \times k}$, where $k = n + m$, and*

$$m_{ij} = \begin{cases} t_{ij}, & \text{if } i \leq n, \quad j \leq n \\ t'_{ij}, & \text{if } n < i, \quad n < j \\ 0 & \text{elsewhere} \end{cases}$$

is an idempotent very good matrix.

Theorem 8.29 [86]. *Let $M = (t_{ij})_{n \times n}$, $M' = (t'_{ij})_{m \times m}$ be two idempotent very good matrices. Then $M \boxplus M' = (m_{ij})_{k \times k}$, where $k = n + m$, and*

$$m_{ij} = \begin{cases} t_{ij}, & \text{if } i \leq n, \quad j \leq n \\ t'_{ij}, & \text{if } n < i, \quad n < j \\ 1, & \text{if } i \leq n, \quad j > n \\ 0 & \text{if } n < i, \quad j \leq n \end{cases}$$

is an idempotent very good matrix. Moreover, if M is regular then $M \boxplus M'$ is regular, too.

We conclude this section with the following result.

Theorem 8.30 [86]. *A regular matrix $M = M(\rho)$ is reversible if $M = M^T$.*

Let $m_{i1}, ..., m_{in}$ be the elements on the i'th row. Then the corresponding integer is given by the formula:

$$\sum_{i=1}^{n}\sum_{j=1}^{n} m_{ij} \times 2^{(i-1)n+j-1}$$

From now on, we denote the integer by $I(M)$. Let us see one example. For instance consider the following matrix

$$M = \begin{pmatrix} 1 & 0 & 0 & 1 \\ 1 & 1 & 0 & 0 \\ 0 & 1 & 0 & 1 \\ 0 & 0 & 1 & 1 \end{pmatrix}$$

then the corresponding integer is

$$I(M) = \sum_{i=1}^{4}\sum_{j=1}^{4} m_{ij} \times 2^{4(i-1)+j-1}$$

$$= \sum_{i=1}^{4}\left(m_{i1} \times 2^{4(i-1)} + m_{i2} \times 2^{4(i-1)+1} + m_{i3} \times 2^{4(i-1)+2} + m_{i4} \times 2^{4(i-1)+3}\right)$$

$$= 2^0 + 2^3 + 2^4 + 2^5 + 2^9 + 2^{11} + 2^{14} + 2^{15}$$

$$= 51769.$$

Obviously, there are 2^{n^2} Boolean matrices of order n, which we denote by $M_1, ..., M_{2^{n^2}}$ and all of them can easily be generated by computer. So computation of vg (Number of very good matrices) and vgi (Number of very good idempotent matrices) is straightforward. The tough part is when we want to enumerate all non-isomorphic matrices, and this is where our integer representation of boolean matrices comes to play. Let A and A_2 are initialized as two $1 \times m$ arrays with all entries equal to 0. Once a Boolean matrix is approved to be a very good matrix, we have to decide weather any of its isomorphic copies has already been approved. Let $I(M)$ be the integer representation of the matrix M. If the $I(M)$'th element of array A_2 is already equal to 1, then this is a copy of another matrix that is already considered and we skip it, otherwise it is new and we increment the variable co3 (the nuumber of very good matrices up to isomorphism) by 1. Also we produce all its $n!$ isomorphic copies which are denoted by M_σ where σ is a permutation of $\{1, ..., n\}$ and compute $I(M_\sigma)$ (the set of corresponding integers) and set the corresponding array in A_2 equal to 1.

A similar approach is given in the idempotent case and decide if a very good idempotent matrix is new (up to isomorphism) by looking at corresponding entry in array A and increment the variable co2 (the number of very good idempotent matrices up to isomorphism) if the case is approved to be new (up to isomorphism).

The Main Algorithm is summarized in the following:

```
co2:=0; co3:=0;vgi:=0;vg:=0; rc:=0; rv:=0;
for j=1 to 2^{n^2}
    A(j) := 0; A_2(j) := 0
end
for j=1 to 2^{n^2}
    M := M_j
    if M ≤ M^2 then
        if M satisfies in conditions (1), (2), (3) of Theorem 8.25 then
            if A_2(I(A)) == 0 then
                co3=co3+1;
                for every permutation σ of {1,...,n} set A_2(I(M_σ)) = 1
                if M is regular rc=rc+1;
                if M is reversible rv=rv+1;
            end
            vg=vg+1;
            if M == M^2 then
                vgi=vgi+1;
                if A(I(M)) == 0 then
                    co2=co2+1;
                    for every permutation σ of {1,...,n} set A(I(M_σ)) = 1
                end
            end
        end
    end
end
```

The results of the computation (for $n = 2, 3, 4, 5$ and 6), using a program written in $C\#$, are summarized in the following table:

vg: Number of very good matrices

co3: Number of very good matrices up to isomorphism

vgi: Number of very good idempotent matrices

co2: Number of very good idempotent matrices up to isomorphism

rc: Number of regular matrices up to isomorphism

rv: Number of reversible matrices up to isomorphism

N =	2	3	4	5	6
vg	6	149	9729	2921442	4578277389
co3	4	33	501	26409	6502030
vgi	4	35	559	14962	636217
co2	3	10	44	239	1668
rc	4	32	466	22672	5000329
rv	3	9	31	147	1095

By using the results presented in above table we can provide a lower bound for very good idempotent matrices of higher orders. For example for order 11 the lower bound would be $14962 + 636217 = 651179$.

Chapter 9

Hypergroups and n-ary Relations

In this chapter, a hypergroupoid $(H, \otimes_\rho)$ is associated with an n-ary relation ρ defined on a non-empty set H. It is investigated when it is an H_v-group, a hypergroup or a join space. Then some connections between this hypergroupoid and Rosenberg's hypergroupoid associated with a binary relation are determined.

The results of this chapter are contained in papers [41; 43].

9.1 n-ary relations

First we present some basic notions about the n-ary relations, see [140]. We suppose that $H \neq \emptyset$ is a set, $n \in \mathbb{N}$ a natural number such that $n \geq 2$, and $\rho \subseteq H^n$ is an n-ary relation on H.

Definition 9.1. The relation ρ is said to be

(1) *reflexive* if, for all $x \in H$, the n-tuple $(x, \ldots, x) \in \rho$;

(2) *n-transitive* if it has the following property: if $(x_1, \ldots, x_n) \in \rho$, $(y_1, \ldots, y_n) \in \rho$ hold and if there exist natural numbers $i_0 > j_0$ such that $1 < i_0 \leq n$, $1 \leq j_0 < n$, $x_{i_0} = y_{j_0}$, then the n-tuple $(x_{i_1}, \ldots, x_{i_k}, y_{j_{k+1}}, \ldots, y_{j_n}) \in \rho$ for any natural number $1 \leq k < n$ and $i_1, \ldots, i_k, j_{k+1}, \ldots, j_n$ such that $1 \leq i_1 < \ldots < i_k < i_0$, $j_0 < j_{k+1} < \ldots < j_n \leq n$;

(3) *strongly symmetric* if $(x_1, \ldots, x_n) \in \rho$ implies $(x_{\sigma(1)}, \ldots, x_{\sigma(n)}) \in \rho$ for all permutation σ of the set $\{1, \ldots, n\}$;

(4) *n-ary preordering* on H if it is reflexive and n-transitive;

(5) *n-equivalence* on H if it is reflexive, strongly symmetric and n-transitive.

177

Example 9.1. For $n = 2$, a binary relation is 2-transitive if and only if it is transitive in the usual sense and therefore it is a 2-equivalence if and only if it is an equivalence relation.

Example 9.2. Let $n = 3$. A ternary relation ρ is 3-transitive if and only if it satisfies the following conditions:

(1) If $(x, y, z) \in \rho$, $(y, u, v) \in \rho$, then $(x, u, v) \in \rho$;
(2) If $(x, y, z) \in \rho$, $(z, u, v) \in \rho$, then $(x, y, u) \in \rho$, $(x, y, v) \in \rho$, $(x, u, v) \in \rho$, $(y, u, v) \in \rho$;
(3) If $(x, y, z) \in \rho$, $(u, z, v) \in \rho$, then $(x, y, v) \in \rho$.

Example 9.3. Set $H = \{a_1, a_2, \ldots, a_n\}$ and let $\rho = \{(a_i, \ldots, a_i) \mid 1 \leq i \leq n\}$, where any sequence is formed by n equal symbols, be the diagonal relation on H. Then ρ is an n-equivalence.

Example 9.4. Set $H = \{1, 2, 3\}$ and let $\rho \subseteq H \times H \times H$ be the ternary relation on H defined by $\rho = \{(1, 1, 3), (1, 1, 1), (2, 2, 2), (3, 3, 3), (1, 3, 3)\}$. Then ρ is a 3-ary preordering on H, but it is not a 3-equivalence.

Definition 9.2. Let ρ be an n-ary relation on a set H. We may associate with ρ a binary relation ρ^b as follows. For all $(x, y) \in H^2$, we put $(x, y) \in \rho^b$ if there exist $(x_1, x_2, \ldots, x_n) \in \rho$ and natural numbers i, j such that $1 \leq i < j \leq n$, $x = x_i$, $y = x_j$.

Definition 9.3. Let σ be a binary relation on a set H. We may associate with σ an n-ary relation σ_n on H by putting for all $(x_1, x_2, \ldots, x_n) \in H^n$, $(x_1, x_2, \ldots, x_n) \in \sigma_n$ if $(x_i, x_j) \in \sigma$ for every pair of natural numbers (i, j) such that $1 \leq i < j \leq n$.

Example 9.5. Let ρ be a ternary relation on H. Then the binary relation ρ^b associated with H is defined as follows: $(x, y) \in \rho^b$ if there exists $z \in H$ such that $(x, y, z) \in \rho$ or $(z, x, y) \in \rho$ or $(x, z, y) \in \rho$.

More general, Novak and Novotny [140] proved that:

(1) If ρ is an n-ary preordering on a set H, then ρ^b is a preordering on H.
(2) If ρ be an n-equivalence on a set H, then ρ^b is an equivalence relation on H.
(3) Let σ be a preordering on a set H. Then σ_n is an n-ary preordering on H.

(4) Let σ be an equivalence relation on a set H. Then σ_n is an n-equivalence on H.

(5) If ρ is an n-ary preordering on a set H, then $(\rho^b)_n = \rho$. If σ is a preordering on a set H, then $(\sigma_n)^b = \sigma$.

Corollary 9.1.

(1) *If ρ is an n-ary relation on a set H, then $(\rho^b)_n \supset \rho$, but the equality does not necessarily hold.*

(2) *If σ is a binary relation on a set H, then $(\sigma_n)^b \subset \sigma$, but the equality does not necessarily hold.*

Proof. The first part of the two statements are easily verified. We give only counterexamples for the second part of them.

(1) On the set $H = \{1,2\}$ we consider $\rho = \{(1,2,\ldots,2)\}$. Then, we have $\rho^b = \{(1,2),(2,2)\}$ and thus $(\rho^b)_n = \{(1,2,\ldots,2),(2,\ldots,2)\} \supsetneq \rho$.

(2) On the set $H = \{1,2,3\}$ we take the binary relation $\sigma = \{(1,2),(2,1),(2,3),(3,3)\}$. Then, we have $\sigma_n = \{(2,3\ldots,3),(3,\ldots,3)\}$ and therefore $(\sigma_n)^b = \{(2,3),(3,3)\} \subsetneq \sigma$. $\qquad\square$

Theorem 9.1.

(1) *If ρ_1 and ρ_2 are two n-ary relations on a set H such that $\rho_1 \subset \rho_2$, then $\rho_1^b \subset \rho_2^b$.*

(2) *For all n-ary relation ρ on a set H, it follows $((\rho^b)_n)^b = \rho^b$.*

Proof. (1) Set $(x,y) \in \rho_1^b$; then there exist $(x_1,x_2,\ldots,x_n) \in \rho_1$ and natural numbers i,j such that $1 \leq i < j \leq n$, $x = x_i$, $y = x_j$. Since $\rho_1 \subset \rho_2$, it follows that there exist $(x_1,x_2,\ldots,x_n) \in \rho_2$ and natural numbers i,j such that $1 \leq i < j \leq n$, $x = x_i$, $y = x_j$, that is $(x,y) \in \rho_2^b$.

(2) For the proof of the direct inclusion we denote ρ^b by σ. We have $(\sigma_n)^b \subset \sigma$, that is $((\rho^b)_n)^b \subset \rho^b$. Conversely, for an n-ary relation ρ, we have $\rho \subset (\rho^b)_n$ and by the item (1) of this proposition, it follows $\rho^b \subset ((\rho^b)_n)^b$, which concludes the proof. $\qquad\square$

Definition 9.4. Let ρ be an n-ary relation on a non-empty set H and $k < n$. The $(i_1,\ldots,i_k)$-*projection* of ρ, denoted by $\rho_{i_1,\ldots,i_k}$, is a k-ary relation on H defined by: if $(a_1,\ldots,a_n) \in \rho$, then $(a_{i_1},\ldots,a_{i_k}) \in \rho_{i_1,\ldots,i_k}$.

Definition 9.5. Let ρ be an m-ary relation on a non-empty set H, λ an n-ary relation on the same set H. The *join* of ρ and λ, denoted by $J_p(\rho,\lambda)$, where $1 \leq p < m, 1 \leq p < n$, is an $(m+n-p)$-relation on H that consists of all $(m + n - p)$-tuples $(a_1, \ldots, a_{m-p}, c_1, \ldots, c_p, b_1, \ldots, b_{n-p})$ such that $(a_1, \ldots, a_{m-p}, c_1, \ldots, c_p) \in \rho$ and $(c_1, \ldots, c_p, b_1, \ldots, b_{n-p}) \in \lambda$.

Example 9.6. On $H = \{1, 2, 3, 4, 5, 6\}$ we define the 5-ary relation ρ and the ternary relation λ:

$$\rho = \{(1, 1, 3, 4, 1), \ (2, 3, 5, 2, 1), \ (3, 3, 4, 5, 1)\},$$
$$\lambda = \{(4, 1, 6), \ (5, 1, 2), \ (3, 4, 3), \ (2, 2, 2)\}.$$

The $\rho_{2,4,5}$ projection is the ternary relation

$$\rho_{2,4,5} = \{(1, 4, 1), \ (3, 2, 1), \ (3, 5, 1)\};$$

the join $J_2(\rho, \lambda)$ is the 6-ary relation

$$J_2(\rho, \lambda) = \{(1, 1, 3, 4, 1, 6), \ (3, 3, 4, 5, 1, 2)\}.$$

Lemma 9.1. *If ρ is an n-ary preordering, then any projection $\rho_{1,i,n}$, $2 \leq i \leq n - 1$, is a 3-ary preordering.*

Proof. The reflexivity is immediately. Set an arbitrary i in $\{2, \ldots, n-1\}$. We prove that $\rho_{1,i,n}$ is 3-transitive in three steps:

(1) If $(x, y, z) \in \rho_{1,i,n}$ and $(y, u, v) \in \rho_{1,i,n}$, then there exist $a_1, \ldots, a_{n-3}$, $b_1, \ldots, b_{n-3} \in H$ such that $(x, a_1, \ldots, a_{i-2}, y, a_{i-1}, \ldots, a_{n-3}, z) \in \rho$ and $(y, b_1, \ldots, b_{i-2}, u, b_{i-1}, \ldots, b_{n-3}, v) \in \rho$; by the n-transitivity of the relation ρ it follows that $(x, b_1, \ldots, b_{i-2}, u, b_{i-1}, \ldots, b_{n-3}, v) \in \rho$, that is $(x, u, v) \in \rho_{1,i,n}$.

(2) If $(x, y, z) \in \rho_{1,i,n}$ and $(z, u, v) \in \rho_{1,i,n}$, then there exist $a_1, \ldots, a_{n-3}$, $b_1, \ldots, b_{n-3} \in H$ such that $(x, a_1, \ldots, a_{i-2}, y, a_{i-1}, \ldots, a_{n-3}, z) \in \rho$ and $(z, b_1, \ldots, b_{i-2}, u, b_{i-1}, \ldots, b_{n-3}, v) \in \rho$; by the n-transitivity of the relation ρ it follows that:

$$(x, b_1, \ldots, b_{i-2}, u, b_{i-1}, \ldots, b_{n-3}, v) \in \rho, \text{ that is } (x, u, v) \in \rho_{1,i,n},$$
$$(y, b_1, \ldots, b_{i-2}, u, b_{i-1}, \ldots, b_{n-3}, v) \in \rho, \text{ that is } (y, u, v) \in \rho_{1,i,n},$$
$$(x, a_1, \ldots, a_{i-2}, y, b_{i-1}, \ldots, b_{n-3}, v) \in \rho, \text{ that is } (x, y, v) \in \rho_{1,i,n},$$
$$(x, a_1, \ldots, a_{i-2}, y, a_{i-1}, \ldots, a_{n-3}, u) \in \rho, \text{ that is } (x, y, u) \in \rho_{1,i,n}.$$

(3) If $(x, y, z) \in \rho_{1,i,n}$ and $(u, z, v) \in \rho_{1,i,n}$, then there exist $a_1, \ldots, a_{n-3}$, $b_1, \ldots, b_{n-3} \in H$ such that $(x, a_1, \ldots, a_{i-2}, y, a_{i-1}, \ldots, a_{n-3}, z) \in \rho$ and $(u, b_1, \ldots, b_{i-2}, z, b_{i-1}, \ldots, b_{n-3}, v) \in \rho$; by the n-transitivity it follows that: $(x, a_1, \ldots, a_{i-2}, y, a_{i-1}, \ldots, a_{n-3}, v) \in \rho$, that is $(x, y, v) \in \rho_{1,i,n}$. $\square$

Lemma 9.2. *Let ρ be an n-ary relation on H such that $\rho_{1,n} = H \times H$. If ρ is strongly symmetric and n-transitive, then ρ is an n-equivalence on H.*

Proof. Set x arbitrary in H. Since $\rho_{1,n} = H \times H$, there exist $u_1, \ldots, u_{n-2} \in H$ such that $(x, u_1, \ldots, u_{n-2}, x) \in \rho$ and by the symmetry it follows that $(u_1, \ldots, u_{n-2}, x, x) \in \rho$. Using the n-transitivity, $(x, u_2, \ldots, u_{n-2}, x, x) \in \rho$. Again by the symmetry, we obtain that $(u_2, \ldots, u_{n-2}, x, x, x) \in \rho$ and therefore, by the n-transitivity, that $(x, u_3, \ldots, u_{n-2}, x, x, x) \in \rho$ and so on; finally it results that $(x, x, \ldots, x) \in \rho$, for all $x \in H$, so ρ is an n-equivalence on H. $\qquad\square$

Corollary 9.2. *If ρ is an n-ary equivalence, then any projection $\rho_{1,i,n}$, $2 \leq i \leq n-1$, is a 3-equivalence.*

Corollary 9.3. *If ρ is a strongly symmetric n-ary relation, then, for all $i \neq j \in \{2, \ldots, n-1\}$, the following equivalence holds:*

$$(x, y, z) \in \rho_{1,i,n} \Leftrightarrow (x, y, z) \in \rho_{1,j,n}.$$

9.2 Hypergroupoids associated with n-ary relations

Let ρ be an n-ary relation on a set H. In what follows we use the notations:

$$L(\rho) = \{x \in H \mid \exists u_2, u_3, \ldots, u_n \in H : (x, u_2, \ldots, u_n) \in \rho\}$$
$$R(\rho) = \{x \in H \mid \exists u_1, u_2, \ldots, u_{n-1} \in H : (u_1, u_2, \ldots, u_{n-1}, x) \in \rho\}.$$

Moreover, for all $x \in H$, set:

$$L(x) = \{y \in H \mid \exists u_1, \ldots, u_{n-2} \in H : (y, x, u_1, \ldots, u_{n-2}) \in \rho \ \vee$$
$$(u_1, \ldots, u_{n-2}, y, x) \in \rho \vee (u_1, \ldots, u_k, y, x, u_{k+1}, \ldots, u_{n-2}) \in \rho,$$
$$\text{for any } k \in \{1, \ldots, n-3\}\}$$

and similarly

$$R(x) = \{y \in H \mid \exists u_1, \ldots, u_{n-2} \in H : (x, y, u_1, \ldots, u_{n-2}) \in \rho \ \vee$$
$$(u_1, \ldots, u_{n-2}, x, y) \in \rho \vee (u_1, \ldots, u_k, x, y, u_{k+1}, \ldots, u_{n-2}) \in \rho,$$
$$\text{for any } k \in \{1, \ldots, n-3\}\}.$$

Remark 9.1. We have

(1) $y \in L(x)$ if and only if $x \in R(y)$, for all $(x, y) \in H^2$;

(2) $\bigcup_{x \in H} L(x) \neq H$ if and only if there exists $y \in H$ such that $R(y) = \emptyset$;

(3) $\bigcup_{x \in H} R(x) \neq H$ if and only if there exists $y \in H$ such that $L(y) = \emptyset$.

Indeed, $\bigcup_{x \in H} L(x) \neq H$ if and only if there exists $y \in H$ such that $y \notin \bigcup_{x \in H} L(x)$, which is equivalent with the fact there exists $y \in H$ such that $y \notin L(x)$, for any $x \in H$, that is there exists $y \in H$ such that $x \notin R(y)$, for all $x \in H$, equivalent with the fact there exists $y \in H$ such that $R(y) = \emptyset$.

Let ρ be an n-ary relation on a non-empty set H. We define on H the following hyperoperation

$$x \otimes_\rho y = L(x) \cup R(y) \tag{9.1}$$

and we notice that if $(H, \otimes_\rho)$ is a hypergroupoid then, for all $x \in H$, $L(x) \neq \emptyset$ or $R(x) \neq \emptyset$. The converse is not true as the following counter example shows: let $H = \{x, y, z, t\}$ and $\rho = \{(x, y, z), (x, z, t), (x, y, t)\}$. We have $L(x) = R(t) = \emptyset$ and so $x \otimes_\rho t = \emptyset$, that is $(H, \otimes_\rho)$ is not a hypergroupoid.

Lemma 9.3. *The hypergroupoid $(H, \otimes_\rho)$ is a quasihypergroup if and only if, for all $x \in H$, $L(x) \neq \emptyset$ and $R(x) \neq \emptyset$.*

Proof. The reproducibility law means: for all $x \in H$, $x \otimes_\rho H = H \otimes_\rho x = H$, that is, for all $x, y \in H$, there exist $z_1, z_2 \in H$ such that $y \in [L(x) \cup R(z_1)] \cap [L(z_2) \cup R(x)]$.

Suppose that for all $x \in H$, $L(x) \neq \emptyset$ and $R(x) \neq \emptyset$. Then $\bigcup_{x \in H} L(x) = H = \bigcup_{x \in H} R(x)$ and therefore, for all $y \in H$, there exist $z_1, z_2 \in H$ such that $y \in R(z_1) \cap L(z_2)$; thus $(H, \otimes_\rho)$ is reproductive, so it is a quasihypergroup.

Now we consider $(H, \otimes_\rho)$ a quasihypergroup and we suppose that there exists $y \in H$ such that $L(y) = \emptyset$ or $R(y) = \emptyset$. If $R(y) = \emptyset$ then, for all $x \in H$, $y \notin L(x)$ and thus $L(x) \cup R(y) = x \otimes_\rho y = L(x) \not\ni y$, for all $x \in H$, and then $H \otimes_\rho y \not\ni y$; so $H \otimes_\rho y \neq H$ which is a contradiction. Similarly, if there exists $y \in H$ such that $L(y) = \emptyset$, then $y \otimes_\rho H \neq H$ and again we obtain a contradiction. $\square$

Theorem 9.2. *Let ρ be an n-ary relation on a set H. Then $(H, \otimes_\rho)$ is an H_v-group if and only if for all $x \in H$, $L(x) \neq \emptyset$ and $R(x) \neq \emptyset$.*

Proof. If $(H, \otimes_\rho)$ is an H_v-group, then it is a quasihypergroup, it follows that $L(x) \neq \emptyset$ and $R(x) \neq \emptyset$, for all $x \in H$.

Conversely, suppose that for all $x \in H$, $L(x) \neq \emptyset$ and $R(x) \neq \emptyset$. It follows that $(H, \otimes_\rho)$ is a quasihypergroup. It remains to prove that the hyperoperation "$\otimes_\rho$" is weak associative. We show that for all $x, y, z \in H$,

$$(x \otimes_\rho y) \otimes_\rho z \cap x \otimes_\rho (y \otimes_\rho z) \ni y.$$

Since

$$(x \otimes_\rho y) \otimes_\rho z = \{L(u) \cup R(z) \mid u \in L(x) \cup R(y)\}$$
$$\supseteq \{L(u) \mid u \in R(y)\} = \{L(u) \mid y \in L(u)\} \ni y$$

and

$$x \otimes_\rho (y \otimes_\rho z) = \{L(x) \cup R(v) \mid v \in L(y) \cup R(z)\}$$
$$\supseteq \{R(v) \mid v \in L(y)\} = \{R(v) \mid y \in R(v)\} \ni y$$

it follows that $(H, \otimes_\rho)$ is an H_v-group. $\qquad \square$

Corollary 9.4. *Let ρ be an n-ary relation on a set H. Then $(H, \otimes_\rho)$ is an H_v-group if and only if, it is a quasihypergroup.*

Theorem 9.3. *Let $\rho = \{(x, x, \ldots, x) \mid x \in H\}$ be the diagonal n-ary relation on a set H. Then $(H, \otimes_\rho)$ is a join space.*

Proof. For all $x \in H$ we obtain $L(x) = R(x) = \{x\}$ and so for all $x, y \in H$, it follows that

$$x \otimes_\rho y = y \otimes_\rho x = \{x, y\}$$

and then $x \otimes_\rho H = H \otimes_\rho = H$, for all $x \in H$. Moreover, for all $(x, y, z) \in H^3$, we obtain

$$(x \otimes_\rho y) \otimes_\rho z = x \otimes_\rho (y \otimes_\rho z) = \{x, y, z\},$$

so $(H, \otimes_\rho)$ is a commutative hypergroup.

Finally, for all $a, b, c, d \in H$ such that $a/b \cap c/d \neq \emptyset$ it follows that $a \otimes_\rho d \cap b \otimes_\rho c \neq \emptyset$. We find that $a/a = \{x \in H \mid a \in x \otimes_\rho a\} = H$ and for $a \neq b \in H$, $a/b = \{x \in H \mid a \in x \otimes_\rho b\} = \{a\}$.

Let $a, b, c, d \in H$ such that $a/b \cap c/d \neq \emptyset$. If $a = b$ or $c = d$ then $a \otimes_\rho d \cap b \otimes_\rho c \ni a$ or $a \otimes_\rho d \cap b \otimes_\rho c \ni d$. If $a \neq b$ and $c \neq d$, then $a/b \cap c/d \neq \emptyset$ if and only if $a = c$ and thus $a \otimes_\rho d \cap b \otimes_\rho c \ni a$. In both cases $a \otimes_\rho d \cap b \otimes_\rho c \neq \emptyset$ and therefore $(H, \otimes_\rho)$ is a join space. $\qquad \square$

Lemma 9.4. *If ρ is an n-ary preordering on a set H, then for all $a, x, u \in H$ such that $a \in L(u)$ and $u \in L(x)$, it follows that $a \in L(x)$.*

Proof. Let a, x, u be in H such that $a \in L(u)$ and $u \in L(x)$. Then there exist $a_1, \ldots, a_{n-2} \in H$, $b_1, \ldots, b_{n-2} \in H$ such that

$$(a, u, a_1, \ldots, a_{n-2}) \in \rho \vee (a_1, \ldots, a_{n-2}, a, u)$$
$$\in \rho \vee (a_1, \ldots, a_k, a, u, a_{k+1}, \ldots, a_{n-2})$$
$$\in \rho$$

and

$$(u, x, b_1, \ldots, b_{n-2}) \in \rho \vee (b_1, \ldots, b_{n-2}, u, x)$$
$$\in \rho \vee (b_1, \ldots, b_k, u, x, b_{k+1}, \ldots, b_{n-2})$$
$$\in \rho$$

with $k \in \{1, \ldots, n - 3\}$. We distinguish the following situations:
  (1) If

$$(a, u, a_1, \ldots, a_{n-2}) \in \rho \vee (a_1, \ldots, a_{n-2}, a, u)$$
$$\in \rho \vee (a_1, \ldots, a_k, a, u, a_{k+1}, \ldots, a_{n-2})$$
$$\in \rho$$

and $(u, x, b_1, \ldots, b_{n-2}) \in \rho$, then by the n-transitivity, it results that $(a, x, b_1, \ldots, b_{n-2}) \in \rho$, that is $a \in L(x)$.
  (2) If

$$(b_1, \ldots, b_{n-2}, u, x) \in \rho \vee (b_1, \ldots, b_k, u, x, b_{k+1}, \ldots, b_{n-2})$$
$$\in \rho$$

with $k \in \{1, \ldots, n - 3\}$ and since $(x, \ldots, x) \in \rho$ by the reflexivity, then $(u, x, \ldots, x) \in \rho$ by the n-transitivity. Now, if

$$(a, u, a_1, \ldots, a_{n-2}) \in \rho \vee (a_1, \ldots, a_{n-2}, a, u)$$
$$\in \rho \vee (a_1, \ldots, a_k, a, u, a_{k+1}, \ldots, a_{n-2})$$
$$\in \rho,$$

with $k \in \{1, \ldots, n - 3\}$, and since $(u, x, \ldots, x) \in \rho$ it follows that $(a, x, \ldots, x) \in \rho$, again by the n-transitivity. Therefore $a \in L(x)$. □

Theorem 9.4. *If ρ is an n-ary preordering on H such that $L(x) = R(x)$ for all $x \in H$, then $(H, \otimes_\rho)$ is a join space.*

Proof.   Since ρ is reflexive, it follows that $(H, \otimes_\rho)$ is a quasihypergroup. Moreover, since $L(x) = R(x)$, for all $x \in H$, it follows that

$$x \otimes_\rho y = y \otimes_\rho x = L(x) \cup L(y),$$

for all $x, y \in H$, and therefore $(H, \otimes_\rho)$ is commutative.

Now we prove that the hyperoperation "$\otimes_\rho$" is associative. For all $a \in (x \otimes_\rho y) \otimes_\rho z$, there exists $u \in L(x) \cup L(y)$ such that $a \in L(u) \cup L(z)$. We distinguish the following cases:

Case 1: If $a \in L(z)$, we take $v = z \in L(y) \cup L(z)$ and then $a \in L(x) \cup L(z) = L(x) \cup L(v)$, thus $a \in x \otimes_\rho (y \otimes_\rho z)$.

Case 2: If $a \in L(u)$ with $u \in L(x)$, then, by the Lemma 16, it follows that $a \in L(x)$; therefore there exists $v \in L(y) \cup L(z)$ (for example $v = y$) such that $a \in L(x) \cup L(v)$, so $a \in x \otimes_\rho (y \otimes_\rho z)$.

Case 3: If $a \in L(u)$ with $u \in L(y)$, there exists $v = u \in L(y) \cup L(z)$ such that $a \in L(x) \cup L(u) = L(x) \cup L(v)$, and again $a \in x \otimes_\rho (y \otimes_\rho z)$.

We have proved that, for all $x, y, z \in H$, we have

$$(x \otimes_\rho y) \otimes_\rho z \subset x \otimes_\rho (y \otimes_\rho z).$$

Similarly we can show the other inclusion

$$x \otimes_\rho (y \otimes_\rho z) \subset (x \otimes_\rho y) \otimes_\rho z.$$

It remains to check the condition of join space. Set $a, b, c, d \in H$ such that $a/b \cap c/d \neq \emptyset$; then there exists $x \in a/b \cap c/d$, that is $a \in x \otimes_\rho b = L(x) \cup L(b)$ and $c \in x \otimes_\rho d = L(x) \cup L(d)$. We consider the situations:

(1) If $a \in L(x)$ and $c \in L(x)$ then $x \in R(a) = L(a)$, $x \in R(c) = L(c)$ and therefore $x \in [L(a) \cup L(d)] \cap [L(b) \cup L(c)] = a \otimes_\rho d \cap b \otimes_\rho c$.
(2) If $c \in L(x)$ and $a \notin L(x)$ then $a \in L(b)$ and, and since $a \in L(a)$ (by the reflexivity), it follows $a \in [L(a) \cup L(d)] \cap [L(b) \cup L(c)] = a \otimes_\rho d \cap b \otimes_\rho c$.
(3) If $c \notin L(x)$ then $c \in L(d)$, and since $c \in L(c)$ (by the reflexivity), it follows $c \in [L(a) \cup L(d)] \cap [L(b) \cup L(c)] = a \otimes_\rho d \cap b \otimes_\rho c$.

We can conclude that $a \otimes_\rho d \cap b \otimes_\rho c \neq \emptyset$, so $(H, \otimes_\rho)$ is a join space. □

Corollary 9.5. *If ρ is an n-ary equivalence on a set H, then $(H, \otimes_\rho)$ is a join space.*

Remark 9.2. Let ρ be an n-ary preordering on a set H. The condition $L(x) = R(x)$, for all $x \in H$, is a sufficient condition, but not a necessary one such that the hyperoperation "$\otimes_\rho$" is associative, as we can see in the next examples.

Example 9.7. On the set $H = \{1, 2, 3\}$ we consider the ternary relation $\rho = \{(1,1,1), (2,2,2), (3,3,3), (1,3,3), (1,1,3)\}$ which is a 3-preordering. It is obvious that $L(1) = \{1\} \neq \{1,3\} = R(1)$, $L(2) = \{2\} = R(2)$ and $L(3) = \{1,3\} \neq \{3\} = R(3)$. Moreover, for all $x, y, z \in H$, $(x \otimes_\rho y) \otimes_\rho z = x \otimes_\rho (y \otimes_\rho z)$.

Example 9.8. On the set $H = \{1, 2, 3\}$ we consider the 3-preordering $\rho = \{(1,1,1), (2,2,2), (3,3,3), (1,3,3), (1,1,3), (2,3,3), (2,2,3)\}$, where $L(1) = \{1\} \neq \{1,3\} = R(1)$, $L(2) = \{2\} \neq \{2,3\} = R(2)$, $L(3) = \{1,2,3\} \neq \{3\} = R(3)$. Since $(2 \otimes_\rho 2) \otimes_\rho 3 = \{1,2,3\} \neq \{2,3\} = 2 \otimes_\rho (2 \otimes_\rho 3)$, it follows that "$\otimes_\rho$" is not associative.

Remark 9.3. Let ρ be an n-ary reflexive relation on a set H. If ρ is also n-transitive, then ρ satisfies the property

(T) for all $a, x, u \in H$ such that $a \in L(u)$ and $u \in L(x)$ it follows that $a \in L(x)$.

Moreover, for n-ary reflexive relations ρ on H such that $L(x) = R(x)$, for all $x \in H$, the property (T) does not imply the n-transitivity, also if $L(x) = R(x)$, for all $x \in H$, as we can see in the following example.

Example 9.9. On the set $H = \{1, 2, 3\}$ let us consider the following n-relation $\rho = \{(1, \ldots, 1), (2, \ldots, 2), (3, \ldots, 3), (2, 1, \ldots, 1, 2)\}$. We find $L(1) = \{1, 2\} = R(1)$, $L(2) = \{1, 2\} = R(2)$, $L(3) = \{3\} = R(3)$, and easily the property (T) is satisfied, but ρ is not n-transitive, because we have $(2, \ldots, 2) \in \rho$, $(2, 1, \ldots, 1, 2) \in \rho$, but $(2, \ldots, 2, 1) \notin \rho$.

Theorem 9.5. *Let ρ be an n-ary relation on H such that $x \in L(x) = R(x)$, for all $x \in H$. If ρ satisfies the property (T), then "$\otimes_\rho$" is associative and therefore $(H, \otimes_\rho)$ is a join space.*

Proof. We have to check only the associativity law. Suppose that "$\otimes_\rho$" is not associative. Then there exist $x, y, z \in H$ such that $(x \otimes_\rho y) \otimes_\rho z \neq x \otimes_\rho (y \otimes_\rho z)$. Thus there exists $u \in (x \otimes_\rho y) \otimes_\rho z$ such that $u \notin x \otimes_\rho (y \otimes_\rho z)$ or vice versa. We consider the first situation. It follows that there exists $v \in L(x) \cup L(y)$ such that $u \in L(v) \cup L(z)$, and for all $t \in L(y) \cup L(z)$, $u \notin L(x) \cup L(t)$. If $u \in L(v)$ with $v \in L(x)$, by the property (T) it follows that $u \in L(x)$, which is false, and similarly, if $u \in L(v)$ with $v \in L(y)$, it follows $u \in L(y)$, again false. If $u \in L(z)$ we obtain a contradiction with $u \notin L(x) \cup L(t)$, for all $t \in L(y) \cup L(z)$. $\qquad\square$

Theorem 9.6. *Let ρ be an n-ary relation on a set H, with $|H| \geq 3$ such that $x \notin L(x)$, $|L(x)| = 1 = |R(x)|$, for all $x \in H$. Then "$\otimes_\rho$" is not associative.*

Proof. Since $|L(x)| = 1 = |R(x)|$, for all $x \in H$, it follows that $L(x) \neq \emptyset$ and $R(x) \neq \emptyset$, for all $x \in H$, that is $\bigcup_{x \in H} L(x) = H = \bigcup_{x \in H} R(x)$. Moreover, there exists a unique $y_x \in H \setminus \{x\}$ such that $x \in L(y_x)$ (if there exist $y_x \neq y'_x \in H \setminus \{x\}$ such that $x \in L(y_x) \cap L(y'_x)$, then $y_x, y'_x \in R(x)$, so $|R(x)| \geq 2$, which is false) and similarly there exists a unique $z_x \in H \setminus \{x\}$ such that $x \in R(z_x)$. We distinguish the following situations:

(1) If $L(x) \cap R(x) = \emptyset$, for all $x \in H$, then it is clear that $y_x \neq z_x$ and $L(y_x) = R(z_x) = \{x\}$, $R(x) = \{y_x\}$, $L(x) = \{z_x\}$. Now it follows

$$(x \otimes_\rho x) \otimes_\rho x = \{y_x, z_x\} \otimes_\rho x = L(y_x) \cup L(z_x) \cup R(x)$$
$$= \{x, y_x\} \cup L(z_x) \not\ni z_x;$$

$$x \otimes_\rho (x \otimes_\rho x) = x \otimes_\rho \{y_x, z_x\} = L(x) \cup R(y_x) \cup R(z_x)$$
$$= \{z_x, x\} \cup R(y_x) \ni z_x,$$

so the hyperoperation "$\otimes_\rho$" is not associative, it is only weak associative.

(2) There exists $\bar{x} \in H$ such that $L(\bar{x}) = R(\bar{x}) = \{y_{\bar{x}}\}$. Then

$$(\bar{x} \otimes_\rho \bar{x}) \otimes_\rho y_{\bar{x}} = y_{\bar{x}} \otimes_\rho y_{\bar{x}} = L(y_{\bar{x}}) \cup R(y_{\bar{x}}) \not\ni y_{\bar{x}},$$

$$\bar{x} \otimes_\rho (\bar{x} \otimes_\rho y_{\bar{x}}) = \bar{x} \otimes_\rho (\{y_{\bar{x}}\} \cup R(y_{\bar{x}})) = L(\bar{x}) \cup R(y_{\bar{x}}) \cup R(u) \ni y_{\bar{x}},$$

where $u \in R(y_{\bar{x}})$, which means that the hyperoperation "$\otimes_\rho$" is not associative. $\square$

9.3 Hypergroupoids associated with ternary relations

In [172] M. Stefanescu presented another approach of the connections between hypergroups and ordered sets: given a hypergroupoid $(H, \circ)$, we can consider the ternary relation ρ on H associated with the hyperoperation:

$$(a, b, c) \in \rho \text{ if and only if } c \in a \circ b.$$

If $(H, \circ)$ is a hypergroup, then Stefanescu [172] showed that ρ satisfies the following three conditions:

(1) For all $a, b \in H$, there exists at least one element $c \in H$, such that $(a, b, c) \in \rho$;

(2) If for $a, b, c, z \in H$ there exists $x \in H$ such that $(a, b, x), (x, c, z) \in \rho$, then there exists $y \in H$ such that $(a, y, z), (b, c, y) \in \rho$ and conversely;

(3) For all $a, b \in H$ there exist $x, y \in H$ such that $(a, x, b) \in \rho$ and $(y, a, b) \in \rho$.

Conversely, if ρ is a ternary relation on a non-empty set H such that the conditions (1), (2) and (3) are satisfied, then considering the hyperoperation

$$x \circ y = \{a \in H \mid (x, y, a) \in \rho\},$$

$(H, \circ)$ is a hypergroup.

With any binary relation σ on a set H we associate a ternary relation denoted by $\sigma_t \subseteq H \times H \times H$ as follows

$$(x, y, z) \in \sigma_t \Longleftrightarrow (x, y) \in \sigma \wedge (y, z) \in \sigma \wedge (x, z) \in \sigma. \tag{9.2}$$

Theorem 9.7. *The unique ternary relation σ_t obtained from a binary relation σ using (9.2) and such that*

$$\forall (a,b) \in H^2, \exists c \in H : (a,b,c) \in \sigma_t \tag{9.3}$$

is the total relation $\sigma_t = H \times H \times H$.

Proof. The condition (9.3) is equivalent with the following one: for all $(a,b) \in H^2, (a,b) \in \sigma$, so the relation σ is the total relation $H \times H$ and thus $\sigma_t = H \times H \times H$. $\qquad\square$

Moreover, the hypergroupoid obtained from σ_t considering

$$x \circ y = \{z \in H \mid (x,y,z) \in \sigma_t\}$$

is the total hypergroup on H.

Conversely, with any ternary relation ρ on H we associate a binary relation $\rho^b \subseteq H \times H$ as follows:

$$(x,y) \in \rho^b \Leftrightarrow \exists z \in H : (x,y,z) \in \rho. \tag{9.4}$$

Let $(H, \circ)$ be an arbitrary hypergroupoid which determines the ternary relation ρ defined by

$$(x,y,z) \in \rho \Leftrightarrow z \in x \circ y.$$

Since $(H, \circ)$ is a hypergroupoid, it follows that for all $(x,y) \in H^2$, there exists $z \in H$ such that $z \in x \circ y$, that is $(x,y,z) \in \rho$; therefore for all $(x,y) \in H^2$, we obtain $(x,y) \in \rho^b$, that is $\rho^b = H \times H$. So we have proved the next result.

Theorem 9.8. *The unique binary relation ρ^b obtained, using (9.4), from the ternary relation ρ associated with any hypergroupoid $(H, \circ)$ as follows*

$$(x,y,z) \in \rho \Leftrightarrow z \in x \circ y,$$

is the total relation $H \times H$.

9.4 Connections with Rosenberg's hypergroupoid

Let ρ be a binary relation on a non-empty set H. We denote by

$$D(\rho) = \{x \in H \mid (x,y) \in \rho, \text{ for some } y \in H\},$$
$$R(\rho) = \{x \in H \mid (y,x) \in \rho, \text{ for some } y \in H\},$$

the domain and the range of ρ. Recall the definition of Rosenberg hyper-operation on H. For all $x \in H$, set $U_x = \{y \in H \mid (x,y) \in \rho\}$ and for all $x, y \in H$,

$$x \circ y = U_x \cup U_y.$$

Here we are interested only in weak associativity: for all $x, y, z \in H$, there exists $a \in H$ such that

$$a \in (x \circ y) \circ z \cap x \circ (y \circ z),$$

which is equivalent with: for all $x, y, z \in H$, there exist $a \in H$ and $b \in U_x \cup U_y$ such that $a \in U_b \cup U_z$ and there exists $c \in U_y \cup U_z$ such that $a \in U_x \cup U_c$. This means that there exists $a \in H$ such that

$$(x,a) \in \rho^2 \vee (y,a) \in \rho^2 \vee (z,a) \in \rho, (x,a) \in \rho \vee (y,a) \in \rho^2 \vee (z,a) \in \rho^2.$$

Theorem 9.9. *If ρ is a binary relation on a set H, with full domain and full range, then IH_ρ is an H_v-group.*

Proof. Since $D(\rho) = H$, it follows that for all $x \in H$, there exists $z \in H$ such that $(x,z) \in \rho$ and there exists $y \in H$ such that $(z,y) \in \rho$; so, for all $x \in H$, there exists $y \in H$ such that $(x,y) \in \rho^2$; thus $D(\rho^2) = H$. Hence, for all $y \in H$ there exists $a \in H$ such that $(y,a) \in \rho^2$ and then $a \in (x \circ y) \circ z \cap x \circ (y \circ z)$, which means that IH_ρ is an H_v-group. $\square$

Let ρ be an n-ary relation on H such that, for all $x \in H$, $L(x) \neq \emptyset$ and $R(x) \neq \emptyset$. It follows that the hypergroupoid $(H, \otimes_\rho)$ is an H_v-group. Then the binary relation ρ^b associated with ρ ($(x,y) \in \rho^b$ if there exist $(x_1, x_2, \ldots, x_n) \in \rho$ and natural numbers i,j such that $1 \leq i < j \leq n$, $x = x_i$, $y = x_j$), has full domain and full range, so the Rosenberg's hypergroupoid IH_{ρ^b} is an H_v-group. Therefore, we obtain the next result:

Theorem 9.10. *If ρ is an n-ary relation on H such that, for all $x \in H$ $L(x) \neq \emptyset$ and $R(x) \neq \emptyset$, then the hypergroupoids $(H, \otimes_\rho)$ and $\mathbb{H}_{\rho^b}$ are H_v-groups.*

Conversely, let σ be a binary relation on H, with full domain and full range. Then the n-ary relation σ_n associated with σ ($(x_1, \ldots, x_n) \in \sigma_n$ if $(x_i, x_j) \in \sigma$, for all i,j, $1 \leq i < j \leq n$) has the property that, for all $x \in H$, $L(x) \neq \emptyset$ and $R(x) \neq \emptyset$. Indeed, for all $x \in H$, there exists $y \in H$ such that $(x,y) \in \sigma$; then $y \in R(x)$. Similarly, there exists $z \in H$ such that $(z,x) \in \sigma$, thus $z \in L(x)$.

We obtain the following result.

Theorem 9.11. *If σ is a binary relation on H, with full domain and full range, then the hypergroupoids $\mathbb{H}_\sigma$ and $(H, \otimes_{\sigma_n})$ are H_v-groups.*

9.5   Hypergroups associated with n-ary relations

Let ρ be a ternary relation on H. For all $x, y \in H$, we define the hyperoperation

$$x \circ_\rho y = \{z \mid z \in H, (x, z, y) \in \rho\}.$$

We notice that $(H, \circ_\rho)$ is a hypergroupoid if and only if for all $x, y \in H$ there exists $z \in H$ such that $(x, z, y) \in \rho$, that is the projection $\rho_{1,3}$ is the total relation, i.e., $\rho_{1,3} = H \times H$.

Now we generalize the definition of the hyperproduct associated with a ternary relation to the case of n-ary relation, $n \geq 3$, defined on a non-empty set H.

Definition 9.6. Let ρ be an n-ary relation on H, for all $i \in \{2, \ldots, n-1\}$, using the projections $\rho_{1,i,n}$, we define the hyperproducts:

$$x \circ_i y = \{z \in H \mid (x, z, y) \in \rho_{1,i,n}\}$$

and

$$x \circ_\rho y = \{z \in H \mid (x, z, y) \in \bigcup_{i=2}^{n-1} \rho_{1,i,n}\} = \bigcup_{i=2}^{n-1} x \circ_i y.$$

We notice that $(H, \circ_\rho)$ is a hypergroupoid if and only if the projection $\rho_{1,n}$ is the total relation, i.e., $\rho_{1,n} = H \times H$.

In what follows we determine neccessary conditions such that the hypergroupoid $(H, \circ_\rho)$ is a hypergroup.

Theorem 9.12. *Let ρ be an n-ary relation on H. Then $(H, \circ_\rho)$ is a quasihypergroup if and only if $\rho_{1,n} = H \times H$ and there exist i, j, with $2 \leq i, j \leq n-1$, such that $\rho_{1,i} = \rho_{j,n} = H \times H$.*

Proof.   We know that a hypergroupoid $(H, \circ)$ is a quasihypergroup if and only if, for all $x \in H$, $x \circ H = H = H \circ x$. Therefore $(H, \circ_\rho)$ is a quasihypergroup if and only if, for all $x, y \in H$, there exist $z, t \in H$, such that $y \in x \circ_\rho z \cap t \circ_\rho x$, that is $(x, y, z) \in \bigcup_{i=2}^{n-1} \rho_{1,i,n} \ni (t, y, x)$. It follows that $(H, \circ_\rho)$ is a quasihypergroup if and only if there exist i, j, with $2 \leq i, j \leq n-1$, such that $(x, y, z) \in \rho_{1,i,n}$ and $(t, y, x) \in \rho_{1,j,n}$, so $\rho_{1,i} = \rho_{j,n} = H \times H$. $\square$

Theorem 9.13. *Let ρ be an n-ary relation on H such that $\rho_{1,n} = H \times H$. If ρ is a preordering, then $(H, \circ_\rho)$ is the total hypergroup.*

Proof. We have to prove that, for all $x, y, z \in H$, $z \in x \circ_\rho y$. Set $x, y, x \in H$. Since $\rho_{1,n} = H \times H$, there exist $a_1, a_2, \ldots, a_{n-2} \in H$, such that $(z, a_1, \ldots, a_{n-2}, y) \in \rho$. By the reflexivity, $(z, z, \ldots, z) \in \rho$ and then, by the n-transitivity, it follows that $(z, \ldots, z, y) \in \rho$. Again since $\rho_{1,n} = H \times H$, there exist $c_1, \ldots, c_{n-2} \in H$ such that $(x, c_1, \ldots, c_{n-2}, z) \in \rho$. Using the n-transitivity for $(z, \ldots, z, y) \in \rho$ and $(x, c_1, \ldots, c_{n-2}, z) \in \rho$, we obtain that $(x, z, \ldots, z, y) \in \rho$, so $z \in x \circ_\rho y$. □

Theorem 9.14. *Let ρ be an n-ary relation on H such that $\rho_{1,n} = H \times H$. If ρ is n-ary transitive and strongly symmetric, then $(H, \circ_\rho)$ is the total hypergroup.*

Proof. ρ is an n-equivalence on H so $(H, \circ_\rho)$ is the total hypergroup. □

Theorem 9.15. *Let ρ be a ternary relation on H such that $\rho_{1,3} = \rho_{1,2} = H \times H$ or $\rho_{1,3} = \rho_{2,3} = H \times H$. If ρ is 3-transitive, then $(H, \circ_\rho)$ is the total hypergroup.*

Proof. We prove that ρ is reflexive and then we obtain the conclusion. We suppose that $\rho_{1,3} = \rho_{1,2} = H \times H$; thus, for all $x \in H$, there exist $a_x, c_x \in H$ such that $(x, a_x, x) \in \rho$ and $(a_x, x, c_x) \in \rho$. Then by the 3-transitivity, $(x, x, c_x) \in \rho$. Again, since $\rho_{1,3} = H \times H$ it follows that there exists $u_x \in H$ such that $(c_x, u_x, x) \in \rho$. Using the 3-transitivity for $(x, x, c_x), (c_x, u_x, x) \in \rho$, we get $(x, x, x) \in \rho$, that is ρ is a 3-preordering on H. □

Corollary 9.6. *Let ρ be an n-ary relation on H ($n \geq 4$) such that there exist i, j, with $2 \leq i, j \leq n - 1$, such that $\rho_{1,i} = \rho_{j,n} = \rho_{1,n} = H \times H$. If ρ is n-transitive, then $(H, \circ_\rho)$ is the total hypergroup.*

Proof. If ρ is n-transitive, then the ternary relations $\lambda = \rho_{1,i,n}$ and $\alpha = \rho_{1,j,n}$ are 3-transitive such that $\lambda_{1,2} = \lambda_{1,3} = H \times H$ and $\alpha_{2,3} = \alpha_{1,3} = H \times H$. It follows that the hypergroupoids $(H, \circ_i)$ and $(H, \circ_j)$ are total hypergroups. Therefore, for all $x, y \in H$, we have $x \circ_\rho y = \bigcup_{l=2}^{n-1} x \circ_{\rho_{1,l,n}} y \supseteq x \circ_{\rho_{1,i,n}} y = H$, so $x \circ_\rho y = H$, that is, $(H, \circ_\rho)$ is the total hypergroup. □

Let ρ be a ternary relation on H. We denote the join relation $J_2(\rho, \rho)$ by α.

We recall that $J_2(\rho, \rho)$ is a 4-relation such that

$$(x, y, z, t) \in J_2(\rho, \rho) \Leftrightarrow (x, y, z), (y, z, t) \in \rho.$$

Using projections of $J_2(\rho, \rho)$, we give a neccessary condition for a hyper-groupoid $(H, \circ_\rho)$ to be a semihypergroup.

Theorem 9.16. *Let ρ be a reflexive and symmetric ternary relation on H. If $\rho \not\subset \alpha_{1,2,4}$ or $\rho \not\subset \alpha_{1,3,4}$, then the hyperoperation "$\circ_\rho$" is not associative.*

Proof. We suppose that $\rho \not\subset \alpha_{1,2,4}$; then there exists $(x, y, z) \in \rho \setminus \alpha_{1,2,4}$ and thus, for all $a \in H$,

$$(x, y, a, z) \notin J_2(\rho, \rho). \tag{9.5}$$

Since ρ is reflexive and $(x, y, z) \in \rho$, it follows that $y \in y \circ_\rho y \subset (x \circ_\rho z) \circ_\rho y$. To prove that "$\circ_\rho$" is not associative, it is enough to show that $y \notin x \circ_\rho (z \circ_\rho y)$, that is, for all $a \in z \circ_\rho y$, $y \notin x \circ_\rho a$.

Suppose that there exists $\bar{a} \in z \circ_\rho y$, with $y \in x \circ_\rho \bar{a}$. Then, $(z, \bar{a}, y) \in \rho$ (and by the symmetry $(y, \bar{a}, z) \in \rho$) and $(x, y, \bar{a}) \in \rho$. Therefore it results that $(x, y, \bar{a}, z) \in J_2(\rho, \rho)$, which is a contradiction with (1). Similarly if we suppose $\rho \not\subset \alpha_{1,3,4}$. $\square$

Corollary 9.7. *Let ρ be a reflexive and symmetric ternary relation on H. If $(H, \circ_\rho)$ is a semihypergroup, then $\rho \subset \alpha_{1,2,4} \cap \alpha_{1,3,4}$.*

In what follows we give a similar result for the n-ary relations. Let ρ be an n-ary relation on H and α be the join relation $J_2(\rho, \rho)$; then α is a $(2n - 2)$-relation on H.

Theorem 9.17. *Let ρ be a reflexive and symmetric n-ary relation on H which satisfies the condition:*

$$(S) : (x, a_1, \ldots, a_{n-2}, y) \in \rho \Leftrightarrow (x, a_{\sigma(1)}, \ldots, a_{\sigma(n-2)}, y) \in \rho,$$

for all permutation σ of the set $\{1, 2, \ldots, n - 2\}$.

If there exists $j \in \{2, \ldots, n-1\}$ such that $\rho_{1,j,n} \not\subset \alpha_{1,n-1,2n-2}$ or $\rho_{1,j,n} \not\subset \alpha_{1,n,2n-2}$, then the hyperoperation "$\circ_\rho$" is not associative.

Proof. First we suppose there exists $j \in \{2, \ldots, n-1\}$ such that $\rho_{1,j,n} \not\subset \alpha_{1,n,2n-2}$. Then there exists $(x, y, z) \in \rho_{1,j,n}$ such that $(x, y, z) \notin \alpha_{1,n,2n-2}$. Therefore, $y \in x \circ_\rho z$ and since ρ is reflexive, it follows that $y \in y \circ_\rho (x \circ_\rho z)$. We prove that $y \notin (y \circ_\rho x) \circ_\rho z$, and thus the hyperoperation "$\circ_\rho$" is not associative.

Suppose that $y \in (y \circ_\rho x) \circ_\rho z$; then there exists $a \in y \circ_\rho x$ such that $y \in a \circ_\rho z$. It follows that $(y, \ldots, a, \ldots, x) \in \rho$ and $(a, \ldots, y, \ldots, z) \in \rho$. By the symmetry and the condition (S), we obtain that $(x, \ldots, a, y) \in \rho$

and $(a, y, \ldots, z) \in \rho$ and therefore $(x, \ldots, a, y, \ldots, z) \in J_2(\rho, \rho)$, that is $(x, y, z) \in \alpha_{1,n,2n-2}$ which is a contradiction.

Similarly, if we suppose there exists $j \in \{2, \ldots, n-1\}$ such that $\rho_{1,j,n} \not\subseteq \alpha_{1,n-1,2n-2}$ one can prove that $y \in (x \circ_\rho z) \circ_\rho y$, but $y \notin x \circ_\rho (z \circ_\rho y)$, so the hyperoperation "$\circ_\rho$" is not associative. $\qquad\square$

9.6 An example of reduced hypergroup

Let H be a non-empty set such that $H = \bigcup_{i \in I} H_i$, where $\{H_i\}_{i \in I}$ is a family of subsets of H indexed by a set I. We define on H the following binary relation:

$$(x, y) \in \rho^b \Leftrightarrow \exists i \in I : x, y \in H_i.$$

We obtain the ternary relation ρ defined as:

$$\rho = J_1(\rho^b, \rho^b) = \{(x, z, y) \mid (x, z) \in \rho^b \wedge (z, y) \in \rho^b\}$$
$$= \{(x, z, y) \mid \exists i, j \in I : x, z \in H_i \wedge z, y \in H_j\}.$$

Consider the hyperproduct on H defined by

$$x \circ_\rho y = \{z \mid (x, z, y) \in \rho\}. \tag{9.6}$$

We have

$$x \circ_\rho y = x \circ_\rho x \cap y \circ_\rho y$$

and if we denote $x \circ_\rho x$ by H_x, then

$$x \circ_\rho y = H_x \cap H_y.$$

Notice that $(H, \circ_\rho)$ is a partial hypergroupoid and it is a hypergroupoid if and only if for all $i \neq j \in I$, $H_i \cap H_j \neq \emptyset$.

Theorem 9.18. *Let $(H, \circ_\rho)$ be a partial hypergroupoid associated with a ternary relation as in (9.6). For all $x, y \in H$, $x \sim_e y$ if and only if $H_x = H_y$.*

Proof. First we notice that, for all $a, b \in H$, $a \in H_b$ if and only if $b \in H_a$.

Let x, y be in H such that $x \sim_o y$, that is $x \circ_\rho a = y \circ_\rho a$, for all $a \in H$. Thus $x \circ_\rho x = x \circ_\rho y = y \circ_\rho x = y \circ_\rho y$ and then $H_x = H_y$.

Conversely, set $x, y \in H$ such that $H_x = H_y$. For all $a \in H$, we have $H_x \cap H_a = H_y \cap H_a$, therefore $x \circ_\rho a = y \circ_\rho a$, so $x \sim_o y$.

Now set $x, y \in H$ such that $x \sim_i y$, thus $x \in a \circ_\rho b$ if and only if $y \in a \circ_\rho b$. Let $z \in H_x$, then $x \in H_z = z \circ_\rho z$ and since $x \sim_i y$, it follows that $y \in H_z$, thus $z \in H_y$. So $H_x \subset H_y$ and similarly one can prove that $H_y \subset H_x$.

Conversely, set $x, y \in H$ such that $H_x = H_y$. If $x \in a \circ_\rho b = H_a \cap H_b$, with $a, b \in H$, then $a, b \in H_x = H_y$ and therefore $y \in H_a \cap H_b = a \circ_\rho b$, so $x \sim_i y$. $\qquad\qquad\square$

Theorem 9.19. *Let $\{H_i\}_{i \in I}$ be a partition of a non-empty set H. Then the family of the equivalence classes respected to the equivalence $\sim_e$ on $(H, \circ_\rho)$ coincides with the family $\{H_i\}_{i \in I}$. Moreover, the partial hypergroupoid $(H, \circ_\rho)$ is reduced if and only if, for all $i \in I$, $\mid H_i \mid = 1$.*

Proof. Let $\{H_i\}_{i \in I}$ be a partition of H. for all $x \neq y \in H$, there exist and are unique $i_0, j_0 \in I$ such that $x \in H_{i_0}$ and $y \in H_{j_0}$. Since $\{H_i\}_{i \in I}$ is a partition of H, it follows that $H_x = H_{i_0}$ and $H_y = H_{j_0}$ and then

$$x \sim_e y \Leftrightarrow H_x = H_y \Leftrightarrow \exists! i_0 \in I : x, y \in H_{i_0},$$

so the family of the equivalence classes respected to "$\sim_e$" is $\{H_i\}_{i \in I}$.

Moreover, $(H, \circ_\rho)$ is reduced if and only if, for all $x \neq y \in H$, $x \not\sim_e y$, which means that $H_x = \{x\}$, for all $x \in H$, so $\mid H_i \mid = 1$, for all $i \in I$. $\quad\square$

Chapter 10

Approximations in Hypergroups

Rough set theory was proposed by Pawlak for knowledge discovery in databases and experimental data sets. The purpose of this chapter is to introduce and discuss the concept of lower and upper approximations in a hypergroup. We consider the fundamental relation $\beta^\star$ defined on a hypergroup H and interpret the lower and upper approximations as subsets of the group $H/\beta^\star$ and give some properties of such subsets. Also, connections between hypergroups and approximation operators are studied. The results of this chapter are contained in papers [44; 46; 56].

10.1   Pawlak approximations

Suppose that U is a non-empty set. A partition or classification of U is a family $\mathcal{P}$ of non-empty subsets of U such that each element of U is contained in exactly one element of $\mathcal{P}$. Recall that an equivalence relation ρ on a set U is a reflexive, symmetric, and transitive binary relation on U. Each partition $\mathcal{P}$ induces an equivalence relation ρ on U by setting

$$x\rho y \Leftrightarrow x \text{ and } y \text{ are in the same class of } \mathcal{P}.$$

Conversely, each equivalence relation ρ on U induces a partition $\mathcal{P}$ of U of which classes have the form $\rho(x) = \{y \in U \mid x\rho y\}$.

The following notation will be used. Given a non-empty universe U, by $\mathcal{P}(U)$ we denote a power-set on U. If ρ is an equivalence relation on U then for every $x \in U$, $\rho(x)$ stands for the equivalence class of ρ with the representative x. For any $X \subseteq U$, we write X^c to denote the complement of X in U, that is the set $U \backslash X$.

Definition 10.1. A pair (U, ρ) where $U \neq \emptyset$ and ρ is an equivalence relation on U, is called an *approximation space*.

Definition 10.2. For an approximation space (U, ρ), by a rough approximation in (U, ρ) we mean a mapping $app : \mathcal{P}(U) \to \mathcal{P}(U) \times \mathcal{P}(U)$ defined by for every $X \in \mathcal{P}(U)$,

$$app(X) = \big(\underline{app}(X), \overline{app}(X)\big),$$

where

$$\underline{app}(X) = \{x \in X \mid \rho(x) \subseteq X\}, \quad \overline{app}(X) = \{x \in X \mid \rho(x) \cap X \neq \emptyset\}.$$

$\underline{app}(X)$ is called the *lower approximation* of X in (U, ρ), where as $\overline{app}(X)$ is called the *upper approximation* of X in (U, ρ).

Definition 10.3. Given an approximation space (U, ρ), a pair $(A, B) \in \mathcal{P}(U) \times \mathcal{P}(U)$ is called a *rough set* in (U, ρ) if and only if $(A, B) = app(X)$ for some $X \in \mathcal{P}(U)$.

The reader can find in [147; 148] a deep discussion of rough set theory.

Definition 10.4. A subset X of U is called *definable* if $\underline{app}(X) = \overline{app}(X)$.

The properties of rough sets can be examined via either partition or equivalence classes. The objects of the given universe U can be divided into three classes with respect to any subset X of U, see Figure 10.1. Indeed, if $X \subseteq U$ is given by a predicate P and $x \in U$, then

(1) $x \in \underline{app}(X)$ means that x certainly has property P;

(2) $x \in \overline{app}(X)$ means that x possibly has property P;

(3) $x \in U \backslash \overline{app}(X)$ means that x definitely does not have property P.

Definition 10.5. Let $app(A) = \big(\underline{app}(A), \overline{app}(A)\big)$ and $app(B) = \big(\underline{app}(B), \overline{app}(B)\big)$ be any two rough sets in the approximation space (U, ρ). Then

(1) $app(A) \sqcup app(B) = \big(\underline{app}(A) \cup \underline{app}(B), \overline{app}(A) \cup \overline{app}(B)\big)$;

(2) $app(A) \sqcap app(B) = \big(\underline{app}(A) \cap \underline{app}(B), \overline{app}(A) \cap \overline{app}(B)\big)$;

(3) $app(A) \sqsubseteq app(B) \Leftrightarrow app(A) \cap app(B) = app(A)$.

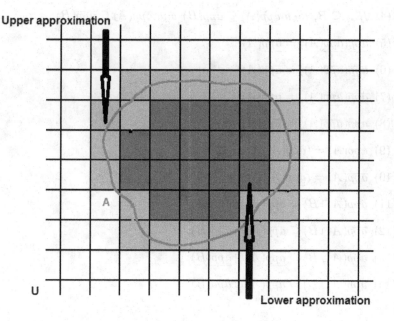

Fig. 10.1   Lower and upper approximations

When $app(A) \sqsubseteq app(B)$, we say that $app(A)$ is a rough subset of $app(B)$. Thus in the case of rough sets $app(A)$ and $app(B)$,

$$app(A) \sqsubseteq app(B) \text{ if and only if } \underline{app}(A) \subseteq \underline{app}(B) \text{ and } \overline{app}(A) \subseteq \overline{app}(B).$$

This property of rough inclusion has all the properties of set inclusion. The rough complement of $app(A)$ denoted by $app^C(A)$ is defined by

$$app^C(A) = \big(U \backslash \overline{app}(A), \ U \backslash \underline{app}(A)\big).$$

Also, we can define $app(A) \backslash app(B)$ as follows:

$$app(A) \backslash app(B) = app(A) \cap app^C(B) = \big(\underline{app}(A) \backslash \overline{app}(B), \ \overline{app}(A) \backslash \underline{app}(B)\big).$$

Theorem 10.1. *We have*

(1) $\underline{app}(A) \subseteq A \subseteq \overline{app}(A)$;

(2) $\underline{app}(\emptyset) = \emptyset = \overline{app}(\emptyset)$;

(3) $\underline{app}(H) = H = \overline{app}(H)$;

(4) *If $A \subseteq B$, then $\underline{app}(A) \subseteq \underline{app}(B)$ and $\overline{app}(A) \subseteq \overline{app}(B)$;*

(5) $\underline{app}(\underline{app}(A)) = \underline{app}(A)$;

(6) $\overline{app}(\overline{app}(A)) = \overline{app}(A)$;

(7) $\overline{app}(\underline{app}(A)) = \underline{app}(A)$;

(8) $\underline{app}(\overline{app}(A)) = \overline{app}(A)$;

(9) $\underline{app}(A) = (\overline{app}(A^C))^C$;

(10) $\overline{app}(A) = (\underline{app}(A^C))^C$;

(11) $\underline{app}(A \cap B) = \underline{app}(A) \cap \underline{app}(B)$;

(12) $\overline{app}(A \cap B) \subseteq \overline{app}(A) \cap \overline{app}(B)$;

(13) $\underline{app}(A \cup B) \supseteq \underline{app}(A) \cup \underline{app}(B)$;

(14) $\overline{app}(A \cup B) = \overline{app}(A) \cup \overline{app}(B)$.

10.2   Lower and upper approximations in a hypergroup

Let $(H, \circ)$ be a hypergroup. For a subset $A \subseteq H$ we define two approximations of A respect to the fundamental equivalence relation β^*:

$$app_{\beta^*}(A) = \{x \in R \mid \beta^*(x) \subseteq A\} \text{ and } \overline{app}_{\beta^*}(A) = \{x \in R \mid \beta^*(x) \cap A \neq \emptyset\}.$$

The set $app_{\beta^*}(A)$ is called the β^*-*lower approximation* of A, and the set $\overline{app}_{\beta^*}(A)$ is called the β^*-*upper approximation* of A. We collect in the following lemma the basic properties of approximations of A, which follow directly from their definitions.

Lemma 10.1.

(1) $app_{\beta^*}(A) \subseteq A \subseteq \overline{app}_{\beta^*}(A)$;

(2) $app_{\beta^*}(app_{\beta^*}(A)) = app_{\beta^*}(A)$ and $\overline{app}_{\beta^*}(\overline{app}_{\beta^*}(A)) = \overline{app}_{\beta^*}(A)$.

The difference $\widehat{\beta^*(A)} = \overline{app}_{\beta^*}(A) \backslash app_{\beta^*}(A)$ is called the β^*-*boundary region* of A. In the case when $\widehat{\beta^*(A)} = \emptyset$ the set A is said to be β^*-*exact*; otherwise A is β^*-*rough*.

Theorem 10.2. *If A and B are two non-empty subsets of H, then*

$$\overline{app}_{\beta^*}(A) \circ \overline{app}_{\beta^*}(B) \subseteq \overline{app}_{\beta^*}(A \circ B).$$

Proof. Let c be any element of $\overline{app_{\beta^*}}(A) \circ \overline{app_{\beta^*}}(B)$. Then $c \in a \circ b$ with $a \in \overline{app_{\beta^*}}(A)$ and $b \in \overline{app_{\beta^*}}(B)$. Thus, there exist elements $x, y \in H$ such that $x \in \beta^*(a) \cap A$ and $y \in \beta^*(b) \cap B$. Therefore, we conclude that

$$x \circ y \subseteq \beta^*(a) \circ \beta^*(b) \subseteq \beta^*(a \circ b).$$

Since $x \circ y \subseteq A \circ B$, we have $x \circ y \subseteq \beta^*(a \circ b) \cap (A \circ B)$ and so $\beta^*(a \circ b) \cap (A \circ B) \neq \emptyset$. Therefore, for every $c \in a \circ b$ we have $\beta^*(c) \cap (A \circ B) \neq \emptyset$ which implies that $c \in \overline{app_{\beta^*}}(A \circ B)$. Therefore, we have $a \circ b \subseteq \overline{app_{\beta^*}}(A \circ B)$. This completes the proof. □

For any $A \subseteq H$, we define a *rough membership function* as follows:

$$\mu_A(x) = \frac{|A \cap \beta^*(x)|}{|\beta^*(x)|},$$

where $|\ |$ denotes the cardinality of a set. By definition, elements in the same equivalence class have the same degree of membership. One can see the similarity between rough membership function and conditional probability. The rough membership value $\mu_A(x)$ may be interpreted as the probability of x belonging to A given that x belongs to an equivalence class.

The following theorem collects the basic properties of the rough membership functions.

Theorem 10.3. *The rough membership functions of the form μ_A have the following properties:*

(1) $\mu_A(x) = 1$ *if and only if* $x \in \underline{app_{\beta^*}}(A)$;

(2) $\mu_A(x) = 0$ *if and only if* $x \in \underline{app_{\beta^*}}(A^c)$;

(3) $0 < \mu_A(x) < 1$ *if and only if* $x \in \widehat{\beta^*(A)}$;

(4) $\mu_A(x) = 1 - \mu_{A^c}(x)$;

(5) $\mu_{A \cup B}(x) \geq \max\{\mu_A(x), \mu_B(x)\}$;

(6) $\mu_{A \cap B}(x) \geq \min\{\mu_A(x), \mu_B(x)\}$.

Proof. It is straightforward. □

10.3 Rough subhypergroups

Let $(H, \circ)$ be a hypergroup. The lower and upper approximations can be presented in an equivalent form as shown below.

Definition 10.6. Let A be a non-empty subsets of H. Then

$$\beta^*(A) = \{\beta^*(x) \in R/\beta^* \mid \beta^*(x) \subseteq A\}$$

and

$$\overline{\beta^*(A)} = \{\beta^*(x) \in R/\beta^* \mid \beta^*(x) \cap A \neq \emptyset\}.$$

Now, we discuss these sets as subsets of the fundamental group H/β^*.

Theorem 10.4. *Let A and B be two non-empty subsets of H. Then the following statements hold:*

(1) $\overline{\beta^*(A \cup B)} = \overline{\beta^*(A)} \cup \overline{\beta^*(B)}$;

(2) $\beta^*(A \cap B) = \beta^*(A) \cap \beta^*(B)$;

(3) $A \subseteq B$ *implies* $\overline{\beta^*(A)} \subseteq \overline{\beta^*(B)}$;

(4) $A \subseteq B$ *implies* $\beta^*(A) \subseteq \beta^*(B)$;

(5) $\beta^*(A) \cup \beta^*(B) \subseteq \beta^*(A \cup B)$;

(6) $\overline{\beta^*(A \cap B)} \subseteq \overline{\beta^*(A)} \cap \overline{\beta^*(B)}$.

Proof. (1) We can write

$$\begin{aligned}
\beta^*(x) \in \overline{\beta^*(X \cup Y)} &\Leftrightarrow \beta^*(x) \cap (X \cup Y) \neq \emptyset \\
&\Leftrightarrow (\beta^*(x) \cap X) \cup (\beta^*(x) \cap Y) \neq \emptyset \\
&\Leftrightarrow \beta^*(x) \cap X \neq \emptyset \;\; \text{or} \;\; \beta^*(x) \cap Y) \neq \emptyset \\
&\Leftrightarrow \beta^*(x) \in \overline{\beta^*(X)} \;\; \text{or} \;\; \beta^*(x) \in \overline{\beta^*(Y)} \\
&\Leftrightarrow \beta^*(x) \in \overline{\beta^*(X)} \cup \overline{\beta^*(Y)}.
\end{aligned}$$

Thus $\overline{\beta^*(X \cup Y)} = \overline{\beta^*(X)} \cup \overline{\beta^*(Y)}$.

(2) We have

$$\begin{aligned}
\beta^*(x) \in \beta^*(X \cap Y) &\Leftrightarrow \beta^*(x) \subseteq X \cap Y \\
&\Leftrightarrow \beta^*(x) \subseteq X \;\; \text{and} \;\; \beta^*(x) \subseteq Y \\
&\Leftrightarrow \beta^*(x) \in \beta^*(X) \;\; \text{and} \;\; \beta^*(x) \in \beta^*(Y) \\
&\Leftrightarrow \beta^*(x) \in \beta^*(X) \cap \beta^*(Y).
\end{aligned}$$

Thus $\beta^*(X \cap Y) = \beta^*(X) \cap \beta^*(Y)$.

(3) Since $X \subseteq Y$ if and only if $X \cap Y = X$, by (2) we have

$$\beta^*(X) = \beta^*(X \cap Y) = \beta^*(X) \cap \beta^*(Y).$$

This implies that $\beta^*(X) \subseteq \beta^*(Y)$.

(4) Since $X \subseteq Y$ if and only if $X \cup Y = Y$, by (1) we have

$$\overline{\beta^*(Y)} = \overline{\beta^*(X \cup Y)} = \overline{\beta^*(X)} \cup \overline{\beta^*(Y)}.$$

This implies that $\overline{\beta^*(X)} \subseteq \overline{\beta^*(Y)}$.

(5) Since $X \subseteq X \cup Y$ and $Y \subseteq X \cup Y$, by (3) we obtain

$$\underline{\beta^*(X) \subseteq \beta^*(X \cup Y)} \quad \text{and} \quad \underline{\beta^*(Y) \subseteq \beta^*(X \cup Y)},$$

which yields that

$$\underline{\beta^*(X) \cup \beta^*(Y) \subseteq \beta^*(X \cup Y)}.$$

(6) Since $X \cap Y \subseteq X$ and $X \cap Y \subseteq Y$, by (4) we have

$$\underline{\overline{\beta^*(X \cap Y)} \subseteq \overline{\beta^*(X)}} \quad \text{and} \quad \underline{\overline{\beta^*(X \cap Y)} \subseteq \overline{\beta^*(Y)}},$$

which yields that

$$\overline{\beta^*(X \cap Y)} \subseteq \overline{\beta^*(X)} \cap \overline{\beta^*(Y)}. \qquad \square$$

Theorem 10.5. *If X is a subhypergroup of H, then $\overline{\beta^*(X)}$ is a subgroup of $(H/\beta^*, \odot)$.*

Proof. The kernel of the canonical map $\varphi : H \longrightarrow H/\beta^*$ is called the core of H and is denoted by ω_H. Here we also denote by ω_H the unit element of H/β^*.

First we show that $\omega_H \in \overline{\beta^*(X)}$. Since X is a subhypergroup of $(H, \circ)$, it follows that for every $a \in X$ we have $a \circ X = X$. Therefore, $a \in a \circ X$ and so there exists $b \in X$ such that $a \in a \circ b$ which implies $\beta^*(a) = \beta^*(a \circ b) = \beta^*(a) \odot \beta^*(b)$. Therefore, $\beta^*(b) = \omega_H$ and so $b \in \omega_H \cap X$ which implies that $\omega_H \cap X \neq \emptyset$. Therefore, $\omega_H \in \overline{\beta^*(X)}$.

Now, suppose that $\beta^*(x), \beta^*(y) \in \overline{\beta^*(X)}$, we show that $\beta^*(x) \odot \beta^*(y) \in H/\beta^*$. We have $\beta^*(x) \cap X \neq \emptyset$ and $\beta^*(y) \cap X \neq \emptyset$ then there exist $a \in \beta^*(x) \cap X$ and $b \in \beta^*(y) \cap X$. Thus $a \in \beta^*(x)$, $a \in X$, $b \in \beta^*(y)$, $b \in X$ and so

$$a \circ b \subseteq \beta^*(x) \circ \beta^*(y) \subseteq \beta^*(x \circ y) = \beta^*(x) \odot \beta^*(y).$$

For every $c \in x \circ y$ we have $\beta^*(c) = \beta^*(x) \circ \beta^*(y)$. Therefore, we get $a \circ b \subseteq \beta^*(c)$ and $a \circ b \subseteq X$.

Therefore, $\beta^*(c) \cap X \neq \emptyset$ which yields $\beta^*(c) \in \overline{\beta^*(X)}$ or $\beta^*(x) \odot \beta^*(y) \in \overline{\beta^*(X)}$.

Finally, if $\beta^*(x) \in \overline{\beta^*(X)}$ then we show that $\beta^*(x)^{-1} \in \overline{\beta^*(X)}$. Since $\omega_H \cap X \neq \emptyset$ then there exists $h \in \omega_H \cap X$ and since $\beta^*(x) \cap X \neq \emptyset$ then there exists $y \in \beta^*(x) \cap X$. By the reproduction axiom we get $h \in y \circ X$ then there exists $a \in X$ such that $h \in y \circ a$ which implies $\beta^*(h) = \beta^*(y) \odot \beta^*(a)$. Since $h \in \omega_H$ then $\beta^*(h) = \omega_H$. Therefore, $\omega_H = \beta^*(y) \odot \beta^*(a)$ or $\omega_H = \beta^*(x) \odot \beta^*(a)$ which yields that $\beta^*(a) = \beta^*(x)^{-1}$. Since $a \in X$ and $a \in \beta^*(a)$ then $\beta^*(a) \cap X \neq \emptyset$ and so $\beta^*(a) \in \overline{\beta^*(X)}$. Therefore, $\overline{\beta^*(X)}$ is a subgroup of $(H/\beta^*, \odot)$. $\qquad \square$

Theorem 10.6. *If X and Y are non-empty subsets of H, then*

$$\overline{\beta^*(X)} \odot \overline{\beta^*(Y)} \subseteq \overline{\beta^*(X \circ Y)}.$$

Proof. We have

$$\overline{\beta^*(X)} \odot \overline{\beta^*(Y)} = \{\beta^*(a) \odot \beta^*(b) \mid \beta^*(a) \in \overline{\beta^*(X)}, \ \beta^*(b) \in \overline{\beta^*(Y)}\}$$
$$= \{\beta^*(a) \odot \beta^*(b) \mid \beta^*(a) \cap X \neq \emptyset, \ \beta^*(b) \cap Y \neq \emptyset\}.$$

Therefore, $(\beta^*(a) \circ \beta^*(b)) \cap (X \circ Y) \neq \emptyset$. Since $\beta^*(a) \circ \beta^*(b) \subseteq \beta^*(a \circ b)$, it follows that $\beta^*(a \circ b) \cap (X \circ Y) \neq \emptyset$. Thus $\beta^*(a \circ b) = \beta^*(a) \odot \beta^*(b) \in \overline{\beta^*(X \circ Y)}$ and so $\overline{\beta^*(X)} \odot \overline{\beta^*(Y)} \subseteq \overline{\beta^*(X \circ Y)}$. $\quad\square$

Theorem 10.7. *Let X and Y be two subhypergroups of H and let $f : X \to Y$ be a strong homomorphism. Then f induces a homomorphism $F : \overline{\beta^*(X)} \to \overline{\beta^*(Y)}$ by setting*

$$F(\beta^*(x)) = \beta^*(f(x)), \text{ for all } x \in X.$$

Proof. First, we prove that F is well-defined. Suppose that $\beta^*(a) = \beta^*(b)$. Then there exist $x_1, \ldots, x_{m+1} \in H$ with $x_1 = a$, $x_{m+1} = b$ and $u_1, \ldots, u_m \in \mathcal{U}$ such that $\{x_i, x_{i+1}\} \subseteq u_i$ $(i = 1, \ldots, m)$ which implies that $\{f(x_i), f(x_{i+1})\} \subseteq f(u_i)$ $(i = 1, \ldots, m)$. Since f is a strong homomorphism and $u_i \in \mathcal{U}$, it follows that $f(u_i) \in \mathcal{U}$. This yields that $f(a)\beta^* f(b)$ or $F(\beta^*(a)) = F(\beta^*(b))$. On the other hand, if $\beta^*(a) \in \overline{\beta^*(X)}$ then $\beta^*(a) \cap X \neq \emptyset$ and so there exists $b \in \beta^*(a) \cap X$. Thus, $b\beta^* a$ and $b \in X$ which yield that $f(b)\beta^* f(a)$ and $f(b) \in Y$. So $f(b) \in \beta^*(f(a))$ and $f(b) \in Y$. Hence, $\beta^*(f(a)) \cap Y \neq \emptyset$ and consequently we have $\beta^*(f(a)) \in \overline{\beta^*(Y)}$ or $F(\beta^*(a)) \in \overline{\beta^*(Y)}$. Thus F is well-defined. Now we have

$$F(\beta^*(a) \odot \beta^*(b)) = F(\beta^*(a \circ b))$$
$$= \beta^*(f(a \circ b))$$
$$= \beta^*(f(a) \circ f(b))$$
$$= \beta^*(f(a)) \odot \beta^*(f(b))$$
$$= F(\beta^*(a)) \odot F(\beta^*(b)).$$

Therefore, F is a homomorphism. $\quad\square$

10.4   Neighborhood operators

Throughout this section $(H, \circ)$ is a hypergroup.

If $\mathcal{U}$ denotes the set of all finite products of elements of H, then the relation β is as follows: for x and y in H we write $x\beta y$ if and only if

$\{x, y\} \subseteq u$, for some $u \in \mathcal{U}$. For the relation β on H and a positive integer k, we now define a new notion of binary relation β^k called the k-*step-relation* of β as follows:

$$\beta^1 = \beta,$$
$$\beta^k = \{(x, y) \in H \times H \mid \text{there exist } y_1, y_2, \ldots, y_i \in H, \ 1 \leq i \leq k-1,$$
$$\text{such that } x\beta y_1, \ y_1\beta y_2, \ \ldots, \ y_i\beta y\} \cup \beta^1, \ k \geq 2.$$

It is easy to see that

$$\beta^{k+1} = \beta^k \cup \{(x, y) \in H \times H \mid \text{there exist } y_1, \ldots, y_k \in H, \text{ such that}$$
$$x\beta y_1, \ y_1\beta y_2, \ldots, y_k\beta y\}.$$

Obviously, $\beta^k \subseteq \beta^{k+1}$, and there exists $n \in \mathbb{N}$ such that $\beta^k = \beta^n$ for all $k \geq n$. (In fact $\beta^n = \beta^*$ is nothing else but the transitive closure of β). Of course β^* is transitive. The relation β^k can be conveniently expressed as a mapping from H to $\mathcal{P}(H)$, $N_k(x) = \{y \in H \mid x\beta^k y\}$ by collecting all β^k-related elements for each element $x \in H$. The set $N_k(x)$ may be viewed as a β^k-*neighborhood* of x defined by the binary relation β^k.

Lemma 10.2. *The β^k-neighborhood $N_k(x)$ becomes an equivalence class containing x.*

Proof. It is straightforward. $\qquad\qquad\qquad\qquad\qquad\qquad\qquad\qquad\square$

Based on the relation β^k on H, we can obtain a *neighborhood system* for each element x: $\{N_k(x) \mid k \geq 1\}$. This neighborhood system is monotonically increasing with respect to k. We can also observe that

$$N_k(x) = \{y \in H \mid \text{there exist } y_1, y_2, \ldots, y_i \in H \text{ such that } x\beta y_1,$$
$$y_1\beta y_2, \ldots, y_i\beta y, 1 \leq i \leq k-1, \text{ or } x\beta y\}.$$

Theorem 10.8. *For each $a, b \in H$ and natural numbers k, l we have*

$$N_k(a) \circ N_l(b) \subseteq N_{k+l-1}(y), \text{ for all } y \in a \circ b.$$

Proof. Suppose that $x \in N_k(a) \circ N_l(b)$, then there exist $a' \in N_k(a)$ and $b' \in N_l(b)$ such that $x \in a' \circ b'$. Since $a' \in N_k(a)$, then $a'\beta^k a$ and so there exist $\{x_1, \ldots, x_{k+1}\} \subseteq H$ with $x_1 = a'$, $x_{k+1} = a$ and $\{u_1, \ldots, u_k\} \subseteq \mathcal{U}$ such that $\{x_i, x_{i+1}\} \subseteq u_i$, $i = 1, \ldots, k$; and since $b' \in N_l(b)$, then $b'\beta^l b$ and so there exist $\{y_1, \ldots, y_{l+1}\} \subseteq H$ with $y_1 = b'$, $y_{l+1} = b$ and $\{v_1, \ldots, v_l\} \subseteq \mathcal{U}$ such that $\{y_j, y_{j+1}\} \subseteq v_j$, $j = 1, \ldots, l$. Therefore we obtain

$$\{x_i, x_{i+1}\} \circ y_i \subseteq u_i \circ v_i, \qquad i = 1, \ldots k-1,$$
$$x_{k+1} \circ \{y_j, y_{j+1}\} \subseteq u_k \circ v_j, j = 1, \ldots, l.$$

If we set

$$u_i \circ v_i = t_i, \qquad i = 1, \ldots, k-1,$$
$$u_k \circ v_j = t_{k+j-1}, \quad j = 1, \ldots, l,$$

and pick up any elements $z_1, \ldots z_{k+l}$ such that

$$z_i \in x_i \circ y_i, \qquad i = 1, \ldots k,$$
$$z_{k+j} \in x_{k+1} \circ y_{j+1}, \quad j = 1, \ldots, l,$$

then we have $\{z_m, z_{m+1}\} \subseteq t_m, m = 1, \ldots, k+l-1$. So for $x \in a' \circ b' = x_1 \circ y_1$ we have $x \beta^{k+l-1} y$ for every element $y \in a \circ b = x_{k+1} \circ y_{k+1}$. Therefore $x \in N_{k+l-1}(y)$. $\qquad\square$

For a neighborhood operator N_k on H, we can extend N_k from $\mathcal{P}(H)$ to $\mathcal{P}(H)$ by:

$$N_k(X) = \bigcup_{x \in X} N_k(x), \text{ for all } X \subseteq H.$$

So we can directly deduce that:

Corollary 10.1. *We have*

(1) $A \subseteq B \ \Rightarrow \ N_k(A) \subseteq N_k(B)$;

(2) *for all* $k, l \geq 1$, *we have* $N_l(N_k(x)) \subseteq N_{l+k}(x)$.

Corollary 10.2. *For each* $a, b \in H$ *and natural numbers* k, l *we have*

$$N_k(a) \circ N_l(b) \subseteq N_{k+l-1}(a \circ b).$$

Definition 10.7. For the relation β, by substituting equivalence class $\beta^*(x)$ with β^k-neighborhood $N_k(x)$ in Definition 10.6, we can define a pair of lower and upper *approximation operators* with respect to N_k as follows:

$$\underline{apr}_k(A) = \{x \in H \mid N_k(x) \subseteq A\} \text{ and } \overline{apr}_k(A) = \{x \in H \mid N_k(x) \cap A \neq \emptyset\}.$$

The set $\underline{apr}_k(A)$ consists of those elements whose β^k-neighborhoods are contained in A, and $\overline{apr}_k(A)$ consists of those elements of which β^k-neighborhoods have a non-empty intersection with A.

Theorem 10.9. *If A is a non-empty subset of H, then we have*

(1) $\underline{apr}_{k+1}(A) \subseteq \underline{apr}_k(A)$;

(2) $\overline{apr}_k(A) \subseteq \overline{apr}_{k+1}(A)$.

Therefore:

Corollary 10.3. *We have*

$$\bigcup\{x \mid \beta^*(x) \in \underline{\beta^*(A)}\} = \bigcap_k \underline{apr}_k(A) \text{ and } \bigcup\{x \mid \beta^*(x) \in \overline{\beta^*(A)}\} = \bigcup_k \overline{apr}_k(A).$$

Theorem 10.10. *If A and B are non-empty subsets of H, then the pair of approximation operators satisfies the following properties:*

(1) $\underline{apr}_k(A) \subseteq A \subseteq \overline{apr}_k(A)$;

(2) $\underline{apr}_k(A) = (\overline{apr}_k(A^c))^c$;

(3) $\overline{apr}_k(A) = (\underline{apr}_k(A^c))^c$;

(4) $\underline{apr}_k(A \cap B) = \underline{apr}_k(A) \cap \underline{apr}_k(B)$;

(5) $\overline{apr}_k(A \cup B) = \overline{apr}_k(A) \cup \overline{apr}_k(B)$;

(6) $\underline{apr}_k(A \cup B) \supseteq \underline{apr}_k(A) \cup \underline{apr}_k(B)$;

(7) $\overline{apr}_k(A \cap B) \subseteq \overline{apr}_k(A) \cap \overline{apr}_k(B)$;

(8) $A \subseteq B \implies \underline{apr}_k(A) \subseteq \underline{apr}_k(B)$;

(9) $A \subseteq B \implies \overline{apr}_k(A) \subseteq \overline{apr}_k(B)$.

Theorem 10.11. *Let A be a non-empty subset of H. For all $k \geq l \geq 1$, we have*

(1) $A \subseteq \underline{apr}_l(\overline{apr}_k(A))$;

(2) $\overline{apr}_l(\underline{apr}_k(A)) \subseteq A$.

Proof. (1) Assume that $a \in A$, if $N_l(a) \neq \emptyset$, then it is clear that $N_l(a) \subseteq \overline{apr}_k(A)$, which implies that $a \in \underline{apr}_l(\overline{apr}_k(A))$, and so $A \subseteq \underline{apr}_l(\overline{apr}_k(A))$. If $N_l(a) \neq \emptyset$, then for each $b \in N_l(a)$, we have $a \in N_l(b)$. Hence $N_l(b) \cap A \neq \emptyset$. Now, we have $b \in \overline{apr}_l(A)$, and by Theorem 10.9, we obtain $b \in \overline{apr}_k(A)$. Therefore $N_l(a) \subseteq \overline{apr}_k(A)$, which implies that $a \in \underline{apr}_l(\overline{apr}_k(A))$, and so $A \subseteq \underline{apr}_l(\overline{apr}_k(A))$.

(2) Assume that $a \in \overline{apr}_l(\underline{apr}_k(A))$, then we have $N_l(a) \cap \underline{apr}_k(A) \neq \emptyset$, and so there exists $b \in N_l(a) \cap \underline{apr}_k(A)$. Therefore, $a \in N_l(b)$ and $N_k(b) \subseteq A$. Hence, we have $a \in N_l(b) \subseteq N_k(b) \subseteq A$, and so we conclude that $\overline{apr}_l(\underline{apr}_k(A)) \subseteq A$. $\square$

Theorem 10.12. *For all $k, l \geq 1$ and $A \subseteq H$, we have*

(1) $\underline{apr}_{l+k}(A) \subseteq \underline{apr}_l(\underline{apr}_k(A))$;

(2) $\overline{apr}_{l+k}(A) \supseteq \overline{apr}_l(\overline{apr}_k(A))$.

Proof. (1) Assume that $a \in \underline{apr}_{l+k}(A)$, then $N_{l+k}(a) \subseteq A$. Using Corollary 10.1, we have $N_k(N_l(a)) \subseteq N_{k+l}(a) \subseteq A$, which implies that $N_l(a) \subseteq \underline{apr}_k(A)$. Therefore $a \in \underline{apr}_l(\underline{apr}_k(A))$.

(2) Assume that $a \in \overline{apr}_l(\overline{apr}_k(A))$, then $N_l(a) \cap \overline{apr}_k(A) \neq \emptyset$, and so there exists $b \in N_l(a) \cap \overline{apr}_k(A)$. Since $b \in \overline{apr}_k(A)$, then $N_k(b) \cap A \neq \emptyset$. Now, we have

$$\emptyset \neq N_k(b) \cap A \subseteq N_k(N_l(a)) \cap A \subseteq N_{l+k}(a) \cap A,$$

and so $N_{l+k}(a) \cap A \neq \emptyset$, which implies that $a \in \overline{apr}_{l+k}(A)$. $\square$

Theorem 10.13. *If A, B are non-empty subsets of H, then*

$$\overline{apr}_k(A) \circ \overline{apr}_l(B) \subseteq \overline{apr}_{k+l-1}(A \circ B).$$

Proof.   Let z be any element of $\overline{apr}_k(A) \circ \overline{apr}_l(B)$. Then there exist $x \in \overline{apr}_k(A)$ and $y \in \overline{apr}_l(B)$ such that $z \in x \circ y$. Since $x \in \overline{apr}_k(A)$ and $y \in \overline{apr}_l(B)$, then there exist $a, b \in H$ such that $a \in N_k(x) \cap A$ and $b \in N_l(y) \cap B$. So $a \in N_k(x)$ and $b \in N_l(y)$. Now, by Theorem 10.8, we have $N_k(x) \circ N_l(y) \subseteq N_{k+l-1}(z)$. Since $a \circ b \subseteq A \circ B$, we obtain $a \circ b \subseteq N_{k+l-1}(z) \cap A \circ B$, and so $z \in \overline{apr}_{k+l-1}(A \circ B)$. This complete the proof. $\square$

Corollary 10.4. *Let H be a hypergroup and A be a closed subset of H, i.e., $A \circ A \subseteq A$, then $\overline{apr}_1(A)$ is also a closed subset of H.*

Theorem 10.14. *Let $\beta^n = \beta^*$ for some $n \in \mathbb{N}$. If A is an H_v-subgroup of H, then $\overline{apr}_n(A)$ is also an H_v-subgroup of H.*

Proof.   Suppose that $x, y \in \overline{apr}_n(A)$ then $N_n(x) \cap A \neq \emptyset$ and $N_n(y) \cap A \neq \emptyset$. Hence, there exist $a \in N_n(x) \cap A$ and $b \in N_n(y) \cap A$. Since $N_n = N_k$ for all $k \geq n$, it follows that

$$a \circ b \subseteq N_n(x) \circ N_n(y) \subseteq N_{2n-1}(z) - N_n(z), \text{ for all } z \in x \circ y.$$

On the other hand, since A is a subhypergroup, it follows that $a \circ b \subseteq A$. Thus $N_n(z) \cap A \neq \emptyset$ which implies that $z \in \overline{apr}_n(A)$, so $x \circ y \subseteq \overline{apr}_n(A)$. Therefore for every $x \in \overline{apr}_n(A)$, we have $x \circ \overline{apr}_n(A) \subseteq \overline{apr}_n(A)$.

Now, we show that $\overline{apr}_n(A) \subseteq x \circ \overline{apr}_n(A)$. Assume that $y \in \overline{apr}_n(A)$. Since $x, y \in H$, then by reproduction axiom, there exists $z \in H$ such that $y \in x \circ z$, and so it is enough to show that $z \in \overline{apr}_n(A)$. By using the definition, we have $\overline{apr}_n(A) = \cup\{\beta^*(x) \mid \beta^*(x) \in \overline{\beta^*(x)}\}$. From $y \in x \circ z$, we obtain $\beta^*(y) = \beta^*(x) \odot \beta^*(z)$. Clearly, $\beta^*(x), \beta^*(y) \in \overline{\beta^*(A)}$. Since $\overline{\beta^*(A)}$ is a subgroup of H/β^*, it follows that $\beta^*(z) \in \overline{\beta^*(A)}$. This yields that $\beta^*(z) \cap A \neq \emptyset$ and so $z \in \overline{apr}_n(A)$. This complete the proof. $\square$

Chapter 11

Links between Hypergraphs and Hypergroups

Hypergraphs are a generalization of graphs in which an edge (hyperedge) is connected to a subset of nodes rather than just two of them. Connections between hypergraphs and hypergroups was studied by Corsini and later, they are studied by Leoreanu-Fotea, Davvaz, Farshi, Nikkhah, Maryati, and many others. In this chapter, we introduce some important and interesting results.

11.1 Hypergraphs

We begin with the definition of a hypergraph.

Definition 11.1. A *hypergraph* is a pair $\Gamma = (V, E)$, where V is a finite set of vertices and $E = \{E_1, \ldots, E_m\}$ is a set of *hyperedges* which are non-empty subsets of V such that $\bigcup_{i=1}^{m} E_i = V$.

The order of the hypergraph $\Gamma = (V, E)$ is the cardinality of V; its size is the cardinality of E.

Example 11.1. Let M be a mathematics meeting with $k \geq 1$ sessions: S_1, $S_2, \ldots, S_k$. Let V be the set of people at this meeting. Assume that each session is attended by one person at least. We can build a hypergraph in the following way:

- The set of vertices is the set of people who attend the meeting;
- The family of hyperedges $(E_i)_{i \in \{1, \ldots, k\}}$ is built in the following way: E_i (for $i = 1, \ldots, k$) is the subset of people who attend the meeting S_i.

Example 11.2. The *Fano plane* is the finite projective plane of order 2, which have the smallest possible number of points and lines, 7 points with 3 points on every line and 3 lines through every point, see Figure 11.1. To a Fano plane we can associate a hypergraph called *Fano hypergraph*:

- The set of vertices is $V = \{0, 1, 2, 3, 4, 5, 6\}$;
- The set of hyperedges is $E = \{013, 045, 026, 124, 346, 235, 156\}$.

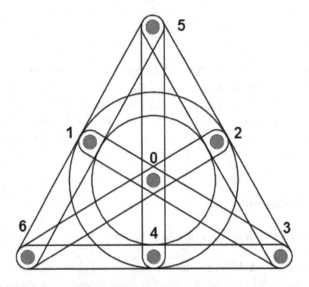

Fig. 11.1　Fano hypergraph

Definition 11.2. Let $\Gamma = (V, E)$ and $\Gamma' = (V', E')$ be two hypergraphs. We say that Γ' is a *subhypergraph* of Γ if $V' \subseteq V$ and $E' \subseteq E$.

In a hypergraph Γ, two vertices x and y are called *connected* if Γ contains a path from x to y. If two vertices are connected by a path of length 1, i.e., by a single hyperedge, the vertices are called *adjacent*. We use the notation $x \sim y$ to denote the adjacency of vertices x and y.

Definition 11.3. A hypergraph is said to be *connected* if every pair of vertices in the hypergraph is connected. A *connected component* of a hypergraph is any maximal set of vertices which are pairwise connected by a path.

Definition 11.4. The length of shortest path between vertices x and y is denoted by $\operatorname{dist}(x, y)$ and the *diameter* of Γ is defined as follows:

$$d := \operatorname{diam}(\Gamma) = \begin{cases} \max\{\operatorname{dist}(x, y) \mid x, y \in H\} & \text{if } \Gamma \text{ is connected,} \\ \infty & \text{otherwise.} \end{cases}$$

In [28], Corsini considered a hypergraph $\Gamma = (H, \{E_i\}_i)$ and defined a hyperoperation $\circ$ on H as follows:

$$x \circ y = E(x) \cup E(y),$$

for all $x, y \in H^2$, where $E(x) = \bigcup_{x \in E_i} E_i$. The hypergroupoid $H_\Gamma = (H, \circ)$ is called a *hypergraph hypergroupoid* or an *h.g. hypergroupoid*. An associative h.g. hypergroupoid is called an *h.g. hypergroup*.

11.2 Hyperoperations associated with hypergraphs and with binary relations

In what follows , we consider the following hyperoperation.

Let $(H, \circ)$ be a hypergroupoid. For all x, y of H, define

$$x \star y = x \circ x \cup y \circ y.$$

Moreover, suppose that "$\circ$" satisfies the following conditions:

(i) for all x of H, we have $x \in x \circ x$;
(ii) for all x, u of H, we have:

$$u \in x \circ x \Leftrightarrow x \in u \circ u.$$

We consider the following particular case.

Let $\Gamma = (H, \{A_i\}_{i \in I})$ be a hypergraph, that is for all $i \in I$, $A_i \neq \emptyset$ and $\bigcup_{i \in I} A_i = H$. For all $x \in H$, consider a hyperoperation, such that:

$$x \circ x = \bigcup_{x \in A_i} A_i.$$

Then "$\circ$" satisfies the conditions (i) and (ii). Indeed, if $u \in x \circ x$, then there is $i_0 \in I$, such that $\{u, x\} \subset A_{i_0}$, hence $x \in \bigcup_{u \in A_i} A_i = u \circ u$. Conversely, it is similar. Therefore, $u \in x \circ x \Leftrightarrow x \in u \circ u$. Conversely, if $(H, \circ)$ is a hypergroupoid, that satisfies (i) and (ii), then there exists a hypergraph $\Gamma = (H, \{x, y\}_{x \in H, y \in x \circ x})$, such that

$$x \circ x = \bigcup_{y \in x \circ x} \{x, y\}.$$

Therefore,

(1) for any hypergraph on H we can consider a hyperoperation on H, that satisfies (i) and (ii).

(2) for any hyperoperation on H, that satisfies (i) and (ii), we can associate a hypergraph on H.

Theorem 11.1. *The hyperoperation "$\star$" is associative if and only if the following condition holds:*

$$\begin{cases} \forall (x, z) \in H^2, \\ \bigcup_{x \in u \circ u} u \circ u \setminus x \circ x \subset \bigcup_{z \in u \circ u} u \circ u. \end{cases} \tag{11.1}$$

Proof. Suppose the condition (11.1) holds. We check that "$\star$" is associative. Let x, y, z be arbitrary elements of H. We have

$$(x \star y) \star z = \left[x \circ x \cup y \circ y \right] \star z$$

$$= z \circ z \cup \bigcup_{u \in x \circ x \cup y \circ y} u \circ u$$

$$= z \circ z \cup \bigcup_{x \in u \circ u} u \circ u \cup \bigcup_{y \in u \circ u} u \circ u.$$

Similarly, we obtain

$$x \star (y \star z) = x \circ x \cup \bigcup_{z \in u \circ u} u \circ u \cup \bigcup_{y \in u \circ u} u \circ u.$$

Denote

$$P = z \circ z \cup \bigcup_{x \in u \circ u} u \circ u \quad \text{and}$$

$$Q = x \circ x \cup \bigcup_{z \in u \circ u} u \circ u.$$

We show that $P = Q$. According to (11.1), we obtain:

$$P = z \circ z \cup x \circ x \cup \left[\bigcup_{x \in u \circ u} u \circ u \setminus x \circ x \right]$$

$$\subset z \circ z \cup x \circ x \cup \bigcup_{z \in u \circ u} u \circ u = Q.$$

Similarly, we obtain $Q \subset P$. Hence, $P = Q$, whence we get that "$\star$" is associative.

Conversely, suppose the hyperoperation "$\star$" is associative. Then, for all x, z of H, we have $(x \star x) \star z = x \star (x \star z)$. On the other hand, we have

$$(x \star x) \star z = z \circ z \cup \bigcup_{x \in u \circ u} u \circ u$$

and

$$x \star (x \star z) = x \circ x \cup \bigcup_{x \in u \circ u} u \circ u \cup \bigcup_{z \in u \circ u} u \circ u$$

$$= \bigcup_{x \in u \circ u} u \circ u \cup \bigcup_{z \in u \circ u} u \circ u.$$

Since $(x \star x) \star z = x \star (x \star z)$, it follows

$$\bigcup_{z \in u \circ u} u \circ u \backslash z \circ z \subset \bigcup_{x \in u \circ u} u \circ u,$$

that is the condition (11.1) holds. □

Corollary 11.1. *If, for all x of H, we have*

$$\bigcup_{x \in u \circ u} u \circ u = x \circ x,$$

then the hyperoperation "$\star$" is associative.

Theorem 11.2. *Let "$\circ$" be a hyperoperation on H, that satisfies* (i) *and* (ii). *The hypergroupoid $(H, \star)$ is a join space if and only if the condition* (11.1) *holds.*

Proof. According to the above theorem, it is sufficient to check that if the condition (11.1) holds, then the following implication holds:

$$x/y \cap z/w \neq \emptyset \Rightarrow x \star w \cap z \star y \neq \emptyset.$$

Let $u \in H$ be such that $x \in u \star y = u \circ u \cup y \circ y$ and $z \in u \star w = u \circ u \cup w \circ w$. We distinguish the following situations:

(1) if $x, z \in u \circ u$, then $u \in x \circ x \cap z \circ z$,
    hence $x \star w \cap z \star y \neq \emptyset$.
(2) if $x \in u \circ u$ and $z \in w \circ w$, then $w \in z \circ z$,
    hence $w \in x \star w \cap y \star z$.
(3) if $x \in y \circ y$ and $z \in u \circ u$, then $y \in x \circ x$,
    hence $y \in x \star w \cap y \star z$.
(4) if $x \in y \circ y$ and $z \in w \circ w$, then $w \in z \circ z$,
    hence $w \in x \star w \cap y \star z$.

Therefore, in all cases we have $x \star w \cap z \star y \neq \emptyset$. □

Let us define inductively the following hyperoperations on H:

Let "$\circ$" be a hyperoperation on H, that satisfies (i) and (ii). For all $x \in H$, define:

$$\begin{cases} x \star_0 x = x \star x = x \circ x; \\[2mm] x \star_1 x = \displaystyle\bigcup_{u \in x \circ x} u \circ u = \bigcup_{u \in x \star_0 x} u \star_0 u; \\[2mm] x \star_2 x = \displaystyle\bigcup_{u \in x \star_1 x} u \star_1 u; ...; x \star_{k+1} x = \bigcup_{u \in x \star_k x} u \star_k u; ...; \quad (11.2) \\[2mm] \text{and for all } x, y \text{ of } H \text{ and } k \in \mathbb{N}, \text{ define} \\[2mm] x \star_k y = x \star_k x \cup y \star_k y. \end{cases}$$

Theorem 11.3. *For all x, y of H and $k \in \mathbb{N}$, we have*

(1) $x \in x \star_k x$;

(2) $x \star_k x \subset x \star_{k+1} x$;

(3) $y \in x \star_k x \Leftrightarrow x \in y \star_k y$.

Proof. (1) By (i) it follows $x \in x \star x$. Suppose $x \in x \star_k x$ and check that $x \in x \star_{k+1} x$. We have $x \in x \star_k x \subset \bigcup_{u \in x \star_k x} u \star_k u = x \star_{k+1} x$. Hence, for all $k \in \mathbb{N}$, we have $x \in x \star_k x$.

(2) Let $u \in x \star_k x$. Then $u \in u \star_k u \subset \bigcup_{u \in x \star_k x} u \star_k u = x \star_{k+1} x$. Hence $x \star_k x \subset x \star_{k+1} x$.

(3) We check by induction on k.

For $k = 0$, it follows by (ii).

Suppose $y \in x \star_k x \Leftrightarrow x \in y \star_k y$ for all x, y of H and check that $b \in a \star_{k+1} a \Leftrightarrow a \in b \star_{k+1} b$ for all a, b of H.

Let $b \in a \star_{k+1} a = \bigcup_{u \in a \star_k a} u \star_k u$. Then there is $u \in a \star_k a$, such that $b \in u \star_k u$. By the induction hypothesis, we get $a \in u \star_k u$ and $u \in b \star_k b$. Hence $a \in \bigcup_{u \in b \star_k b} u \star_k u = b \star_{k+1} b$. $\square$

Theorem 11.4. *If there exists $k \in \mathbb{N}$, such that the hyperoperations "$\star_k$" and "$\star_{k+1}$", defined by (11.2), coincide, then $(H, \star_k)$ is a join space.*

Proof. Notice that by Theorem 11.3 (2) and by (11.2), it follows that, for all $m \geq k$, the hyperoperations "$\star_m$" and "$\star_k$" coincide.

Since $x \star_{k+1} x = x \star_k x$, it follows that

$$\bigcup_{u \in x \star_k x} u \star_k u = x \star_k x. \tag{11.3}$$

Moreover, f_k satisfies the conditions (i) and (ii). Indeed, by Theorem 11.3 (1), we have $x \in x \star_k x$. On the other hand, by Theorem 11.3 (3), we have

$$x \in u \star_k u \Leftrightarrow u \in x \star_k x.$$

According to Corollary 11.1, it follows that the hyperoperation "$\star_k$" is associative. Moreover, by Theorem 11.2, it follows that $(H, \star_k)$ is a join space, as desired. $\square$

Let us reconsider now the Rosenberg hyperoperation for a reflexive and symmetric binary relation R be on H.

$$x_1 \circ x_2 = \{y \in H \mid (x_1, y) \in R \text{ or } (x_2, y) \in R\}.$$

Clearly, we have $x_1 \circ x_2 = x_1 \circ x_1 \cup x_2 \circ x_2$. For all $x \in H$, we have $x \circ x = \{y \in H \mid (x, y) \in R\}$. Notice that "$\circ$" satisfies the conditions (i) and (ii). Indeed, for all $x \in H$, we have $x \in x \circ x$, since R is reflexive. Moreover, we have $x \in u \circ u \Leftrightarrow (u, x) \in R \Leftrightarrow (x, u) \in R \Leftrightarrow u \in x \circ x$, since R is symmetric.

Let us analyze the sequence (11.2) of hyperoperations associated with R. We have

$$x \star_0 x = x \circ x = \{y \mid (x, y) \in R\} = x \star x;$$

$$x \star_1 x = \bigcup_{u \in x \star_0 x} u \star_0 u = \bigcup_{(x, u) \in R} \{v \mid (u, v) \in R\}$$

$$= \{v \mid (u, v) \in R, \ (x, u) \in R\} = \{v \mid (x, v) \in R^2\}.$$

Suppose that $x \star_k x = \{t \mid (x, t) \in R^{2^k}\}$.

Check that $x \star_{k+1} x = \{s \mid (x, s) \in R^{2^{k+1}}\}$. We have:

$$x \star_{k+1} x = \bigcup_{u \in x \star_k x} u \star_k u = \bigcup_{(x, u) \in R^{2^k}} \{t \mid (u, t) \in R^{2^k}\}$$

$$= \{t \mid (u, t) \in R^{2^k}, \ (x, u) \in R^{2^k}\} = \{t \mid (x, t) \in R^{2^{k+1}}\}.$$

Theorem 11.5. *If R is an equivalence relation on H, then $(H, \star)$ is a join space, where for all x, y of H,*

$$x \star y = x \star x \cup y \star y \ and \ x \star x = \{z \mid (x, z) \in R\}.$$

Proof. It follows by Corollary 11.1 and Theorem 11.2. $\qquad\square$

Theorem 11.6. *If R is a reflexive and symmetric relation on H, such that there is $k \in \mathbb{N}$, for which $R^{2^k} = R^{2^{k+1}}$, then $(H, \star_k)$ is a join space.*

Proof. It follows by Theorem 11.4. $\qquad\square$

Theorem 11.7. *If R is a reflexive and symmetric relation on H and for all x, y of H we define*

$$x \sqcup y = \bigcup_{k \geq 0} x \star_k y,$$

then $(H, \sqcup)$ is a join space.

Proof. For all $x \in H$, we have

$$x \sqcup x = \left\{ z \mid (x, z) \in \bigcup_{k \geq 0} R^{2^k} \right\} = \left\{ z \mid (x, z) \in \bigcup_{n \geq 1} R^n \right\},$$

since R is reflexive. Notice that $\bigcup_{n \geq 1} R^n$ is an equivalence relation and by Theorem 11.5, it follows that $(H, \sqcup)$ is a joint space. $\qquad\square$

Theorem 11.8. *Let R be a reflexive and symmetric relation on H, and let $|H| = n$. Then $(H, \star_{k+1})$ is a join space, where $2^k \leq n - 1 < 2^{k+1}$.*

Proof. By hypothesis, it follows that the transitive closure of R is R^{n-1}, which is an equivalence relation on H. Since $2^k \leq n - 1 < 2^{k+1}$, it follows that $R^k \subset R^{n-1} \subset R^{2^{k+1}}$. We have $R^{n-1} = R^{2^{k+1}}$, hence $(H, \star_{k+1})$ is a join space, according to Theorem 11.5. $\qquad\square$

11.3 Relationship between hypergraphs and hypergroups by using an equivalence relation

By using an equivalence relation, Farshi et al. [71] connected hypergraphs to hypergroups. This equivalence relation is called *special relation*. More exactly, starting with a hypergraph and using the special relation, they constructed a hypergroup. In this section we review their results.

Definition 11.5. Let $\Gamma = (H, \{E_i\}_i)$ be a hypergraph. Define the special relation ρ on H as follows:

$$x \rho y \text{ if } \{E_i \mid x \in E_i\} = \{E_i \mid y \in E_i\}.$$

Clearly, the relation ρ is an equivalence relation. We define the hyperoperation $\circ_\rho$ on H as follows:

$$x \circ_\rho y = \rho(x) \cup \rho(y),$$

for all $(x, y) \in H^2$, where $\rho(x)$ denotes the equivalence class of element x of H.

This hyperoperation is a particular case of Rosenberg hyperoperation, which is associated with an equivalence relation.

Theorem 11.9. *Let H be the vertex set of hypergraph Γ. Then, the hypergroupoid $H_\rho = (H, \circ_\rho)$ has the following properties for each $(x, y) \in H^2$:*

(1) $x \circ_\rho y = x \circ_\rho x \cup y \circ_\rho y$;
(2) $x \in x \circ_\rho x$;
(3) $y \in x \circ_\rho x \Leftrightarrow x \in y \circ_\rho y$.

Proof. It is straightforward. $\qquad\square$

One easily checks that for every $x \in H$, we have $x \circ_\rho x \circ_\rho x = x \circ_\rho x$, and so by Theorem 11.2, H_ρ is a hypergroup. H_ρ is called the ρ-hypergroup *induced* by Γ. By Theorem 11.2, H_ρ is a join space.

Let ρ_Γ and $\rho_{\Gamma'}$ be special relations on hypergraphs Γ and Γ' respectively. An *isomorphism* from Γ to Γ' is a bijection f from the vertex set of Γ to that of Γ' such that

$$x\rho_\Gamma y \Leftrightarrow f(x)\rho_{\Gamma'}f(y).$$

We say that Γ and Γ' are isomorphic (written $\Gamma \cong \Gamma'$) if there is an isomorphism between them.

Theorem 11.10. *If H and H' are two ρ-hypergroups induced by hypergraphs Γ and Γ' respectively, then $\Gamma \cong \Gamma'$ if and only if $H \cong H'$.*

Proof. Suppose that $f : \Gamma \to \Gamma'$ be an isomorphism. We show that $f(x \circ_\rho y) = f(x) \circ_\rho f(y)$ or equivalently we show that $f(\rho_\Gamma(x) \cup \rho_\Gamma(y)) = \rho_{\Gamma'}(f(x)) \cup \rho_{\Gamma'}(f(y))$. Let $t \in f(\rho_\Gamma(x) \cup \rho_\Gamma(y))$ be an arbitrary element. Then, there exists $a \in \rho_\Gamma(x) \cup \rho_\Gamma(y)$ such that $t = f(a)$. If $a \in \rho_\Gamma(x)$, then we have $x\rho_\Gamma a$ and by hypothesis we have $f(x)\rho_{\Gamma'}f(a)$ which implies that $t = f(a) \in \rho_{\Gamma'}(f(x))$. If $a \in \rho_\Gamma(y)$, then in a similar way we conclude that $t = f(a) \in \rho_{\Gamma'}(f(y))$. Therefore, in each case we have $t \in \rho_{\Gamma'}(f(x)) \cup \rho_{\Gamma'}(f(y))$. Now, let $t \in \rho_{\Gamma'}(f(x)) \cup \rho_{\Gamma'}(f(y))$ be an arbitrary element. Since f is onto there exists an element "a" in vertex set of Γ such that $t = f(a)$. If $t \in \rho_{\Gamma'}(f(x))$, then we have $f(a)\rho_{\Gamma'}f(x)$ and by hypothesis we have $a\rho_\Gamma x$ which implies that $a \in \rho_\Gamma(x)$ and so in this case we have $t \in f(\rho_\Gamma(x) \cup \rho_\Gamma(y))$. Similarly if $t \in \rho_{\Gamma'}(f(y))$, we have $t \in f(\rho_\Gamma(x) \cup \rho_\Gamma(y))$.

Conversely, suppose that $f : H \to H'$ be an isomorphism. Then, for every $x, y \in H$ we have $f(\rho_\Gamma(x) \cup \rho_\Gamma(y)) = \rho_{\Gamma'}(f(x)) \cup \rho_{\Gamma'}(f(y))$. This implies that for every $x \in H$ we have $f(\rho_\Gamma(x)) = \rho_{\Gamma'}(f(x))$. If $x\rho_\Gamma y$, then we have $\rho_\Gamma(x) = \rho_\Gamma(y)$ and so we have $f(\rho_\Gamma(x)) = f(\rho_\Gamma(y)) = \rho_{\Gamma'}(f(x)) \cup \rho_{\Gamma'}(f(y))$ which implies that $\rho_{\Gamma'}(f(x)) = \rho_{\Gamma'}(f(y)) = \rho_{\Gamma'}(f(x)) \cup \rho_{\Gamma'}(f(y))$ and so we have $\rho_{\Gamma'}(f(x)) = \rho_{\Gamma'}(f(y))$, i.e., we have $f(x)\rho_{\Gamma'}f(y)$. Now, if $f(x)\rho_{\Gamma'}f(y)$, then we have $\rho_{\Gamma'}(f(x)) = \rho_{\Gamma'}(f(y))$ which implies that $f(\rho_\Gamma(x)) = f(\rho_\Gamma(y))$ and so we have $f(\rho_\Gamma(x) \cup \rho_\Gamma(y)) = f(\rho_\Gamma(x)) = f(\rho_\Gamma(y))$. This implies that $\rho_\Gamma(x) = \rho_\Gamma(y)$, i.e., we have $x\rho_\Gamma y$. $\square$

Definition 11.6. Let $(H, \circ)$ be a hypergroup and $a \in H$. Then, the hypergraph $\Gamma_a = (H, \{a \circ x\}_{x \in H})$ is called the hypergraph *extracted* from H. Also, we set the hypergraph $\Gamma_\Delta = (H, \{x \circ x\}_{x \in H})$.

Theorem 11.11. *Let Γ be a hypergraph and H_ρ be the ρ-hypergroup induced by Γ. Then, for every $a \in H$ we have $\Gamma_\Delta \subset \Gamma_a$, which means that every hyperedge of Γ_Δ is a subset of some hyperedges of Γ_a.*

Theorem 11.12. *Let Γ be a hypergraph and $H_\rho = (H, \circ_\rho)$ be the ρ-hypergroup induced by Γ. Then, H_ρ is an h.g. hypergroup.*

Proof. Since $\Gamma_\Delta = (H, \{x \circ x\}_{x \in H})$ is a hypergraph and for every $x \in H$ we have $E(x) = x \circ_\rho x$, then $H_\Gamma = H_\rho$ and H_ρ is an h.g. hypergroup. $\square$

Example 11.3. There exists an h.g. hypergroup such that it is not a ρ-hypergroup induced by Γ. Let $\Gamma = (H, E)$ such that $H = \{1, 2, 3, 4\}$ and $E = \{\{1, 2, 3\}, \{2, 3, 4\}\}$. Then, in the h.g. hypergraph H_Γ we have $1 \circ 1 = \{1, 2, 3\}$ and $3 \circ 3 = H$ but in the every ρ-hypergroup induced by every hypergraph Γ we have $1 \circ_\rho 1 = 3 \circ_\rho 3$ or $1 \circ_\rho 1 \cap 3 \circ_\rho 3 = \emptyset$.

Theorem 11.13. *Let Γ be a hypergraph, $H_\Gamma = (H, \circ)$ be h.g. hypergroupoid and $H_\rho = (H, \circ_\rho)$ be the ρ-hypergroup induced by Γ. Then, for every $x \in H$ we have $x \circ_\rho x \subset x \circ x$ and $\{y \circ_\rho y\}_{y \in x \circ x}$ is a partition of $x \circ x$.*

Theorem 11.14. *Let Γ be a hypergraph and $H_\rho = (H, \circ_\rho)$ be the ρ-hypergroup induced by Γ. Then, for every $a \in H$ we have $\Gamma_a \cong \Gamma$, where Γ_a is the hypergraph extracted from H_ρ.*

Proof. Suppose that $a \in H$ is an arbitrary element and $f : H \to H$ is the identity map. We show that

$$x\rho_\Gamma y \Leftrightarrow f(x)\rho_{\Gamma_a} f(y), \text{ for all } (x, y) \in H^2.$$

Let's prove the implication " $\Longrightarrow$ ". Let $x, y \in H$ and $x\rho_\Gamma y$. In order to verify that $x\rho_{\Gamma_a} y$ we must show that $x \in a \circ_\rho t$ if and only if $y \in a \circ_\rho t$, where t is an arbitrary element of H. If $x \in a \circ_\rho t$, then $x \in \rho_\Gamma(a) \cup \rho_\Gamma(t)$ and so $\rho_\Gamma(x) = \rho_\Gamma(a)$ or $\rho_\Gamma(x) = \rho_\Gamma(t)$. Since $\rho_\Gamma(x) = \rho_\Gamma(y)$ we have $y \in \rho_\Gamma(a) \cup \rho_\Gamma(t)$ which implies that $y \in a \circ_\rho t$. Similarly, if $y \in a \circ_\rho t$ we can conclude that $x \in a \circ_\rho t$.

Now, we prove the implication " $\Longleftarrow$ ". Let $x, y \in H$ and $x\rho_{\Gamma_a} y$. Then, for every $t \in H$ we have

$$x \in a \circ_\rho t \Leftrightarrow y \in a \circ_\rho t.$$

We know that $x \in a \circ_\rho x$, thus we have $y \in a \circ_\rho x$, i.e., $y \in \rho_\Gamma(a) \cup \rho_\Gamma(x)$. If $y \in \rho_\Gamma(x)$, then we have $x\rho_\Gamma y$ and the proof completes. But if $y \in \rho_\Gamma(a)$, then we have $y \in a \circ_\rho a$ and by assumption we have $x \in a \circ_\rho a$ which implies that $\rho_\Gamma(x) = \rho_\Gamma(a) = \rho_\Gamma(y)$ and so in this case we have $x\rho_\Gamma y$. $\square$

By Theorem 11.14, if we induce a hypergroup H by a hypergraph Γ, then all hypergraphs extracted from H are isomorphic. Next example shows that if we start with a hypergroup H and extract a hypergraph Γ_a and H_ρ be the ρ-hypergroup induced by Γ_a, then H and H_ρ may be non-isomorphic.

Theorem 11.15. *Let $(H, \circ)$ be a hypergroup and $a \in H$ and Γ_a be the hypergraph extracted from H. If $H_\rho \cong H$ where H_ρ is the ρ-hypergroup induced by Γ_a, then for every $x, y \in H$ we have $\{x, y\} \subset x \circ y$.*

Proof. Suppose that $f : H \to H_\rho$ is an isomorphism and x, y are two arbitrary elements of H. Then, we have $f(x \circ y) = \rho_{\Gamma_a}(f(x)) \cup \rho_{\Gamma_a}(f(y))$. Since $\{f(x), f(y)\} \subset \rho_{\Gamma_a}(f(x)) \cup \rho_{\Gamma_a}(f(y))$, there exist $x', y' \in x \circ y$ such that $f(x) = f(x')$ and $f(y) = f(y')$. Therefore, we have $x = x'$ and $y = y'$ which implies that $\{x, y\} \subset x \circ y$. $\qquad\square$

Corollary 11.2. *Let $(H, \circ)$ be a hypergroup and $a \in H$ and Γ_a be the hypergraph extracted from H. If there exist $x, y \in H$ such that $x \neq y$ and $|x \circ y| = 1$, then $H_\rho \not\cong H$ where H_ρ is the ρ-hypergroup induced by Γ_a.*

Corollary 11.3. *Let $(H, \circ)$ be a hypergroup and there exist $x, y \in H$ such that $\{x, y\} \not\subset x \circ y$. Then, H is not a ρ-hypergroup.*

Proof. Suppose that H is a ρ-hypergroup induced by Γ and $a \in H$. Then, by Theorem 11.14, $\Gamma_a \cong \Gamma$. Now if H_ρ be the hypergroup induced by Γ_a, then by Theorem 11.38 we have $H \cong H_\rho$ and by Theorem 11.15 we have $\{x, y\} \subset x \circ y$ which is a contradiction. $\qquad\square$

Theorem 11.16. *Let $H_\rho = (H, \circ_\rho)$ be a ρ-hypergroup induced by hypergraph Γ. If $(H', *)$ is a hypergroup and $H' \cong H_\rho$, then there exists a hypergraph Γ' such that $\Gamma' \cong \Gamma$ and H' is the ρ-hypergroup induced by Γ'.*

Proof. Suppose that $\Gamma = (H, \{E_i\}_i)$ and $f : H_\rho \to H'$ is an isomorphism. If $\Gamma' = (H', \{f(E_i)\}_i)$, then for every $x, y \in H_\rho$ we have

$$x \rho_\Gamma y \Leftrightarrow \{E_i \mid x \in E_i\} = \{E_i \mid y \in E_i\}$$
$$\Leftrightarrow \{f(E_i) \mid f(x) \in f(E_i)\} = \{f(E_i) \mid f(y) \in f(E_i)\}$$
$$\Leftrightarrow f(x) \rho_{\Gamma'} f(y),$$

and so $\Gamma \cong \Gamma'$. It is sufficient to show that H' is a ρ-hypergroup induced by Γ'. For every $f(x), f(y) \in H'$ we have

$$f(x) * f(y) = f(x \circ_\rho y) = f(\rho_\Gamma(x) \cup \rho_\Gamma(y)) = \rho_{\Gamma'}(f(x)) \cup \rho_{\Gamma'}(f(y)).$$

This completes the proof. $\qquad\square$

Corollary 11.4. *Let $(H, \circ)$ be a hypergroup, $a \in H$ and H_ρ be the ρ-hypergroup induced by Γ_a, where Γ_a is the hypergraph extracted from H. Then, $H \cong H_\rho$ if and only if H is the ρ-hypergroup induced by a hypergraph Γ.*

Let X, X' be vertex sets of hypergraphs Γ, Γ', respectively. We say that Γ' is a *special subhypergraph* of Γ if

$$\rho_{\Gamma'}(x) = \rho_\Gamma(x), \text{ for all } x \in X'.$$

Therefore, if Γ' is a special subhypergraph of Γ, then we have $X' \subset X$.

Lemma 11.1. *Let $(H, \circ_\rho)$ be the ρ-hypergroup induced by a hypergraph Γ and let H' be a subhypergroup of H. Then,*

(1) $H' = \bigcup_{x \in H'} \rho_\Gamma(x)$;

(2) H' *is the hypergroup induced by the hypergraph* $\Gamma' = (H', \{\rho_\Gamma(x)\}_{x \in H'})$;

(3) Γ' *is a special subhypergraph of Γ.*

Proof.   The assertions (2) and (3) follow from (1) and so we prove (1). Obviously we have $H' \subset \bigcup_{x \in H'} \rho_\Gamma(x)$. Let $y \in \bigcup_{x \in H'} \rho_\Gamma(x)$ be an arbitrary element. Then, there exists $x \in H'$ such that $y \in \rho_\Gamma(x)$. Since H' is a subhypergroup of H we have $x \circ_\rho H' = H'$ and so we have $x \circ_\rho x \subset H'$ which implies that $\rho_\Gamma(x) \subset H'$. Therefore, we have $y \in H'$.   □

Theorem 11.17. *Let H, H' be the ρ-hypergroups induced by hypergraphs Γ, Γ' respectively. Then, H' is a subhypergroup of H if and only if Γ' is a special subhypergraph of Γ.*

Proof.   Suppose that H' is a subhypergroup of H. Since H' is the ρ-hypergroup induced by Γ', for every $x \in H'$ we have $x \circ_\rho x = \rho_{\Gamma'}(x)$. On the other hand, since H' is a subhypergroup of H we have $x \circ_\rho x = \rho_\Gamma(x)$ and so for every $x \in H'$ we have $\rho_{\Gamma'}(x) = \rho_\Gamma(x)$, i.e., Γ' is a special subhypergraph of Γ. Conversely, let Γ' be a special subhypergraph of Γ and let $x \in H'$ be an arbitrary element. We show that $x \circ_\rho H' = H'$. For every $y \in H'$ we have

$$x \circ_\rho y = \rho_{\Gamma'}(x) \cup \rho_{\Gamma'}(y) = \rho_\Gamma(x) \cup \rho_\Gamma(y).$$

By Lemma 11.1 (1) we have $\rho_\Gamma(x) \cup \rho_\Gamma(y) \subset H'$ and so we have $x \circ_\rho H' \subset H'$. On the other hand, we have $y \in x \circ_\rho y$ which implies that $y \in x \circ_\rho H'$. Therefore, $H' \subset x \circ_\rho H'$.   □

Corollary 11.5. *Let H be the ρ-hypergroup induced by hypergraph Γ. Then, the number of subhypergroups of H is equal to $2^n - 1$ where $n = |H/\rho_\Gamma|$.*

Now, we relate the product of hypergraphs to the product of hypergroups. Let $\Gamma = (H, E)$ and $\Gamma' = (H', E')$ be two hypergraphs where $E = \{E_1, \ldots, E_m\}$ and $E' = \{E'_1, \ldots, E'_n\}$. We define their product to be the hypergraph $\Gamma \times \Gamma'$ whose vertices set is $H \times H'$ and whose hyperedges are the sets $E_i \times E'_j$ with $1 \leq i \leq m$, $1 \leq j \leq n$.

Lemma 11.2. *Let $\Gamma = (H, E)$ and $\Gamma' = (H', E')$ be two hypergraphs and $(x_1, y_1), (x_2, y_2) \in H \times H'$. Then,*

$$(x_1, y_1)\rho_{\Gamma \times \Gamma'}(x_2, y_2) \Leftrightarrow x_1 \rho_\Gamma x_2 \text{ and } y_1 \rho_{\Gamma'} y_2.$$

Proof. It is obvious. □

Let H_ρ and H'_ρ be ρ-hypergroups induced by two hypergraphs $\Gamma = (H, E)$ and $\Gamma' = (H', E')$ respectively. We define the hyperoperation $\otimes$ on the cartesian product $H_\rho \times H'_\rho$ as follows:

$$(x_1, y_1) \otimes (x_2, y_2) = \{(x, y) \mid x\rho_\Gamma x_1, \ y\rho_{\Gamma'} y_1 \quad \text{or} \quad x\rho_\Gamma x_2, \ y\rho_{\Gamma'} y_2\}.$$

The hypergroup $(H_\rho \times H'_\rho, \otimes)$ is called *special product* of H_ρ and H'_ρ.

Theorem 11.18. *Let H_ρ, H'_ρ be ρ-hypergroups induced by two hypergraphs $\Gamma = (H, E)$ and $\Gamma' = (H', E')$ respectively. Then, $H_\rho \times H'_\rho$ is the ρ-hypergroup induced by $\Gamma \times \Gamma'$.*

Proof. Suppose that $(H \times H')_\rho$ is the ρ-hypergroup induced by $\Gamma \times \Gamma'$. Then, we have

$$(x_1, y_1) \circ (x_2, y_2) = \rho_{\Gamma \times \Gamma'}(x_1, y_1) \cup \rho_{\Gamma \times \Gamma'}(x_2, y_2)$$
$$= \{(x, y) \mid (x, y)\rho_{\Gamma \times \Gamma'}(x_1, y_1) \quad \text{or} \quad (x, y)\rho_{\Gamma \times \Gamma'}(x_2, y_2)\},$$

for all $(x_1, y_1), (x_2, y_2) \in H \times H'$. Now, by Lemma 11.2, we have

$$(x_1, y_1) \circ (x_2, y_2) = (x_1, y_1) \otimes (x_2, y_2)$$

and so $(H \times H')_\rho = H_\rho \times H'_\rho$. □

Let $(H, \circ)$ be a hypergroup and R be an equivalence relation on H. If A and B are non-empty subsets of H, then $A\overline{R}B$ means that for all $a \in A$, there exists $b \in B$ such that aRb and for all $b' \in B$ there exists $a' \in A$ such that $a'Rb'$. We say that R is *regular* if for all $a \in H$ from xRy, it follows that $a \circ x\overline{R}a \circ y$ and $x \circ a\overline{R}y \circ a$. It is easy to verify that if $H_\rho = (H, \circ)$

be the ρ-hypergroup induced by hypergraph Γ, then special relation ρ_Γ is regular. Therefore, $H/\rho_\Gamma = \{\rho_\Gamma(x) \mid x \in H\}$ is a hypergroup with respect to the following hyperoperation:

$$\rho_\Gamma(x) \odot \rho_\Gamma(y) = \{\rho_\Gamma(z) \mid z \in x \circ y\}.$$

It is not difficult to show that

$$\rho_\Gamma(x) \odot \rho_\Gamma(y) = \{\rho_\Gamma(x), \rho_\Gamma(y)\}.$$

Clearly, H/ρ_Γ is the ρ-hypergroup induced by hypergraph $(\{\rho_\Gamma(x)\}_{x \in H}, \{E_x\}_{x \in H})$ where $E_x = \{\rho_\Gamma(x)\}$.

Theorem 11.19. *Let H_ρ, H'_ρ be two ρ-hypergroups induced by hypergraphs $\Gamma = (H, E)$ and $\Gamma' = (H', E')$ respectively, where $E = \{E_i\}_i$ and $E' = \{E'_j\}_j$. Then,*

$$H/\rho_\Gamma \otimes H'/\rho_{\Gamma'} \cong (H \otimes H')/\rho_{\Gamma \times \Gamma'}.$$

Proof. We define $\varphi : H/\rho_\Gamma \otimes H'/\rho_{\Gamma'} \to (H \otimes H')/\rho_{\Gamma \times \Gamma'}$ by setting

$$\varphi(\rho_\Gamma(x), \rho_{\Gamma'}(y)) = \rho_{\Gamma \times \Gamma'}(x, y),$$

for all $(x, y) \in H \times H'$. We prove firstly, that φ is well defined. Consider $(\rho_\Gamma(x_1), \rho_{\Gamma'}(y_1)) = (\rho_\Gamma(x_2), \rho_{\Gamma'}(y_2))$. Hence, we have $\rho_\Gamma(x_1) = \rho_\Gamma(x_2)$ and $\rho_{\Gamma'}(y_1) = \rho_{\Gamma'}(y_2)$. Since

$$(x_1, y_1) \in E_i \times F_j \Leftrightarrow x_1 \in E_i, y_1 \in F_j \Leftrightarrow x_2 \in E_i, y_2 \in F_j \Leftrightarrow (x_2, y_2) \in E_i \times F_j,$$

We obtain $\rho_{\Gamma \times \Gamma'}(x_1, y_1) = \rho_{\Gamma \times \Gamma'}(x_2, y_2)$, i.e., φ is well defined. Now we check that φ is one to one. Let $\rho_{\Gamma \times \Gamma'}(x_1, y_1) = \rho_{\Gamma \times \Gamma'}(x_2, y_2)$. We have

$$x_1 \in E_i, y_1 \in F_j \Leftrightarrow (x_1, y_1) \in E_i \times F_j \Leftrightarrow (x_2, y_2) \in E_i \times F_j \Leftrightarrow x_2 \in E_i, y_2 \in F_j.$$

This implies that $\rho_\Gamma(x_1) = \rho_\Gamma(x_2)$ and $\rho_{\Gamma'}(y_1) = \rho_{\Gamma'}(y_2)$, i.e., φ is injective. Clearly φ is onto. We need only to show that φ is a homomorphism.

$$\begin{aligned}
\varphi((\rho_\Gamma(x_1), \rho_{\Gamma'}(y_1)) &\otimes (\rho_\Gamma(x_2), \rho_{\Gamma'}(y_2))) \\
&= \varphi(\{(\rho_\Gamma(x_1), \rho_{\Gamma'}(y_1)), (\rho_\Gamma(x_2), \rho_{\Gamma'}(y_2))\}) \\
&= \{\rho_{\Gamma \times \Gamma'}(x_1, y_1), \rho_{\Gamma \times \Gamma'}(x_2, y_2)\} \\
&= \rho_{\Gamma \times \Gamma'}(x_1, y_1) \otimes \rho_{\Gamma \times \Gamma'}(x_2, y_2) \\
&= \varphi((\rho_\Gamma(x_1), \rho_{\Gamma'}(y_1))) \otimes \varphi((\rho_\Gamma(x_2), \rho_{\Gamma'}(y_2))).
\end{aligned}$$

Therefore, φ is an isomorphism. $\qquad\square$

Theorem 11.20. *Let Γ be a hypergraph and $H_\rho = (H, \circ_\rho)$ be the ρ-hypergroup induced by Γ. Then, we have $\beta^* = H^2$ where β^* is the fundamental relation on H_ρ.*

Proof. For every x, y in H we have $\{x, y\} \subset x \circ_\rho y$ and therefore we have $x \beta_2 y$ which implies that $x \beta y$. This means that $H^2 \subset \beta$. Since H_ρ is a hypergroup, we have $\beta^* = \beta$ and so $\beta^* = H^2$. $\qquad \square$

Theorem 11.21. *Let Γ be a hypergraph and $H_\rho = (H, \circ_\rho)$ be the ρ-hypergroup induced by Γ. Then, the complete part of H_ρ is equal to H.*

Proof. Suppose that A is a complete part of H_ρ and $b \in H$ is an arbitrary element. Since for every $a \in A$ we have $A \cap a \circ_\rho b \neq \emptyset$, we have $a \circ_\rho b \subset A$ and so $b \in A$. Therefore, $H \subset A$ which implies that $A = H$. $\qquad \square$

11.4 A special hyperoperation on hypergraphs

In [138], Nikkhah et al. constructed a hypergroupoid by defining a hyperoperation on the set of degrees of vertices of a hypergraph. They showed the constructed hypergroupoid is always an Hv-group. The main reference for this section is [138].

We define a hyperoperation $_n\circ_m$ for all $n, m \in \mathbb{N}$ on H as follows:

$$x \ _n\circ_m \ y = E^n(x) \cup E^m(y),$$

for all $(x, y) \in H^2$, where $E^0(x) = x$, $E(x) = \bigcup_{x \in E_i} E_i$ and $E^n(x) = E^{n-1}(E(x)) = \underbrace{E(E(...(E(x))))}_{n \text{ times}}$. It is clear that for $n \geq m$, $x \ _n\circ_m \ x = E^n(x) \cup E^m(x) = E^n(x)$.

The hypergroupoid $H_\Gamma = (H, _n\circ_m)$ is called a *hypergraph hypergroupoid* or an *h.g. hypergroupoid*.

In the following, we consider three cases:

Case 1: $n \geq d$ or $m \geq d$.

Case 2: $n = m$ $(n < d$ and $m < d)$.

Case 3: $n \neq m$ $(n < d$ and $m < d)$.

For the case 1, we have the following:

Lemma 11.3. *Let Γ be a connected hypergraph. If $n \geqslant d$, then for every $x \in H$, $E^n(x) = H$.*

Proof. By induction the result follows. So, we prove the claim for $n = d$, i.e., we show that $E^d(x) = H$.

By contradiction, suppose that $E^d(x) \neq H$. We know that $|E^d(x)| \geqslant d$. Since $E^d(x) \neq H$, it follows that there exists a vertex in Γ, say x_d, which is not in $E^d(x)$. Obviously, there is a vertex x_i in $E^d(x)$ such that $dist(x_i, x_d) > d$ and this is a contradiction. $\qquad\square$

Remark 11.1. If $n \geq d$ or $m \geq d$, then $x \ _n\circ_m y = H$, for all $(x, y) \in H^2$.

For the case 2, we have (Theorems 11.22, 11.23, 11.24, 11.25, Corollary 11.6, Theorem 11.26).

Theorem 11.22. *For each $(x, y) \in H^2$ and $n \in \mathbb{N}$, the hypergroupoid $H_\Gamma = (H, \ _n\circ_n)$ satisfies the following conditions:*

(1) $x \ _n\circ_n y = x \ _n\circ_n x \cup y \ _n\circ_n y$;

(2) $x \in x \ _n\circ_n x$;

(3) $y \in x \ _n\circ_n x \Leftrightarrow x \in y \ _n\circ_n y$.

Proof. We only prove (3). The proof of other conditions is straightforward Let $y \in x \ _n\circ_n x = E^n(x)$. Then $E(y) \cap E^{n-1}(x) \neq \emptyset$ and so $E^2(y) \cap E^{n-2}(x) \neq \emptyset$, ..., $E^{n-1}(y) \cap E(x) \neq \emptyset$. Hence $E^n(y) \cap x \neq \emptyset$. Therefore, $x \in E^n(y)$. $\qquad\square$

Theorem 11.23. *If the hypergroupoid $H_\Gamma = (H, \ _n\circ_n)$ satisfied the conditions (1), (2), (3) of the Theorem 11.22, then also satisfies the conditions:*

(4) $x \ _n\circ_n y \supset \{x, y\}$,

(5) $x \ _n\circ_n y = y \ _n\circ_n x$,

(6) $H \ _n\circ_n x = x \ _n\circ_n H = H$,

(7) $< H; \{x \ _n\circ_n x\}_{x \in H} >$ is a hypergraph,

(8) $(x \ _n\circ_n x) \ _n\circ_n x = \bigcup\limits_{x \in z \ _n\circ_n z} z \ _n\circ_n z$,

(9) $(x \ _n\circ_n x) \ _n\circ_n (x \ _n\circ_n x) = x \ _n\circ_n x \ _n\circ_n x$.

Proof. It is straightforward. $\qquad\square$

Theorem 11.24. *The hypergroupoid $H_\Gamma = (H, \ _n\circ_m)$ is commutative, if one of the following conditions is satisfied:*

(1) $n = m$;

(2) $n \geq d$ or $m \geq d$.

Proof. (1) It is clear.

(2) It is straightforward, by Lemma 11.3. $\qquad\square$

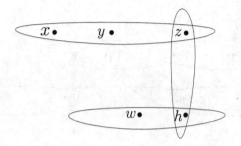

Fig. 11.2   The hypergraph defined in Example 11.6

Example 11.4. Diameter of the hypergraph presented in Figure 11.4 is 3 $(d = 3)$. Suppose that $n = 2$ and $m = 1$. Then, we have

${}_2\circ_1$	x	y	z	w	h
x	$\{x, y, z, h\}$	$\{x, y, z, h\}$	$\{x, y, z, h\}$	H	H
y	$\{x, y, z, h\}$	$\{x, y, z, h\}$	$\{x, y, z, h\}$	H	H
z	H	H	H	H	H
w	H	H	H	$\{w, h, z\}$	$\{w, h, z\}$
h	H	H	H	H	H

It is clear that $(H, {}_2\circ_1)$ is not commutative.

Remark 11.2. By using the conditions (5) and (6) of Theorem 11.23, it is clear that $H_\Gamma = (H, {}_n\circ_n)$ is a commutative quasihypergroup.

Theorem 11.25. *The hypergroupoid $(H, {}_n\circ_n)$ satisfying the conditions (1), (2) and (3) of Theorem 11.22 is a hypergroup if and only if the following condition is valid:*

$$\forall (a, c) \in H^2 \text{ and } n < \lceil \tfrac{d}{2} \rceil, \quad c \,{}_n\circ_n\, c \,{}_n\circ_n\, c - c \,{}_n\circ_n\, c \subset a \,{}_n\circ_n\, a \,{}_n\circ_n\, a.$$

Proof.   It is straightforward.   □

Corollary 11.6. *If the hypergroupoid satisfies the conditions (1), (2) and (3) of Theorem 11.22 and the condition:*

$$n < \lceil \tfrac{d}{2} \rceil, \quad E^{2n}(x) = E^n(x), \text{ for all } x \in H,$$

then it is a hypergroup.

Example 11.5. Suppose that $H = \{v_1, v_2, v_3, v_4, v_5, v_6, v_7, v_8\}$, $E = \{E_1, E_2, E_3, E_4, E_5, E_6, E_7\}$ and $d = 5$, where $E_1 = \{v_1, v_2, v_3\}$, $E_2 =$

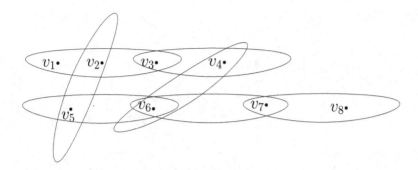

Fig. 11.3   The hypergraph defineded in Example 11.7

$\{v_3, v_4\}$, $E_3 = \{v_2, v_5\}$, $E_4 = \{v_4, v_6\}$, $E_5 = \{v_5, v_6\}$, $E_6 = \{v_6, v_7\}$ and $E_7 = \{v_7, v_8\}$, see Figure 11.3. Then, we have

$_2\circ_2$	v_1	v_2	v_3	v_4	v_5	v_6	v_7	v_8
v_1	$\{v_1,...,v_5\}$	$\{v_1,...,v_6\}$	$\{v_1,...,v_6\}$	$\{v_1,...,v_7\}$	$\{v_1,...,v_7\}$	H	H	H
v_2	$\{v_1,...,v_6\}$	$\{v_1,...,v_6\}$	$\{v_1,...,v_6\}$	$\{v_1,...,v_7\}$	$\{v_1,...,v_7\}$	H	H	H
v_3	$\{v_1,...,v_6\}$	$\{v_1,...,v_6\}$	$\{v_1,...,v_6\}$	$\{v_1,...,v_7\}$	$\{v_1,...,v_7\}$	H	H	H
v_4	$\{v_1,...,v_7\}$	$\{v_1,...,v_7\}$	$\{v_1,...,v_7\}$	$\{v_1,...,v_7\}$	$\{v_1,...,v_7\}$	H	H	H
v_5	$\{v_1,...,v_7\}$	$\{v_1,...,v_7\}$	$\{v_1,...,v_7\}$	$\{v_1,...,v_7\}$	$\{v_1,...,v_7\}$	H	H	H
v_6	H	H	H	H	H	$\{v_2,...,v_8\}$	$\{v_2,...,v_8\}$	$\{v_2,...,v_8\}$
v_7	H	H	H	H	H	$\{v_2,...,v_8\}$	$\{v_4,...,v_8\}$	$\{v_4,...,v_8\}$
v_8	H	H	H	H	H	$\{v_2,...,v_8\}$	$\{v_4,...,v_8\}$	$\{v_6,v_7,v_8\}$

It is clear that $(H, _2\circ_2)$ is a commutative hypergroup.

Theorem 11.26. *If the hypergroup $H_\Gamma = (H, _n\circ_n)$ satisfies the conditions (1), (2) and (3) of Theorem 11.22, then it is a join space.*

Proof. It is sufficient to prove that the following implication:

$$x/y \cap z/w \neq \emptyset \Rightarrow x \,_n\circ_n w \cap y \,_n\circ_n z \neq \emptyset,$$

where $x/y = \{z \mid x \in z \,_n\circ_n y\}$. We have:

$$u \in x/y \cap z/w \Leftrightarrow [x \in u \,_n\circ_n y \text{ and } z \in u \,_n\circ_n w].$$

Moreover, $x \in u \,_n\circ_n y \Leftrightarrow x \in u \,_n\circ_n u \cup y \,_n\circ_n y$ and $z \in u \,_n\circ_n w \Leftrightarrow z \in u \,_n\circ_n u \cup w \,_n\circ_n w$. The following four cases are possible:

(1) If $x \in u \,_n\circ_n u$, $z \in u \,_n\circ_n u$, then $u \in x \,_n\circ_n x \cap z \,_n\circ_n z$ and so $u \in x \,_n\circ_n w \cap y \,_n\circ_n z$.

(2) If $x \in u \,_n\circ_n u$, $z \in w \,_n\circ_n w$, then $w \in z \,_n\circ_n z$. Hence, $w \in x \,_n\circ_n w \cap y \,_n\circ_n z$.

(3) If $x \in y \; {}_n{\circ}_n \; y$, $z \in u \; {}_n{\circ}_n \; u$, then $y \in x \; {}_n{\circ}_n \; x$. This implies that $y \in x \; {}_n{\circ}_n \; w \cap y \; {}_n{\circ}_n \; z$.

(4) If $x \in y \; {}_n{\circ}_n \; y$, $z \in w \; {}_n{\circ}_n \; w$, then $w \in z \; {}_n{\circ}_n \; z$. This implies that $w \in x \; {}_n{\circ}_n \; w \cap y \; {}_n{\circ}_n \; z$. $\qquad\square$

Remark 11.3. If $n = m$ ($n < d$ and $m < d$), then $x \; {}_n{\circ}_m \; y = E^n(x) \cup E^n(y)$, for all $(x, y) \in H^2$.

For the case 3, we have (Theorems 11.27, 11.28, Lemma 11.4, Theorems 11.29):

Theorem 11.27. *For each $(x, y) \in H^2$, $n, m \in \mathbb{N}$ ($n \neq m, n < d$ and $m < d$), the hypergroupoid H_Γ satisfies the following conditions:*

(1) $x \; {}_n{\circ}_m \; y \subset x \; {}_n{\circ}_m \; x \cup y \; {}_n{\circ}_m \; y$. *In the other words:*
$$x \; {}_n{\circ}_m \; y = \begin{cases} x \; {}_n{\circ}_m \; x \cup E^m(y) & \text{if } n > m, \\ E^n(x) \cup y \; {}_n{\circ}_m \; y & \text{if } n < m; \end{cases}$$

(2) $x \in x \; {}_n{\circ}_m \; x$;

(3) $y \in x \; {}_n{\circ}_m \; x \Leftrightarrow x \in y \; {}_n{\circ}_m \; y$.

Proof. We only prove (3). The proof of other items is straightforward. Let $n > m$ and $y \in x \; {}_n{\circ}_m \; x = E^n(x)$. Then $E(y) \cap E^{n-1}(x) \neq \emptyset$ and so $E^2(y) \cap E^{n-2}(x) \neq \emptyset$, ..., $E^{n-1}(y) \cap E(x) \neq \emptyset$. Hence $E^n(y) \cap x \neq \emptyset$. Therefore, $x \in E^n(y)$. $\qquad\square$

Theorem 11.28. *If a hypergroupoid H_Γ satisfied the conditions (1), (2), (3) of Theorem 11.27, then also satisfies the following conditions:*

(4) $x \; {}_n{\circ}_m \; y \supset \{x, y\}$;

(5) $x \; {}_n{\circ}_m \; y \cap y \; {}_n{\circ}_m \; x \neq \emptyset$;

(6) $H \; {}_n{\circ}_m \; x = x \; {}_n{\circ}_m \; H = H$;

(7) $< H; \{x \; {}_n{\circ}_m \; x\}_{x \in H} >$ *is a hypergraph;*

(8) $(x \; {}_n{\circ}_m \; x) \; {}_n{\circ}_m \; x \subset \bigcup_{x \in z \, {}_n{\circ}_m z} z \; {}_n{\circ}_m \; z$.

Proof. It is enough to prove (8). We have $(x \; {}_n{\circ}_m \; x) \; {}_n{\circ}_m \; x = \bigcup_{z \in x \, {}_n{\circ}_m x} z \; {}_n{\circ}_m \; x$. By (1), we obtain
$$x \; {}_n{\circ}_m \; x \; {}_n{\circ}_m \; x \subset \bigcup_{z \in x \, {}_n{\circ}_m x} z \; {}_n{\circ}_m \; z \cup x \; {}_n{\circ}_m \; x.$$

Now, from (2) we have
$$x \; {}_n{\circ}_m \; x \; {}_n{\circ}_m \; x \subset \bigcup_{z \in x \, {}_n{\circ}_m x} z \; {}_n{\circ}_m \; z,$$

and finally from (3) we obtain (8). $\qquad\square$

Remark 11.4. From the condition (5) of Theorem 11.28, it is clear that hyperoperation $_n\circ_m$ is weak commutative and weak associative. In the other words:

$$\forall\, x, y, z \in H, \quad x \,_n\circ_m (y \,_n\circ_m z) \cap (x \,_n\circ_m y) \,_n\circ_m z \neq \emptyset.$$

Corollary 11.7. *From the condition (6) of Theorem 11.28 and Remark 11.4, it is clear that an h.g. hypergroupoid is an H_v-group.*

Lemma 11.4. *Let $(H, {}_n\circ_m)$ is a hypergroupoid. Then:*

$$y \in H, \quad E^n(E^m(x) \cup E^p(y)) = E^{n+m}(x) \cup E^{n+p}(y), \text{ for all } x, y \in H.$$

Proof. Let $a \in E^n(E^m(x) \cup E^p(y))$. Then $a \in E^m(x)$ or $a \in E^p(y)$ or $a \in E^{n+m}(x)$ or $a \in E^{n+p}(y)$. But $E^m(x) \subset E^{n+m}(x)$ and $E^p(y) \subset E^{n+p}(y)$. So $a \in E^{n+m}(x)$ or $a \in E^{n+p}(y)$. Hence $E^n(E^m(x) \cup E^p(y)) \subset E^{n+m}(x) \cup E^{n+p}(y)$.

The converse is straightforward. □

Theorem 11.29. *The hypergroupoid $(H, {}_n\circ_m)$ is a hypergroup if and only if for every $a, b, c \in H$, $n < \lceil \dfrac{d}{2} \rceil$ and $m < \lceil \dfrac{d}{2} \rceil$ one of the following conditions is valid:*

(1) *If $E^{2n}(a) \cup E^m(c) \subset E^n(a) \cup E^{2m}(c)$, then $(E^n(a) \cup E^{2m}(c)) - (E^{2n}(a) \cup E^m(c)) \subset E^{n+m}(b)$;*

(2) *If $E^n(a) \cup E^{2m}(c) \subset E^{2n}(a) \cup E^m(c)$, then $(E^{2n}(a) \cup E^m(c)) - (E^n(a) \cup E^{2m}(c)) \subset E^{n+m}(b)$;*

(3) *If $E^{2n}(a) \cup E^m(c) \not\subset E^n(a) \cup E^{2m}(c)$ and $E^n(a) \cup E^{2m}(c) \not\subset E^{2n}(a) \cup E^m(c)$, then $((E^n(a) \cup E^{2m}(c)) - (E^{2n}(a) \cup E^m(c))) \cup ((E^{2n}(a) \cup E^m(c)) - (E^n(a) \cup E^{2m}(c))) \subset E^{n+m}(b)$.*

Proof. (1) Let $(H, {}_n\circ_m)$ is a hypergroup and suppose $a, b, c \in H$. Then $a \,_n\circ_m (b \,_n\circ_m c) = (a \,_n\circ_m b) \,_n\circ_m c$. So $E^n(a) \cup E^{n+m}(b) \cup E^{2m}(c) = E^{2n}(a) \cup E^{n+m}(b) \cup E^m(c)$ by Lemma 11.4. Since $E^{2n}(a) \cup E^m(c) \subset E^n(a) \cup E^{2m}(c)$, thus $(E^n(a) \cup E^{2m}(c)) - (E^{2n}(a) \cup E^m(c)) \subset E^{n+m}(b)$.

The converse is a routine verification.

(2) It follows directly from (1).

(3) Let $(H, {}_n\circ_m)$ is a hypergroup and suppose $a, b, c \in H$. Then $a \,_n\circ_m (b \,_n\circ_m c) = (a \,_n\circ_m b) \,_n\circ_m c$. So $E^n(a) \cup E^{n+m}(b) \cup E^{2m}(c) = E^{2n}(a) \cup E^{n+m}(b) \cup E^m(c)$ by Lemma 11.4. Since $E^{2n}(a) \cup E^m(c) \not\subset E^n(a) \cup E^{2m}(c)$ and $E^n(a) \cup E^{2m}(c) \not\subset E^{2n}(a) \cup E^m(c)$, thus $((E^n(a) \cup E^{2m}(c)) - (E^{2n}(a) \cup E^m(c))) \cup ((E^{2n}(a) \cup E^m(c)) - (E^n(a) \cup E^{2m}(c))) \subset E^{n+m}(b)$.

The converse is a routine verification. □

Corollary 11.8. *If the hypergroupoid $(H, {}_n\circ_m)$ satisfies the conditions:*

$$\forall a, \begin{cases} E^{2n}(a) = E^n(a) & \text{if } n < \lceil \frac{d}{2} \rceil \\ E^{2m}(a) = E^m(a) & \text{if } m < \lceil \frac{d}{2} \rceil \end{cases}$$

then it is a hypergroup.

Remark 11.5. If $n \neq m$ ($n < d$ and $m < d$), then $x \; {}_n\circ_m \; y = E^n(x) \cup E^m(y)$, for all $(x, y) \in H^2$.

11.5 Rough hypergroups of hypergraphs

In this section, by using the concepts of lower and upper approximations, we define two new hyperoperations on the set of vertices of a given hypergraph. Some examples are presented and some properties of these hyperoperations are discussed. The results of this section are contained in [116].

Definition 11.7. Let $\Gamma = (H, \{E_i\}_i)$ be a hypergraph, $E(x) = \bigcup\limits_{x \in E_i} E_i$ and ρ be an equivalence relation on H. By using the equivalence relation ρ and the notions of lower and upper approximations, we define two hyperoperations on H as follows:

$$x \underline{\circ} y = \underline{app}(E(x)) \cup \underline{app}(E(y)) \cup \{x, y\}$$

and

$$x \overline{\circ} y = \overline{app}(E(x)) \cup \overline{app}(E(y)),$$

for all $x, y \in H$. The hypergroupoid $\mathbb{H}_\rho = (H, \underline{\circ})$ is called the *lower hypergraph hypergroupoid* associated to ρ and the hypergroupoid $\mathbb{H}^\rho = (H, \overline{\circ})$ is called the *upper hypergraph hypergroupoid* associated to ρ

Remark 11.6. Note that $\overline{app}(E(x))$ is a non-empty set, for all $x \in H$. But it is possible to have the case $\underline{app}(E(x)) = \emptyset$, for some $x \in H$.

Example 11.6. Suppose that $H = \{v_1, v_2, v_3, v_4, v_5, v_6\}$ and $E = \{E_1, E_2, E_3, E_4\}$, where $E_1 = \{v_1, v_2, v_3\}$, $E_2 = \{v_2, v_3, v_4\}$, $E_3 = \{v_5, v_6\}$ and $E_4 = \{v_4, v_5\}$, see Figure 11.4. Then, we have:

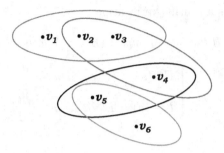

Fig. 11.4   The hypergraph defined in Example 11.6

$$E(v_1) = E_1 = \{v_1, v_2, v_3\},$$
$$E(v_2) = E_1 \cup E_2 = \{v_1, v_2, v_3, v_4\},$$
$$E(v_3) = E_1 \cup E_2 = \{v_1, v_2, v_3, v_4\},$$
$$E(v_4) = E_2 \cup E_4 = \{v_2, v_3, v_4, v_5\},$$
$$E(v_5) = E_3 \cup E_4 = \{v_4, v_5, v_6\},$$
$$E(v_6) = E_3 = \{v_5, v_6\}.$$

Let ρ be an equivalence relation on H that forms a partition on H as follows:

$$\{\{v_1, v_2\},\ \{v_3, v_5\},\ \{v_4\},\ \{v_6\}\}.$$

Then, we have

$$\underline{app}(E(v_1)) = \{v_1, v_2\},$$
$$\underline{app}(E(v_2)) = \{v_1, v_2, v_4\},$$
$$\underline{app}(E(v_3)) = \{v_1, v_2, v_4\},$$
$$\underline{app}(E(v_4)) = \{v_3, v_4, v_5\},$$
$$\underline{app}(E(v_5)) = \{v_4, v_6\},$$
$$\underline{app}(E(v_6)) = \{v_6\}.$$

Therefore, the lower hypergroupoid $\mathbb{H}_\rho = (H, \underline{\circ})$ associated to ρ is as follows:

$\underline{\circ}$	v_1	v_2	v_3	v_4	v_5	v_6
v_1	$\{v_1, v_2\}$	$\{v_1, v_2, v_4\}$	$H \setminus \{v_5, v_6\}$	$H \setminus \{v_6\}$	$H \setminus \{v_3\}$	$\{v_1, v_2, v_6\}$
v_2	$\{v_1, v_2, v_4\}$	$\{v_1, v_2\}$	$H \setminus \{v_5, v_6\}$	$H \setminus \{v_6\}$	$H \setminus \{v_3\}$	$H \setminus \{v_3, v_5\}$
v_3	$H \setminus \{v_5, v_6\}$	$H \setminus \{v_5, v_6\}$	$H \setminus \{v_5, v_6\}$	$H \setminus \{v_6\}$	H	$H \setminus \{v_5\}$
v_4	$H \setminus \{v_6\}$	$H \setminus \{v_6\}$	$H \setminus \{v_6\}$	$\{v_3, v_4, v_5\}$	$H \setminus \{v_1, v_2\}$	$H \setminus \{v_1, v_2\}$
v_5	$H \setminus \{v_3\}$	$H \setminus \{v_3\}$	H	$H \setminus \{v_1, v_2\}$	$\{v_4, v_5, v_6\}$	$\{v_4, v_5, v_6\}$
v_6	$\{v_1, v_2, v_6\}$	$H \setminus \{v_3, v_5\}$	$H \setminus \{v_5\}$	$H \setminus \{v_1, v_2\}$	$\{v_4, v_5, v_6\}$	$\{v_6\}$

Similarly, we obtain

$$\overline{app}(E(v_1)) = \{v_1, v_2, v_3, v_5\},$$

$$\overline{app}(E(v_2)) = \{v_1, v_2, v_3, v_4, v_5\},$$

$$\overline{app}(E(v_3)) = \{v_1, v_2, v_3, v_4, v_5\},$$

$$\overline{app}(E(v_4)) = \{v_1, v_2, v_3, v_4, v_5\},$$

$$\overline{app}(E(v_5)) = \{v_3, v_4, v_5, v_6\},$$

$$\overline{app}(E(v_6)) = \{v_3, v_5, v_6\}.$$

Therefore, the upper hypergroupoid $\mathbb{H}^\rho = (H, \overline{\circ})$ associated to ρ is as follows:

$\overline{\circ}$	v_1	v_2	v_3	v_4	v_5	v_6
v_1	$H \setminus \{v_4, v_6\}$	$H \setminus \{v_6\}$	$H \setminus \{v_6\}$	$H \setminus \{v_6\}$	H	$H \setminus \{v_4\}$
v_2	$H \setminus \{v_6\}$	$H \setminus \{v_6\}$	$H \setminus \{v_6\}$	$H \setminus \{v_6\}$	H	H
v_3	$H \setminus \{v_6\}$	$H \setminus \{v_6\}$	$H \setminus \{v_6\}$	$H \setminus \{v_6\}$	H	H
v_4	$H \setminus \{v_6\}$	$H \setminus \{v_6\}$	$H \setminus \{v_6\}$	$H \setminus \{v_6\}$	H	H
v_5	H	H	H	H	$H \setminus \{v_1, v_2\}$	$H \setminus \{v_1, v_2\}$
v_6	$H \setminus \{v_4\}$	H	H	H	$H \setminus \{v_1, v_2\}$	$H \setminus \{v_1, v_2, v_4\}$

Let $(H, \circ)$ and $(H, \star)$ be two hypergroupoids defined on the same set H. Recall that the hyperoperation $\circ$ is less than $\star$, or $\star$ greater than $\circ$ if there exists $f \in Aut(H)$ such that $x \circ y \subset f(x \star y)$, for all $x, y \in H$. In this case, we write $\circ \leqslant \star$ and we say that $(H, \star)$ contains $(H, \circ)$.

Corollary 11.9. *Let $\Gamma = (H, \{E_i\}_i)$ be a hypergraph and ρ be an equivalence relation on H. If $\circ$ is the hyperoperation defined by Corsini, i.e., $x \circ y = E(x) \cup E(y)$, then*

$$\underline{\circ} \leqslant \circ \leqslant \overline{\circ}.$$

Proof. It follows by Theorem 10.1(1). □

Definition 11.8. We say an equivalence relation ρ on H is $\underline{\rho}$-good if

$$x \in \underline{app}E(y) \iff y \in \underline{app}E(x)$$

and we say ρ is $\overline{\rho}$-good if

$$x \in \overline{app}E(y) \iff y \in \overline{app}E(x).$$

Theorem 11.30. *Let $\Gamma = (H, \{E_i\}_i)$ be a hypergraph and ρ be a $\underline{\rho}$-good equivalence relation on H. Then, $\mathbb{H}_\rho = (H, \underline{\circ})$ satisfies the following conditions:*

(1) $x \underline{\circ} y = x \underline{\circ} x \cup y \underline{\circ} y$,

(2) $x \in x \underline{\circ} x$,

(3) $x \in y \underline{\circ} y \Leftrightarrow y \in x \underline{\circ} x$.

Proof. (1) We have

$$x \underline{\circ} x = \underline{app}(E(x)) \cup \underline{app}(E(x)) \cup \{x\} = \underline{app}(E(x)) \cup \{x\},$$
$$y \underline{\circ} y = \underline{app}(E(y)) \cup \underline{app}(E(y)) \cup \{y\} = \underline{app}(E(y)) \cup \{y\}.$$

Thus, we have

$$x \underline{\circ} y = \underline{app}(E(x)) \cup \underline{app}(E(y)) \cup \{x, y\} = x \underline{\circ} x \cup y \underline{\circ} y.$$

The proofs of (2) and (3) are straightforward. □

Corollary 11.10. $\mathbb{H}_\rho = (H, \underline{\circ})$ *is a commutative H_v-group.*

Proof. Since $\{x, y, z\} \subset (x \underline{\circ} y) \underline{\circ} z \cap x \underline{\circ} (y \underline{\circ} z)$ for all $x, y, z \in H$, it follows that $(x \underline{\circ} y) \underline{\circ} z \cap x \underline{\circ} (y \underline{\circ} z) \neq \emptyset$. Moreover, obviously, we have $x \underline{\circ} H \subset H$, for all $x \in H$. So, we show that $H \subset x \underline{\circ} H$. Suppose that $y \in H$ is an arbitrary element. By Theorem 12.3 (2), we have $y \in y \underline{\circ} y$ and so $y \in x \underline{\circ} x \cup y \underline{\circ} y$. This implies that $y \in x \underline{\circ} y$. Therefore, $y \in x \underline{\circ} H$. □

Corollary 11.11. *If the H_v-group $\mathbb{H}_\rho = (H, \underline{\circ})$ satisfies the following condition*

$$x \underline{\circ} x \underline{\circ} x = x \underline{\circ} x, \text{ for all } x \in H,$$

then $\mathbb{H}_\rho = (H, \underline{\circ})$ is a hypergroup.

Proof. It is straightforward. □

The hypergroup $\mathbb{H}_\rho$ is called the *lower $\underline{\rho}$-hypergroup induced by Γ.* Moreover, $\mathbb{H}_\rho$ is a join space.

Example 11.7. Suppose that $H = \{v_1, v_2, v_3, v_4\}$ and $E = \{E_1, E_2\}$, where $E_1 = \{v_1, v_2, v_3\}$ and $E_2 = \{v_1, v_2, v_4\}$, see Figure 11.5.

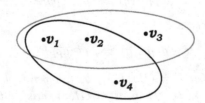

Fig. 11.5   The hypergraph defined in Example 11.7

Then, we have

$$E(v_1) = E_1 \cup E_2 = H,$$
$$E(v_2) = E_1 \cup E_2 = H,$$
$$E(v_3) = E_1 = \{v_1, v_2, v_3\}$$
$$E(v_4) = E_2 = \{v_1, v_2, v_4\}.$$

Let ρ be the following relation on H,

$$\rho = \{(v_1, v_1), (v_2, v_2), (v_3, v_3), (v_4, v_4), (v_1, v_2), (v_2, v_1), (v_3, v_4), (v_4, v_3)\}.$$

Clearly, ρ is an equivalence relation. Moreover, we have

$[v_1]_\rho \subset E(v_1)$	$[v_2]_\rho \subset E(v_1)$	$[v_3]_\rho \subset E(v_1)$	$[v_4]_\rho \subset E(v_1)$
$[v_1]_\rho \subset E(v_2)$	$[v_2]_\rho \subset E(v_2)$	$[v_3]_\rho \subset E(v_2)$	$[v_4]_\rho \subset E(v_2)$
$[v_1]_\rho \subset E(v_3)$	$[v_2]_\rho \subset E(v_3)$	$[v_3]_\rho \not\subset E(v_3)$	$[v_4]_\rho \not\subset E(v_3)$
$[v_1]_\rho \subset E(v_4)$	$[v_2]_\rho \subset E(v_4)$	$[v_3]_\rho \not\subset E(v_4)$	$[v_4]_\rho \not\subset E(v_4)$

This means that $x \in \underline{app}(E(y))$ if and only if $y \in \underline{app}(E(x))$. So, ρ is a ρ-good equivalence relation. Now, we obtain

$$\underline{app}(E(v_1)) = H,$$
$$\underline{app}(E(v_2)) = H,$$
$$\underline{app}(E(v_3)) = \{v_1, v_2\},$$
$$\underline{app}(E(v_4)) = \{v_1, v_2\}.$$

Therefore, the upper ρ-hypergroup $\mathbb{H}_\rho = (H, \underline{\circ})$ induced by Γ is as follows:

$\underline{\circ}$	v_1	v_2	v_3	v_4
v_1	H	H	H	H
v_2	H	H	H	H
v_3	H	H	$\{v_1, v_2\}$	$\{v_1, v_2\}$
v_4	H	H	$\{v_1, v_2\}$	$\{v_1, v_2\}$

Remark 11.7. Note that in Example 11.7, since $x \in v_1 \circ x \cap x \circ v_1$ and $x \in v_2 \circ x \cap x \circ v_2$ for all $x \in H$, it follows that both v_1 and v_2 are identity elements $\mathbb{H}^\rho$.

We recall the special relation ρ on H as follows:

$$x \,\rho\, y \Leftrightarrow \{E_i \mid x \in E_i\} = \{E_j \mid y \in E_j\}.$$

Clearly, ρ is an equivalence relation. Moreover, we have the following two lemmas regarding to this special relation.

Lemma 11.5.

(1) ρ is a $\underline{\rho}$-good equivalence relation.

(2) ρ is a $\overline{\rho}$-good equivalence relation.

Proof. (1) Suppose that $x \in \underline{app}(E(y))$. This implies that $x \in E(y)$. We show that $[y]_\rho \subset E(x)$. Let $z \in [y]_\rho$ be an arbitrary element. Then, we have $z \, \rho \, y$ and so $E(z) = E(y)$. Thus, $x \in E(z)$. This implies that $z \in E(x)$. Therefore, we conclude that $[y]_\rho \subset E(x)$ and so $y \in \underline{app}(E(x))$. The proof of converse is similar.

(2) Suppose that $x \in \overline{app}(E(y))$. This implies that $[x]_\rho \cap E(y) \neq \emptyset$. So, there exists $z \in [x]_\rho \cap E(y)$. Since $z \in E(y)$, it follows that $y \in E(z)$. Since $z \in [x]_\rho$, it follows that $z \, \rho \, x$. Hence, by definition we have $\{E_i \mid x \in E_i\} = \{E_j \mid y \in E_j\}$, and so $E(x) = E(z)$. Thus, we obtain $y \in E(x)$. On the other hand, we have $y \in [y]_\rho$. Therefore, $[y]_\rho \cap E(x) \neq \emptyset$. This implies that $y \in \overline{app}(E(x))$. The proof of the converse is similar. $\square$

Lemma 11.6. *For all $a \in H$, $E(x)$ is a definable set.*

Proof. By Theorem 10.1(1), it is enough to show that $\overline{app}(E(x)) \subset \underline{app}(E(x))$. Suppose that $y \in \overline{app}(E(x))$ is an arbitrary element. Then, $[y]_\rho \cap E(x) \neq \emptyset$. Hence, there exists $a \in [y]_\rho$ and $a \in E(x)$. By the definition of ρ we obtain $E(a) = E(y)$. Since $a \in E(x)$, it follows that $x \in E(a)$.

Now, we show that $[y]_\rho \subset E(x)$. Let $b \in [y]_\rho$ be an arbitrary element. Then, $E(b) = E(y)$, and so we obtain $x \in E(b)$. This implies that $b \in E(x)$. Therefore, $y \in \underline{app}(E(x))$. $\square$

Theorem 11.31. *Let $\Gamma = (H, \{E_i\}_i)$ be a hypergraph and ρ be a $\overline{\rho}$-good equivalence relation on H. Then, $(H, \overline{\circ})$ satisfies the following conditions:*

(1) $x \overline{\circ} y = x \overline{\circ} x \cup y \overline{\circ} y$,

(2) $x \in x \overline{\circ} x$,

(3) $x \in y \overline{\circ} y \iff y \in x \overline{\circ} x$.

Proof. (1) We have
$$x \overline{\circ} x = \overline{app}(E(x)) \cup \overline{app}(E(x)) = \overline{app}(E(x)),$$
$$y \overline{\circ} y = \overline{app}(E(y)) \cup \overline{app}(E(y)) = \overline{app}(E(y)).$$
Thus, we have
$$x \overline{\circ} y = \overline{app}(E(x)) \cup \overline{app}(E(y)) = x \overline{\circ} x \cup y \overline{\circ} y.$$

(2) Since $x \in E(x)$, it follows that $x \in \overline{app}(E(x))$. Thus, we obtain $x \in x \overline{\circ} x$.

(3) It is clear. $\square$

Corollary 11.12. $\mathbb{H}^\rho = (H, \bar{\circ})$ *is a commutative H_v-group.*

Proof. Since $\{x, y, z\} \subset (x\bar{\circ}y)\bar{\circ}z \cap x\bar{\circ}(y\bar{\circ}z)$ for all $x, y, z \in H$, it follows that $(x\bar{\circ}y)\bar{\circ}z \cap x\bar{\circ}(y\bar{\circ}z) \neq \emptyset$. Moreover, obviously, we have $x\bar{\circ}H \subset H$, for all $x \in H$. So, we show that $H \subset x\bar{\circ}H$. Suppose that $y \in H$ is an arbitrary element. By Theorem 12.4(2), we have $y \in y\bar{\circ}y$ and so $y \in x\bar{\circ}x \cup y\bar{\circ}y$. This implies that $y \in x\bar{\circ}y$. Therefore, $y \in x\bar{\circ}H$. $\qquad\square$

Corollary 11.13. *If the H_v-group $\mathbb{H}^\rho = (H, \bar{\circ})$ satisfies the following condition*

$$x\bar{\circ}x\bar{\circ}x = x\bar{\circ}x, \text{ for all } x \in H,$$

then $\mathbb{H}_\rho = (H, \bar{\circ})$ is a hypergroup.

Proof. The proof is clear. $\qquad\square$

The hypergroup $\mathbb{H}^\rho$ is called the *upper $\bar{p}$-hypergroup induced by Γ*. Moreover, it is easy to see that $\mathbb{H}^\rho$ is a join space.

Example 11.8. Suppose that $H = \{v_1, v_2, v_3, v_4, v_5, v_6\}$ and $E = \{E_1, E_2, E_3\}$, where $E_1 = \{v_1, v_2, v_3\}$, $E_2 = \{v_4, v_5\}$ and $E_3 = \{v_5, v_6\}$, see Figure 11.6.

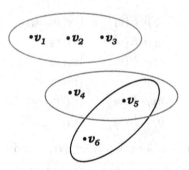

Fig. 11.6   The hypergraph defined in Example 11.8

Then, we have

$$E(v_1) = E_1 = \{v_1, v_2, v_3\},$$
$$E(v_2) = E_1 = \{v_1, v_2, v_3\},$$
$$E(v_3) = E_1 = \{v_1, v_2, v_3\},$$
$$E(v_4) = E_2 = \{v_4, v_5\},$$
$$E(v_5) = E_2 \cup E_3 = \{v_4, v_5, v_6\},$$
$$E(v_6) = E_3 = \{v_5, v_6\}.$$

Let ρ be the following relation on H,

$$\rho = \{(v_1, v_1), (v_2, v_2), (v_3, v_3), (v_4, v_4), (v_5, v_5), (v_6, v_6), (v_1, v_2), (v_2, v_1),$$
$$(v_4, v_6), (v_6, v_4)\}.$$

Clearly, ρ is an equivalence relation. Moreover, we have

$[v_1]_\rho \cap E(v_1) \neq \emptyset$	$[v_2]_\rho \cap E(v_1) \neq \emptyset$	$[v_3]_\rho \cap E(v_1) \neq \emptyset$
$[v_1]_\rho \cap E(v_2) \neq \emptyset$	$[v_2]_\rho \cap E(v_2) \neq \emptyset$	$[v_3]_\rho \cap E(v_2) \neq \emptyset$
$[v_1]_\rho \cap E(v_3) \neq \emptyset$	$[v_2]_\rho \cap E(v_3) \neq \emptyset$	$[v_3]_\rho \cap E(v_3) \neq \emptyset$
$[v_1]_\rho \cap E(v_4) = \emptyset$	$[v_2]_\rho \cap E(v_4) = \emptyset$	$[v_3]_\rho \cap E(v_4) = \emptyset$
$[v_1]_\rho \cap E(v_5) = \emptyset$	$[v_2]_\rho \cap E(v_5) = \emptyset$	$[v_3]_\rho \cap E(v_5) = \emptyset$
$[v_1]_\rho \cap E(v_6) = \emptyset$	$[v_2]_\rho \cap E(v_6) = \emptyset$	$[v_3]_\rho \cap E(v_6) = \emptyset$
$[v_4]_\rho \cap E(v_1) = \emptyset$	$[v_5]_\rho \cap E(v_1) = \emptyset$	$[v_6]_\rho \cap E(v_1) = \emptyset$
$[v_4]_\rho \cap E(v_2) = \emptyset$	$[v_5]_\rho \cap E(v_2) = \emptyset$	$[v_6]_\rho \cap E(v_2) = \emptyset$
$[v_4]_\rho \cap E(v_3) = \emptyset$	$[v_5]_\rho \cap E(v_3) = \emptyset$	$[v_6]_\rho \cap E(v_3) = \emptyset$
$[v_4]_\rho \cap E(v_4) \neq \emptyset$	$[v_5]_\rho \cap E(v_4) \neq \emptyset$	$[v_6]_\rho \cap E(v_4) \neq \emptyset$
$[v_4]_\rho \cap E(v_5) \neq \emptyset$	$[v_5]_\rho \cap E(v_5) \neq \emptyset$	$[v_6]_\rho \cap E(v_5) \neq \emptyset$
$[v_4]_\rho \cap E(v_6) \neq \emptyset$	$[v_5]_\rho \cap E(v_6) \neq \emptyset$	$[v_6]_\rho \cap E(v_6) \neq \emptyset$

Hence, we obtain

$$\overline{app}(E(v_1)) = \{v_1, v_2, v_3\},$$
$$\overline{app}(E(v_2)) = \{v_1, v_2, v_3\},$$
$$\overline{app}(E(v_3)) = \{v_1, v_2, v_3\},$$
$$\overline{app}(E(v_4)) = \{v_4, v_5, v_6\},$$
$$\overline{app}(E(v_5)) = \{v_4, v_5, v_6\},$$
$$\overline{app}(E(v_6)) = \{v_4, v_5, v_6\}.$$

Therefore, the upper $\overline{\rho}$-hypergroup $\mathbb{H}^\rho = (H, \overline{\circ})$ induced by Γ is as follows:

$\overline{\circ}$	v_1	v_2	v_3	v_4	v_5	v_6
v_1	$\{v_1, v_2, v_3\}$	$\{v_1, v_2, v_3\}$	$\{v_1, v_2, v_3\}$	H	H	H
v_2	$\{v_1, v_2, v_3\}$	$\{v_1, v_2, v_3\}$	$\{v_1, v_2, v_3\}$	H	H	H
v_3	$\{v_1, v_2, v_3\}$	$\{v_1, v_2, v_3\}$	$\{v_1, v_2, v_3\}$	H	H	H
v_4	H	H	H	$\{v_4, v_5, v_6\}$	$\{v_4, v_5, v_6\}$	$\{v_4, v_5, v_6\}$
v_5	H	H	H	$\{v_4, v_5, v_6\}$	$\{v_4, v_5, v_6\}$	$\{v_4, v_5, v_6\}$
v_6	H	H	H	$\{v_4, v_5, v_6\}$	$\{v_4, v_5, v_6\}$	$\{v_4, v_5, v_6\}$

Remark 11.8. Note that in Example 11.8, every element is an identity element.

Theorem 11.32. *Let $\Gamma = (H, \{E_i\}_i)$ be a hypergraph and ρ be a $\overline{\rho}$-good equivalence relation on H. If $\mathbb{H}^\rho = (H, \overline{\circ})$ is the upper $\overline{\rho}$-hypergroup induced by Γ, then $\{x, y\} \subset x\overline{\circ}y$.*

Proof. Since $x \in E(x)$ and $y \in E(x)$, it follows that $x \in \overline{app}(E(x))$ and $y \in \overline{app}(E(y))$. Therefore, we obtain $\{x, y\} \subset \overline{app}(E(x)) \cup \overline{app}(E(y)) = x\overline{\circ}y$. $\square$

Theorem 11.33. *Let $\Gamma = (H, \{E_i\}_i)$ be a hypergraph and ρ be a $\overline{\rho}$-good equivalence relation on H. If $\mathbb{H}^\rho = (H, \overline{\circ})$ is the upper $\overline{\circ}$-hypergroup induced by Γ, then $\beta^* = H^2$ where β^* is the fundamental relation on $\mathbb{H}^\rho$.*

Proof. For every x, y in H we have $\{x, y\} \subset x\overline{\circ}y$ and so we have $x\beta_2 y$ which implies that $x\beta y$. This means that $H^2 \subset \beta$. Since $\mathbb{H}^\rho$ is a hypergroup, it follows that $\beta^* = \beta$ and so $\beta^* = H^2$. $\square$

Theorem 11.34. *Let $\Gamma = (H, \{E_i\}_i)$ be a hypergraph and ρ be a ρ-good equivalence relation on H. If $\mathbb{H}_\rho = (H, \underline{\circ})$ is the lower $\underline{\circ}$-hypergroup induced by Γ, then $\beta^* = H^2$ where β^* is the fundamental relation on $\mathbb{H}^\rho$.*

Proof. The proof is similar to the proof of Theorem 11.33. $\square$

Theorem 11.35. *Let $\Gamma = (H, \{E_i\}_i)$ be a hypergraph and ρ be a $\overline{\rho}$-good equivalence relation on H. If $\mathbb{H}^\rho = (H, \overline{\circ})$ is the upper $\overline{\circ}$-hypergroup induced by Γ, then a complete part of $\mathbb{H}^\rho$ is equal to H.*

Proof. Suppose that A is a complete part of $\mathbb{H}^\rho$ and $b \in H$ is an arbitrary element. Let $a \in A$ be an arbitrary. Since $a \in a\overline{\circ}b$, it follows that $A \cap a\overline{\circ}b \neq \emptyset$. Thus, we obtain $a\overline{\circ}b \subset A$ and so $b \in A$. Therefore, $H \subset A$ which implies that $A = H$. $\square$

Theorem 11.36. *Let $\Gamma = (H, \{E_i\}_i)$ be a hypergraph and ρ be a ρ-good equivalence relation on H. If $\mathbb{H}_\rho = (H, \underline{\circ})$ is the lower $\underline{\circ}$-hypergroup induced by Γ, then a complete part of $\mathbb{H}_\rho$ is equal to H.*

Proof. The proof is similar to the proof of Theorem 11.35. $\square$

Definition 11.9. A *hypergraph homomorphism* is a map from the vertex set of one hypergraph to another such that each hyperedge maps to one other hyperedge.

Let ρ_Γ and $\rho_{\Gamma'}$ be $\overline{\rho_\Gamma}$-good and $\overline{\rho_{\Gamma'}}$-good equivalence relations on hypergraphs Γ and Γ' respectively. A *ρ-isomorphism* from Γ to Γ' is a bijective homomorphism f from the vertex set of Γ to that of Γ' such that

$$x\rho_r y \Leftrightarrow f(x)\rho_{r'} f(y).$$

We say that Γ and Γ' are ρ-isomorphic (written $\Gamma \cong_\rho \Gamma'$) if there is an isomorphism between them.

Lemma 11.7. *Let* $\mathbb{H}^\rho = (H, \overline{\circ})$ *and* $\mathbb{H}'^\rho = (H', \overline{\ast})$ *be two upper* $\overline{\rho}$-*hypergroups induced by hypergraphs* $\Gamma = (H, \{E_i\}_i)$ *and* $\Gamma' = (H', \{E'_j\}_j)$ *respectively.*

(1) *If* $f : \Gamma \to \Gamma'$ *is a* ρ-*isomorphism, then* $f(\overline{app}(E(x))) = \overline{app}(E'(f(x)))$.
(2) *If* $f : \mathbb{H}^\rho \to \mathbb{H}'^\rho$ *is an isomorphism, then* $f(\overline{app}(E(x))) = \overline{app}(E'(f(x)))$.

Proof. (1) Suppose that $z \in f(\overline{app}(E(x)))$ be an arbitrary element. Then, there exists $a \in \overline{app}(E(x))$ such that $z = f(a)$. Hence, $[a]_{\rho_r} \cap E(x) \neq \emptyset$. This implies that there exists $b \in [a]_{\rho_r}$ and $b \in E(x)$. So, we have $a \, \rho_r \, b$ and $b \in \bigcup\limits_{x \in E_i} E_i$. Since f is a ρ-isomorphism, we obtain

$$f(a) \, \rho_{\Gamma'} \, f(b) \text{ and } f(b) \in f\left(\bigcup\limits_{x \in E_i} E_i\right),$$

and so

$$f(b) \in [f(a)]_{\rho_{r'}} \text{ and } f(b) \in \bigcup\limits_{f(x) \in f(E_i)} f(E_i).$$

Hence, we obtain $f(b) \in E'(f(x))$. Therefore, $[f(a)]_{\rho_{r'}} \cap E'(f(x)) \neq \emptyset$. This implies that $z = f(a) \in \overline{app}(E'(f(x)))$.

Now, let $z \in \overline{app}(E'(f(x)))$ be an arbitrary element. Then, $[z]_{\rho_{r'}} \cap E'(f(x)) \neq \emptyset$. So, there exists $b \in H'$ such that $b \in [z]_{\rho_{r'}}$ and $b \in E'(f(x))$. Since f is onto, it follows that there exists $a \in H$ such that $b = f(a)$. Thus, we have

$$f(a) \, \rho_{\Gamma'} \, z \Rightarrow a \, \rho_\Gamma \, f^{-1}(z) \Rightarrow a \in [f^{-1}(z)]_{\rho_\Gamma}.$$

On the other hand, we have

$$b \in \bigcup\limits_{f(x) \in E'_j} E'_j \Rightarrow a \in f^{-1}\left(\bigcup\limits_{f(x) \in E'_j} E'_j\right)$$

$$\Rightarrow a \in \bigcup\limits_{x \in f^{-1}(E'_j)} f^{-1}(E'_j)$$

$$\Rightarrow a \in \bigcup\limits_{x \in E_i} E_i$$

$$\Rightarrow a \in E(x).$$

Thus, we conclude that $[f^{-1}(z)]_{\rho_\Gamma} \cap E(x) \neq \emptyset$. This implies that $f^{-1}(z) \in \overline{app}(E(x))$. Therefore, $z \in f(\overline{app}(E(x)))$. This completes the proof.

(2) Let $f : H \to H'$ be an isomorphism. We have $f(x \bar{\circ} y) = f(x) \bar{\star} f(y)$, for all $x, y \in H$. So, $f(x \bar{\circ} x) = f(x) \bar{\star} f(x)$. This implies that $f(\overline{app}(E(x))) = \overline{app}(E'(f(x)))$. □

Theorem 11.37. *Let* $\mathbb{H}^\rho = (H, \bar{\circ})$ *and* $\mathbb{H}'^\rho = (H', \bar{\star})$ *be two upper ρ-hypergroups induced by hypergraphs* $\Gamma = (H, \{E_i\}_i)$ *and* $\Gamma' = (H', \{E'_j\}_j)$ *respectively. If* $\Gamma \cong \Gamma'$, *then* $\mathbb{H}^\rho \cong \mathbb{H}'^\rho$.

Proof. Suppose that $f : \Gamma \to \Gamma'$ is a ρ-isomorphism. We show that $f(x \bar{\circ} y) = f(x) \bar{\star} f(y)$ or

$$f(\overline{app}(E(x)) \cup \overline{app}(E(y))) = \overline{app}(E'(f(x))) \cup \overline{app}(E'(f(y))).$$

Suppose that $z \in f(\overline{app}(E(x)) \cup \overline{app}(E(y)))$. Then, there exists $a \in \overline{app}(E(x)) \cup \overline{app}(E(y))$ such that $z = f(a)$.

If $a \in \overline{app}(E(x))$, then $[a]_{\rho_\Gamma} \cap E(x) \neq \emptyset$. This implies that $[f(x)]_{\rho_{\Gamma'}} \cap E'(f(x)) \neq \emptyset$. So, we obtain $z \in \overline{app}(E'(f(x)))$.

If $a \in \overline{app}(E(y))$, then in a similar way we obtain $z \in \overline{app}(E'(f(y)))$. Therefore,

$$f(\overline{app}(E(x)) \cup \overline{app}(E(y))) \subset \overline{app}(E'(f(x))) \cup \overline{app}(E'(f(y))).$$

For the converse, we have

$$\overline{app}(E(x)) \subset \overline{app}(E(x)) \cup \overline{app}(E(y)) \text{ and } \overline{app}(E(y)) \subset \overline{app}(E(x)) \cup \overline{app}(E(y)).$$

Then, we have

$$f(\overline{app}(E(x))) \subset f(\overline{app}(E(x)) \cup \overline{app}(E(y))) \text{ and } f(\overline{app}(E(y)))$$
$$\subset f(\overline{app}(E(x)) \cup \overline{app}(E(y))).$$

Hence, we obtain

$$f(\overline{app}(E(x))) \cup f(\overline{app}(E(y))) \subset f(\overline{app}(E(x)) \cup \overline{app}(E(y))).$$

Now, by Lemma 11.7(1), we obtain

$$\overline{app}(E'(f(x))) \cup \overline{app}(E'(f(y))) \subset f(\overline{app}(E(x)) \cup \overline{app}(E(y))). \quad \square$$

Lemma 11.8. *Let* $\mathbb{H}_\rho = (H, \underline{\circ})$ *and* $\mathbb{H}'_\rho = (H', \underline{\star})$ *be two lower ρ-hypergroups induced by hypergraphs* $\Gamma = (H, \{E_i\}_i)$ *and* $\Gamma' = (H', \{E'_j\}_j)$ *respectively.*

(1) *If* $\Gamma \cong_\rho \Gamma'$, *then* $f(\underline{app}(E(x))) = \underline{app}(E'(f(x)))$.

(2) *If* $\mathbb{H}_\rho \cong \mathbb{H}'_\rho$, *then* $f(\underline{app}(E(x))) = \underline{app}(E'(f(x)))$.

Proof.   (1) Suppose that $z \in f(\underline{app}(E(x)))$ be an arbitrary element. Then, there exists $a \in \underline{app}(E(x))$ such that $z = f(a)$. Hence, $[a]_{\rho_\Gamma} \subset E(x)$. So, we have $f([a]_{\rho_\Gamma}) \subset f(E(x))$. Therefore, we obtain

$$[f(a)]_{\rho_{\Gamma'}} \subset f\left(\bigcup_{x \in E_i} E_i \right) = \bigcup_{f(x) \in f(E_i)} f(E_i)$$
$$= \bigcup_{f(x) \in E'_j} E'_j = E'_j(f(x)).$$

This implies that $z = f(a) \in \underline{app}(E'(f(x)))$.

Now, assume that $z \in \underline{app}(E'(f(x)))$ be an arbitrary element. Then, $[z]_{\rho_{\Gamma'}} \subset E'(f(x))$. So, $f^{-1}\left([z]_{\rho_{\Gamma'}}\right) \subset f^{-1}(E'(f(x)))$. Then, we conclude that $f^{-1}\left([z]_{\rho_{\Gamma'}}\right) \subset E(x)$. This implies that $f^{-1}(z) \in \underline{app}(E(x))$. Therefore, $z \in f(\underline{app}(E(x)))$.

(2) The proof is similar to the proof of Lemma 11.7(2).   □

Theorem 11.38. *Let* $\mathbb{H}_\rho = (H, \underline{\circ})$ *and* $\mathbb{H}'_\rho = (H', \underline{\star})$ *be two lower ρ-hypergroups induced by hypergraphs* $\Gamma = (H, \{E_i\}_i)$ *and* $\Gamma' = (H', \{E'_j\}_j)$ *respectively. If* $\Gamma \cong_\rho \Gamma'$, *then* $\mathbb{H}_\rho \cong \mathbb{H}'_\rho$.

Proof.   Let $f : \Gamma \to \Gamma'$ be a ρ-isomorphism. We show that $f(x\underline{\circ}y) = f(x)\underline{\star}f(y)$. Since $f(\{x, y\}) = \{f(x), f(y)\}$, it is enough to show that

$$f(\underline{app}(E(x)) \cup \underline{app}(E(y))) = \underline{app}(E'(f(x))) \cup \underline{app}(E'(f(y))).$$

Suppose that $z \in f(\underline{app}(E(x)) \cup \underline{app}(E(y)))$. Then, there exists $a \in \underline{app}(E(x)) \cup \underline{app}(E(y))$ such that $z = f(a)$.

If $a \in \underline{app}(E(x))$, then $[a]_{\rho_\Gamma} \subset E(x)$. This implies that $[f(a)]_{\rho_{\Gamma'}} \subset E'(f(x))$. So, $z \in \underline{app}(E'(f(x)))$. Similarly, if $a \in \underline{app}(E(y))$, then we obtain $z \in \underline{app}(E'(f(y)))$. So, we have

$$f(\underline{app}(E(x)) \cup \underline{app}(E(y))) \subset f(\underline{app}(E(x))) \cup f(\underline{app}(E(y))).$$

Now, by using Lemma 11.8(1), we obtain

$$f(\underline{app}(E(x)) \cup \underline{app}(E(y))) \subset \underline{app}(E'(f(x))) \cup \underline{app}(E'(f(y))).$$

The proof of the converse inclusion is similar.   □

Chapter 12

Topological Hypergroups

In this chapter we study topological hypergroupoids and topological hypergroups. Connections between topology and algebraic hyperstructures were studied by Mittas, Konstantinidou, Hošková-Mayerová, Davvaz, Al Tahan, Heidari, Modarres, Singha, Das, and many others. A topological hypergroup is a topological space with a continuous hypergroup structure.

12.1 Topological spaces

A topological space is a geometrical space in which closure is defined but, generally, can not be measured by a numeric distance. More specifically, a topological space is a set of points, along with a set of neighborhoods for each point, satisfying a set of axioms relating points and neighborhoods.

We begin to present topological spaces.

Definition 12.1. A *topology* on a non-empty set X is a collection $\mathcal{T}$ of subsets of X, called the *open sets*, satisfying:

(1) $\emptyset \in \mathcal{T}$ and $X \in \mathcal{T}$;

(2) If $U_1, \ldots, U_n \in \mathcal{T}$, then $\bigcap_{i=1}^{n} U_i \in \mathcal{T}$ ($\mathcal{T}$ is closed under finite intersections);

(3) If I is an arbitrary index set and $U_i \in \mathcal{T}$ for all $i \in I$, then $\bigcup_{i \in I} U_i \in \mathcal{T}$ ($\mathcal{T}$ is closed under arbitrary unions).

A *topological space* is a pair $(X, \mathcal{T})$, where X is a non-empty set and $\mathcal{T}$ is a topology on X.

The condition (2) that a topology $\mathcal{T}$ must be closed under finite intersections can be replaced by the condition (2') that

(2′) If $U_1, U_2 \in \mathcal{T}$, then $U_1 \cap U_2 \in \mathcal{T}$ (that is, $\mathcal{T}$ is closed under binary intersections).

Example 12.1. If X is any non-empty set, the *discrete topology* on X is $\mathcal{T}_D = \mathcal{P}(X) = \{U \mid U \subset X\}$. Every subset of X is open in the discrete topology (so it is easily confirmed that $\mathcal{T}_D$ is in fact a topology). Any topology is a collection of subsets of X; the discrete topology is the collection of all subsets of X and thus is the largest possible topology on X.

Example 12.2. The indiscrete topology on X is $\mathcal{T}_I = \{\emptyset, X\}$. Thus, the only open sets in the indiscrete topology are those required by the definition of a topology.

Example 12.3. Consider the set $X = \{a, b, c, d\}$ and the collections of subsets

$$\mathcal{T}_1 = \{\emptyset, \{a\}, \{c\}, \{c,d\}, X\},$$
$$\mathcal{T}_2 = \{\emptyset, \{a,b,c\}, \{c\}, \{c,d\}, X\},$$

as shown in Figure 12.1.

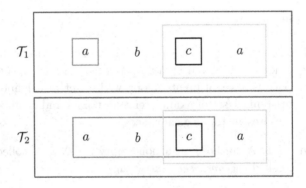

Fig. 12.1 $\mathcal{T}_1$ is not a topology on X but $\mathcal{T}_2$ is a topology on X

The collection $\mathcal{T}_1$ is not a topology on X since it is not closed under arbitrary unions. The sets $\{a\}$ and $\{c\}$ are in $\mathcal{T}_1$, but their union $\{a, c\}$ is not; but the collection $\mathcal{T}_2$ is a topology on X.

Definition 12.2. If $\mathcal{T}$ is a topology on X, a *basis* for $\mathcal{T}$ is a collection $\mathcal{B}$ of open subsets of X such that every open set is a union of elements of $\mathcal{B}$. The open sets in $\mathcal{B}$ are called *basic open sets* with respect to $\mathcal{B}$. If $\mathcal{B}$ is a basis for $\mathcal{T}$, we say $\mathcal{T}$ is generated by $\mathcal{B}$.

Theorem 12.1. *Suppose that $\mathcal{T}$ is a topology on the set X and $\mathcal{B} \subset \mathcal{T}$ is a collection of open sets. The following are equivalent:*

(a) *$\mathcal{B}$ is a basis for $\mathcal{T}$;*
(b) *U is open if and only if U is a union of elements of $\mathcal{B}$;*
(c) *U is open if and only if for every $x \in U$, there exists a basic open set $B \in \mathcal{B}$ which contains x and is contained in U.*

We note that (c) could be formally written as:

(c′) *$U \in \mathcal{T}$ if and only if for all $x \in U$ there exists $B \in \mathcal{B}$ such that $x \in B$ and $B \subset U$.*

Proof. The definition of $\mathcal{B}$ being a basis for $\mathcal{T}$ was that U is open only if U is a union of elements of $\mathcal{B}$. The converse also holds: U is open if U is a union of elements of $\mathcal{B}$. Indeed, since $\mathcal{B} \subset \mathcal{T}$, any union of elements of $\mathcal{B}$ is a union of open sets, which is open. Thus, (a) and (b) are equivalent.

To see (b) and (c) are equivalent, it suffices to show that (b2) U is a union of elements of $\mathcal{B}$ is equivalent to (c2) for every $x \in U$, there exists a basic open set $B \in \mathcal{B}$ which contains x and is contained in U. Suppose (b2). Express U as the union $\bigcup\{B_i \mid i \in I\}$ of basic open sets $B_i \in \mathcal{B}$. For any $x \in U$, there exists $j \in I$ such that $x \in B_j$. Furthermore, $B_j \subset \{B_i \mid i \in I\} = U$, proving (c2). Now suppose (c2). Then, for every $x \in U$, there exists a basic open set $B_x \in \mathcal{B}$ with $\{x\} \subset B_x \subset U$. Taking the union over all $x \in U$ gives $U = \bigcup_{x \in U} \{x\} \subset \bigcup_{x \in U} B_x \subset \bigcup_{x \in U} U = U$, so U is a union of basis elements, proving (b2). □

Definition 12.3. If $(X, \mathcal{T})$ is a topological space and $x \in X$, an *open neighborhood* of x is an open set containing x. A neighborhood of x is a set containing an open neighborhood of x.

We may ask, if we start with a collection of subsets of X, when might it be a basis for some topology on X? Suppose that $\mathcal{B}$ is a basis for a topology $\mathcal{T}$ on X. Since X is an open set, condition (b) of Theorem 12.1 implies that $\bigcup \mathcal{B} = X$. If $B_1, B_2 \in \mathcal{B}$, then B_1 and B_2 are both open, so $B_1 \cap B_2$ is open. If $x \in B_1 \cap B_2$, then the neighborhood $B_1 \cap B_2$ of x must contain a basic neighborhood B_3 of x. That is, there exists $B_3 \in \mathcal{B}$ with $x \in B_3$ and $B_3 \subset B_1 \cap B_2$, as suggested in Figure 12.2. In fact, these properties are all that is needed to guarantee that a collection $\mathcal{B}$ is a basis for some topology on X.

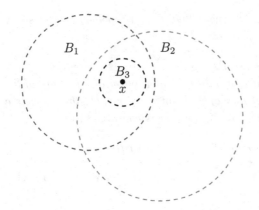

Fig. 12.2 If $B_1, B_2 \in \mathcal{B}$ and $x \in B_1 \cap B_2$, then there exists $B_3 \in \mathcal{B}$ with $x \in B_3 \subset B_1 \cap B_2$

Theorem 12.2. *A collection $\mathcal{B}$ of subsets of X is a basis for some topology on X if and only if*

(1) $\bigcup \mathcal{B} = X$, *and*
(2) *if $B_1, B_2 \in \mathcal{B}$ and $x \in B_1 \cap B_2$, then there exists $B_3 \in \mathcal{B}$ with $x \in B_3 \subset B_1 \cap B_2$.*

Proof. If $\mathcal{B}$ is a basis for some topology on X, then (1) and (2) are satisfied.

Now, assume that $\mathcal{B}$ satisfies (1) and (2) above. If $\mathcal{B}$ is to be a basis for a topology, then that topology must consist of all unions of elements of $\mathcal{B}$. Let $\mathcal{T}_\mathcal{B}$ be the collection of all unions of elements of $\mathcal{B}$. We show that $\mathcal{T}_\mathcal{B}$ is a topology. By condition (1) above, $X \in \mathcal{T}_\mathcal{B}$, and $\emptyset \in \mathcal{T}_\mathcal{B}$ since $\emptyset$ is the union of the empty collection of elements of $\mathcal{B}$. An arbitrary union of elements of $\mathcal{T}_\mathcal{B}$ is a union of unions of elements of $\mathcal{B}$, which is just a larger union of elements of $\mathcal{B}$, so $\mathcal{T}_\mathcal{B}$ is closed under arbitrary unions. For finite intersections, suppose that $U_1, U_2 \in \mathcal{T}_\mathcal{B}$. If $U_1 \cap U_2 = \emptyset$, then $U_1 \cap U_2 \in \mathcal{T}_\mathcal{B}$. Otherwise, suppose that $x \in U_1 \cap U_2$. Since U_1 is a union of sets in $\mathcal{B}$, there exists $B_1 \in \mathcal{B}$ with $\{x\} \subset B_1 \subset U_1$. Similarly, there exists $B_2 \in \mathcal{B}$ with $\{x\} \subset B_2 \subset U_2$. By condition (2), there exists $B_x \in \mathcal{B}$ with $\{x\} \subset B_x \subset B_1 \cap B_2 \subset U_1 \cap U_2$. Now taking the union over all $x \in U_1 \cap U_2$ shows that

$$U_1 \cap U_2 = \bigcup_{x \in U_1 \cap U_2} \{x\} \subset \bigcup_{x \in U_1 \cap U_2} B_x \subset \bigcup_{x \in U_1 \cap U_2} (U_1 \cap U_2) = U_1 \cap U_2,$$

hence $U_1 \cap U_2 = \bigcup\limits_{x \in U_1 \cap U_2} B_x$ is a union of elements of $\mathcal{B}$ and thus is in $\mathcal{T}_\mathcal{B}$.

Hence, the result follows for finite intersections inductively. $\square$

Definition 12.4. Let $(X, \mathcal{T})$ be a topological space.

(1) $(X, \mathcal{T})$ is a T_0-*space* if for all $x \neq y \in X$, there exists $U \in \mathcal{T}$ such that $x \in U$ and y is not in U or $y \in U$ and x is not in U;
(2) $(X, \mathcal{T})$ is a T_1-*space* if for all $x \neq y \in X$, there exist $U, V \in \mathcal{T}$ such that $x \in U$ and y is not in U and $y \in V$ and x is not in V;
(3) $(X, \mathcal{T})$ is a T_2-*space* if for all $x \neq y \in X$, there exist $U, V \in \mathcal{T}$ such that $x \in U$, $y \in V$ and $U \cap V = \emptyset$.

Every T_2-topological space is a T_1-topological space and every T_1-topological space is a T_0-topological space.

Topological groups have the algebraic structure of a group and the topological structure of a topological space and they are linked by the requirement that multiplication and inversion are continuous functions.

Definition 12.5. A *topological group* is a group G together with a topology on G that satisfies the following two properties:

(1) the mapping $p : G \times G \to G$ defined by $p(g, h) = gh$ is continuous when $G \times G$ is endowed with the product topology;
(2) the mapping $inv : G \to G$ defined by $inv(g) = g^{-1}$ is continuous.

We remark that the item (1) is equivalent to the statement that, whenever $U \subset G$ is open, and $g_1 g_2 \in U$, then there exist open sets V_1, V_2 such that $g_1 \in V_1, g_2 \in V_2$ and $V_1 V_2 = \{v_1 v_2 \mid v_1 \in V_1, v_2 \in V_2\} \subset U$. Also, the item (2) is equivalent to showing that whenever $U \subset G$ is open, then $U^{-1} = \{g^{-1} \mid g \in U\}$ is open.

Definition 12.6. Let $(H_1, \circ_1)$, $(H_2, \circ_2)$ be two hypergroupoids and define the topologies $\mathcal{T}, \mathcal{T}'$ on H_1, H_2 respectively. A mapping f from H_1 to H_2 is said to be *good topological homomorphism* if for all $x, y \in H_1$,

(1) $f(a \circ_1 b) = f(a) \circ_2 f(b)$;
(2) f is continuous;
(3) f is open.

A good topological homomorphism is a *topological isomorphism* if f is one to one and onto and we say that H_1 is topologically isomorphic to H_2. Let $(H, \circ)$ be a hypergroupoid and A, B be non empty subsets of H. By $A \approx B$ we mean that $A \cap B \neq \emptyset$.

12.2   Topological hypergroupoids

Hošková-Mayerová in [80] introduced some new definitions inspired by the
definition of a topological groupoid. She investigated various kinds of continuity of hyperoperations, namely pseodocontinuous, strongly pseodocontinuous and continuous hyperoperations. In addition, the relation between
them is studied. Then, Al Tahan, Hošková-Mayerová and Davvaz in [7]
studied examples of topological hypergroupoids. They showed that there
is no relation (in general) between pseudotopolgical and strongly pseudotopolgical hypergroupoids. In particular, they presented a topological hypergroupoid that does not depend on the pseodocontinuity nor on strongly
pseodocontinuity of the hyperoperation. Their results are summarized in
this section.

Definition 12.7. Let $(H, \circ)$ be a hypergroupoid and $(H, \mathcal{T})$ be a topological space. The hyperoperation " $\circ$ " is called:

(1) pseudocontinuous (p-continuous) if for every $O \in \mathcal{T}$, the set $O_\star =
\{(x, y) \in H^2 \mid x \circ y \subset O\}$ is open in $H \times H$.
(2) strongly pseudocontinuous (sp-continuous) if for every $O \in \mathcal{T}$, the set
$O^\star = \{(x, y) \in H^2 \mid x \circ y \approx O\}$ is open in $H \times H$.

A simple way to prove that a hyperoperation " $\circ$ " is p-continuous
(sp-continuous) is to take any open set O in $\mathcal{T}$ and $(x, y) \in H^2$ such that
$x \circ y \subset O$ ($x \circ y \approx O$) and prove that there exist $U, V \in \mathcal{T}$ such that $u.v \subset O$
($u.v \approx O$) for all $(u, v) \in U \times V$.

Definition 12.8. Let $(H, \circ)$ be a hypergroupoid, $(H, \mathcal{T})$ be a topological
space and $\mathcal{T}_\star$ be a topology on $\mathcal{P}^*(H)$. The triple $(H, \circ, \mathcal{T})$ is called a
pseudotopological or strongly pseudotopological hypergroupoid if the hy-
peroperation " $\circ$ " is p-continuous or sp-continuous respectively.

The quadruple $(H, \circ, \mathcal{T}, \mathcal{T}_\star)$ is called $\mathcal{T}_\star$-topological hypergroupoid if the
hyperoperation " $\circ$ " is $\mathcal{T}_\star$-continuous.

Notation 12.1. Let $(H, \mathcal{T})$ be a topological space, $V \in \mathcal{T}$. We define S_V
and I_V as follows:

- $S_V = \{U \in \mathcal{P}^*(H) \mid U \subset V\} = \mathcal{P}^*(V)$.
- $I_V = \{U \in \mathcal{P}^*(H) \mid U \approx V\}$.

Remark 12.1. $S_\emptyset = I_\emptyset = \emptyset$. For all $V \neq \emptyset$, we have
$$S_V = \mathcal{P}^*(V) \text{ and } I_V \supseteq \{H, \mathcal{P}^*(V)\}.$$

Lemma 12.1. *Let $(H, \mathcal{T})$ be a topological space. Then, the family $\mathcal{U} = \{S_V\}_{V \in \mathcal{T}}$ is a base for a topology $\mathcal{T}_{\mathcal{U}}$ on $\mathcal{P}^*(H)$.*

Proof. It is straightforward. $\qquad\qquad\qquad\qquad\qquad\qquad\qquad\qquad\square$

The topology $\mathcal{T}_{\mathcal{U}}$ in Lemma 12.1 is called the *upper topology* induced by the topology $\mathcal{T}$ on H.

Lemma 12.2. *Let $(H, \mathcal{T})$ be a topological space. Then, the family $\mathcal{L} = \{I_V\}_{V \in \mathcal{T}}$ forms a subbase for a topology $\mathcal{T}_{\mathcal{L}}$ on $\mathcal{P}^*(H)$.*

Proof. It is straightforward. $\qquad\qquad\qquad\qquad\qquad\qquad\qquad\qquad\square$

The topology $\mathcal{T}_{\mathcal{L}}$ in Lemma 12.2 is called the *lower topology* induced by the topology $\mathcal{T}$ on H.

Theorem 12.3. *Let $(H, \circ)$ be a hypergroupoid and $(H, \mathcal{T})$ be a topological space. Then the triple $(H, \circ, \mathcal{T})$ is a pseudotopological hypergroupoid if and only if the quadruple $(H, \circ, \mathcal{T}, \mathcal{T}_{\mathcal{U}})$ is a $\mathcal{T}_{\mathcal{U}}$-topological hypergroupoid.*

Proof. Suppose that $\varphi : H \times H \to \mathcal{P}^*(H)$ is the mapping that assigns the $x \circ y$ to the pair (x, y). For an open set $V \in \mathcal{T}$ we can write

$$\varphi^{-1}(S_V) = \{(x, y) \in H \times H \mid x \circ y \in S_V\}$$
$$= \{(x, y) \in H \times H \mid x \circ y \subset V\}$$
$$= V_*.$$

This means that φ is continuous if and only if V_* is open for any $V \in \mathcal{T}$, i.e., $\circ$ is p-continuous. $\qquad\qquad\qquad\qquad\qquad\qquad\qquad\qquad\square$

Theorem 12.4. *Let $(H, \circ)$ be a hypergroupoid and $(H, \mathcal{T})$ be a topological space. Then the triple $(H, \circ, \mathcal{T})$ is a strongly pseudotopological hypergroupoid if and only if the quadruple $(H, \circ, \mathcal{T}, \mathcal{T}_{\mathcal{L}})$ is a $\mathcal{T}_{\mathcal{L}}$-topological hypergroupoid.*

Proof. Suppose that $\varphi : H \times H \to \mathcal{P}^*(H)$ is the mapping that assigns the $x \circ y$ to the pair (x, y). For an open set $V \in \mathcal{T}$ we can write

$$\varphi^{-1}(I_V) = \{(x, y) \in H \times H \mid x \circ y \in I_V\}$$
$$= \{(x, y) \in H \times H \mid x \circ y \approx V\}$$
$$= V^*.$$

This means that φ is continuous if and only if V^* is open for any $V \in \mathcal{T}$, i.e., $\circ$ is sp-continuous. $\qquad\qquad\qquad\qquad\qquad\qquad\qquad\square$

Lemma 12.3. *Let* $(H, \mathcal{T})$ *be a topological space. For any* $U_1, \ldots, U_k$, $k \in \mathbb{N}$ *with* $U_i \in \mathcal{T}$ *for* $i = 1, \ldots, k$, *we denote*

$$\aleph(U_1, \ldots, U_k) = \Big\{ B \in \mathcal{P}^*(H) \mid b \subset \bigcup_{i=1}^{k} U_i \text{ and } B \approx U_i \text{ for } i = 1, \ldots, k \Big\}.$$

The family $\mathcal{B}$ *of all* $\aleph(U_1, \ldots, U_k)$ *forms a base of a topological space* $(\mathcal{P}^*(H), \mathcal{T}_{\aleph})$

Proof. Assume that $O_1, O_2 \in \mathcal{B}$, i.e., $O_1 = \aleph(U_1, \ldots, U_k)$ and $O_2 = \aleph(V_1, \ldots, V_l)$, where $U_1, \ldots, U_k \in \mathcal{T}$ and $V_1, \ldots, V_l \in \mathcal{T}$. If $X \in O_1 \mathcal{O}_\epsilon$, then

$$X \subset \bigcup_{i=1}^{k} U_i \cap \bigcup_{j=1}^{l} V_j = \bigcup_{i,j} U_i \cap V_j.$$

Let us denote by $W_1, \ldots, W_r$ those intersections $U_i \cap V_j$ which are not disjoint with X. Denote $O_3 = \aleph((W_1, \ldots, W_r)$. Evidently $X \in O_3$. Now, we show that $O_3 \subset O_1 \cap O_2$. For any $Y \in O_3$ we have $Y \subset \bigcup_{i=1}^{r} W_i$ and $Y \approx W_i$, $i = 1, \ldots, r$. Therefore, we have $\bigcup_{i=1}^{r} W_i \subset \bigcup_{i=1}^{k} U_i$.

Choose an arbitrary but fixed U_i, $1 \leq i \leq k$. Then, there exists $1 \leq j \leq l$ such that $U_i \cap V_j = W_s$ for an appropriate $1 \leq s \leq r$. In fact, as $X \approx U_i$ there is $a \in X \cap U_i$. Further $V_1, \ldots, V_l$ cover X, so there exists j such that $a \in V_j$. Therefore, $X \approx U_i \cap V_j = W_s$. Now, we have $Y \approx W_s \subset U_i$, so $Y \approx U_i$, which means that $Y \in O_1$, i.e., $O_3 \subset O_1$. Similarly, we have $O_3 \subset O_2$, so $O_3 \subset O_1 \cap O_2$.

In addition, as $H \in \mathcal{T}$ we have $\aleph(H) = \mathcal{P}^*(H)$. $\qquad\square$

The topology $\mathcal{T}_{\aleph}$ in Lemma 12.3 is called the *Vietoris topology* on $\mathcal{P}^*(H)$.

Lemma 12.4. *Vietoris topology* $\mathcal{T}_{\aleph}$ *is the lowest common refinement of upper and lower topologies* $\mathcal{T}_{\mathcal{U}}$ *and* $\mathcal{T}_{\mathcal{L}}$.

Proof. If $\mathcal{T}_1, \mathcal{T}_2$ are topologies on a set H with subbases $\mathcal{B} = \mathcal{B}_1, \mathcal{B}_2$ respectively, then $\mathcal{B}_1 \cup \mathcal{B}_2$ is a subbase of a topology $\mathcal{T}$, which is the lowest common refinement of $\mathcal{T}_1$ and $\mathcal{T}_2$. We apply this result to the (sub)bases $\mathcal{U}$ and $\mathcal{L}$. Denote $\mathcal{C}$ the base corresponding to the subbase $\mathcal{U} \cup \mathcal{L}$. The system $\mathcal{C}$ consists of three types of sets:

(1) $S_{V_1} \cap S_{V_2} \cap \ldots \cap S_{V_k} = S_{V_1 \cap V_2 \cap \ldots \cap V_k} = S_V$, where $V = V_1 \cap V_2 \cap \ldots \cap V_k$ and $V_1, V_2, \ldots, V_k \in \mathcal{T}_{\mathcal{U}}$, $k \in \mathbb{N}$;
(2) $I_{U_1} \cap I_{U_2} \cap \ldots \cap I_{U_l}$, where $U_1, U_l, \ldots, U_l \in \mathcal{T}_{\mathcal{L}}$, $l \in \mathbb{N}$;

(3) $S_{V_1} \cap S_{V_2} \cap \ldots \cap S_{V_k} \cap I_{U_1} \cap I_{U_2} \cap \ldots \cap I_{U_l} = S_V \cap I_{U_1} \cap I_{U_2} \cap \ldots \cap I_{U_l}$,
where $V = V_1 \cap V_2 \cap \ldots \cap V_k$, $V_1, V_2, \ldots, V_k \in \mathcal{T_U}$, $U_1, U_l, \ldots, U_l \in \mathcal{T_L}$
and $k, l \in \mathbb{N}$.

As $H \in \mathcal{T_L} \cap \mathcal{T_U}$ and $S_H = I_H = \mathcal{P}^*(H)$, we can limit only to the type 3.
Then, it is not difficult to see $\mathcal{C} = \mathcal{B}$. □

Theorem 12.5. *Let $(H, \circ)$ be a hypergroupoid and $(H, \mathcal{T})$ be a topological space. Then the triple $(H, \circ, \mathcal{T})$ is a pseudotopological hypergroupoid and strongly pseudotopological hypergroupoid if and only if the quadruple $(H, \circ, \mathcal{T}, \mathcal{T_N})$ is a $\mathcal{T_N}$-topological hypergroupoid.*

Proof. The result follows from Lemma 12.4. □

Now, we will show that there is no relation (in general) between pseudotopolgical and strongly pseudotopolgical hypergroupoids.

Theorem 12.6. *Let $H = \{a, b\}$, $\mathcal{T} = \{\emptyset, \{a\}, H\}$ and define a hyperoperation "$\circ_1$" on H as follows:*

$\circ_1$	a	b
a	b	H
b	H	H

Then $(H, \circ_1, \mathcal{T})$ is a pseudotopological hypergroupoid.

Proof. It is clear that $\mathcal{T}$ is a topology on H. By means of Definition 12.8, it suffices to show that $\circ_1$ is p-continuous. Let $(x, y) \in H^2$ and $O \in \mathcal{T}$ such that $x \circ_1 y \subset O$. Since $x \circ_1 y \neq \emptyset$, it follows that $O \neq \emptyset$. We have the following cases for (x, y):

- Case $x = y = a$. Having $\{b\} = a \circ_1 a = x \circ_1 y \subset O$ and $O \in \mathcal{T}$ implies that $O = H$. Take $U = V = \{a\} \in \mathcal{T}$. For every $(u, v) \in U \times V$, we have that $u \circ_1 v = \{b\} \subset O$.
- Case $x = a, y = b$ or $x = b, y = a$. Having $H = a \circ_1 b = x \circ_1 y \subset O$ and $O \in \mathcal{T}$ implies that $O = H$. Take $U = V = H$. For every $(u, v) \in U \times V$, we have that $u \circ_1 v \subset O$.
- Case $x = y = b$. Having $H = b \circ_1 b = x \circ_1 y \subset O$ and $O \in \mathcal{T}$ implies that $O = H$. Take $U = V = H$. For every $(u, v) \in U \times V$, we have that $u \circ_1 v \subset O$.

Therefore, $(H, \circ_1, \mathcal{T})$ is a pseudotopological hypergroupoid. □

Corollary 12.1. *The quadruple $(H, \circ_1, \mathcal{T}, \mathcal{T_U})$ is a $\mathcal{T_U}$-topological hypergroupoid.*

Proof.   The proof results from Theorems 12.3 and 12.6. One can easily see that $\mathcal{T}_{\mathcal{U}} = \{\emptyset, \{\{a\}\}, \mathcal{P}^*(H)\}$. □

Remark 12.2. $(H, \circ_1, \mathcal{T})$ is not strongly pseudotopological hypergroupoid.

Proof.   By means of Definition 12.8, it suffices to show that $\circ_1$ is not sp-continuous. Let $O = \{a\} \in \mathcal{T}$. We have that $a \circ_1 b \cap O \neq \emptyset$. Suppose, to get a contradiction, that $\circ_1$ is sp-continuous. Then there exist $U, V \in \mathcal{T}$ such that $a \in U$, $b \in V$ and for all $(u, v) \in U \times V$, $u \circ_1 v \approx O$. It is easy to see that $V = H$ and U is either $\{a\}$ or H. By setting $(u, v) = (a, a) \in U \times V$, we get that $\{b\} = a \circ_1 a = u \circ_1 v \approx O$. The latter is impossible since $\{a\} \cap \{b\} = \emptyset$. □

Corollary 12.2. *Not every pseudotopological hypergroupoid is strongly pseudotopological hypergroupoid.*

Proof.   The proof results from Theorem 12.6 and Remark 12.2. □

Theorem 12.7. *Let $H = \{a, b\}$, $\mathcal{T} = \{\emptyset, \{a\}, H\}$ and define a hyperoperation " $\circ_2$ " on H as follows:*

$\circ_2$	a	b
a	H	a
b	a	b

Then $(H, \circ_2, \mathcal{T})$ is a strongly pseudotopological hypergroupoid.

Proof.   It is clear that $\mathcal{T}$ is a topology on H. By means of Definition 12.8, it suffices to show that " $\circ_1$ " is sp-continuous. Let $(x, y) \in H^2$ and $O \in \mathcal{T}$ such that $x \circ_2 y \approx O$. Since $x \circ_2 y \neq \emptyset$, it follows that $O \neq \emptyset$. We have the following cases for (x, y):

- Case $x = y = a$. Having $H = a \circ_2 a = x \circ_2 y$, $x \circ_2 y \approx O$ and $O \in \mathcal{T}$ imply that $O = \{a\}$ or $O = H$. Take $U = V = \{a\} \in \mathcal{T}$. For every $(u, v) \in U \times V$, we have that $u \circ_2 v = H$. Thus, $u \circ_2 v \approx O$.
- Case $x = a, y = b$ or $x = b, y = a$. Having $\{a\} = a \circ_2 b = x \circ_2 y \approx O$ and $O \in \mathcal{T}$ imply that $O = \{a\}$ or $O = H$. Take $U = \{a\}$ and $V = H$. For every $(u, v) \in U \times V$, we have that $u \circ_2 v \approx O$.
- Case $x = y = b$. Having $\{b\} = b \circ_2 b = x \circ_2 y \approx O$ and $O \in \mathcal{T}$ implies that $O = H$. Take $U = V = H$. For every $(u, v) \in U \times V$, we have that $u \circ_2 v \approx O$.

Therefore, $(H, \circ_2, \mathcal{T})$ is a strongly pseudotopological hypergroupoid. □

Corollary 12.3. *The quadruple* $(H, \circ_2, \mathcal{T}, \mathcal{T}_{\mathcal{L}})$ *is a* $\mathcal{T}_{\mathcal{L}}$*-topological hypergroupoid.*

Proof. The proof results from Theorems 12.4 and 12.7. One can easily see that $\mathcal{T}_{\mathcal{L}} = \{\emptyset, \{H, \{a\}\}, \mathcal{P}^*(H)\}$. □

Remark 12.3. $(H, \circ_2, \mathcal{T})$ is not pseudotopological hypergroupoid.

Proof. By means of Definition 12.8, it suffices to show that $\circ_2$ is not p-continuous. Let $O = \{a\} \in \mathcal{T}$. We have that $a \circ_2 b \subset O$. Suppose, to get contradiction, that $\circ_2$ is p-continuous. Then there exist $U, V \in \mathcal{T}$ such that $a \in U$, $b \in V$ and for all $(u, v) \in U \times V$, $u \circ_2 v \subset O$. It is easy to see that $V = H$ and U is either $\{a\}$ or H. By setting $(u, v) = (a, a) \in U \times V$, we get that $\{b\} = a \circ_2 a = u \circ_2 v \subset O$. The latter is impossible since $\{a\} \not\subseteq \{b\}$. □

Corollary 12.4. *Not every strongly pseudotopological hypergroupoid is a pseudotopological hypergroupoid.*

Proof. The proof results from Theorem 12.7 and Remark 12.3. □

Remark 12.4. Let $(H, \circ)$ be a hypergroupoid and $\mathcal{T}$ a topology on H. Then $(H, \circ, \mathcal{T})$ may be neither a pseudotopological hypergroupoid nor a strongly pseudotopological hypergroupoid.

We illustrate Remark 12.4 by the following example.

Example 12.4. Let $H = \{a, b\}$, $\mathcal{T} = \{\emptyset, \{a\}, H\}$ and define a hyperoperation " $\circ_3$ " on H as follows:

$\circ_3$	a	b
a	b	a
b	a	b

It is easy to check, by taking $O = \{a\}$ and $a \circ_3 b \in O$, that $(H, \circ_3, \mathcal{T})$ is neither a pseudotopological hypergroupoid nor a strongly pseudotopological hypergroupoid.

Theorem 12.8. *Let* $(H, \circ, \mathcal{T}, \mathcal{T}_{\mathcal{U}})$ *be a topological hypergroupoid. Then* $(H, \mathcal{T})$ *is the trivial topology if and only if* $(\mathcal{P}^*(H), \mathcal{T}_{\mathcal{U}})$ *is the trivial topology.*

Proof. Let $\mathcal{T} = \{\emptyset, H\}$. It is easy to see that $\mathcal{T}_{\mathcal{U}} = \{\emptyset, S_H\} = \{\emptyset, \mathcal{P}^*(H)\}$.

Let $\mathcal{T}_{\mathcal{U}} = \{\emptyset, \mathcal{P}^*(H)\}$ and let $V \neq \emptyset \in \mathcal{T}$. Then $S_V = \mathcal{P}^*(V) \in \mathcal{T}_{\mathcal{U}}$. The latter implies that $\mathcal{P}^*(V) = \mathcal{P}^*(H)$. Thus, $V = H$. □

Corollary 12.5. *Let* $(H, \circ, \mathcal{T}, \mathcal{T}_\aleph)$ *be a topological hypergroupoid. Then* $(H, \mathcal{T})$ *is the trivial topology if and only if* $(\mathcal{P}^*(H), \mathcal{T}_\aleph)$ *is the trivial topology.*

Proof. The proof is results from Theorems 12.3, 12.5 and 12.8. $\square$

Theorem 12.9. *Let* $(H, \circ, \mathcal{T}, \mathcal{T}_\mathcal{L})$ *be a topological hypergroupoid. Then* $(H, \mathcal{T})$ *is the trivial topology if and only if* $(\mathcal{P}^*(H), \mathcal{T}_\mathcal{L})$ *is the trivial topology.*

Proof. The proof is similar to that of Theorem 12.8. $\square$

Theorem 12.10. *Let* $(H, \circ, \mathcal{T}, \mathcal{T}_\mathcal{U})$ *be a topological hypergroupoid,* $|H| \geq 2$ *and* $(H, \mathcal{T})$ *be the powerset topology. Then* $(\mathcal{P}^*(H), \mathcal{T}_\mathcal{U})$ *is not the powerset topology on* $\mathcal{P}^*(H)$.

Proof. Let $a \neq b \in H$. It is clear that $\{\{a, b\}\}$ is not in $\mathcal{T}_\mathcal{U}$. $\square$

Theorem 12.11. *Let* $(H, \circ, \mathcal{T}, \mathcal{T}_\mathcal{L})$ *be a topological hypergroupoid,* $|H| \geq 2$ *and* $(H, \mathcal{T})$ *be the powerset topology. Then* $(\mathcal{P}^*(H), \mathcal{T}_\mathcal{L})$ *is not the powerset topology on* $\mathcal{P}^*(H)$.

Proof. Let $a \in H$. It is clear that $\{\{a\}\}$ is not in $\mathcal{T}_\mathcal{L}$. $\square$

Theorem 12.12. *Let* $(H_1, \circ_1, \mathcal{T})$ *and* $(H_2, \circ_2, \mathcal{T}')$ *be two topologically isomorphic hypergroupoids. If* $(H_1, \circ_1, \mathcal{T}, \mathcal{T}_\mathcal{U})$ *is a* $\mathcal{T}_\mathcal{U}$*-topological hypergroupoid then* $(H_2, \circ_2, \mathcal{T}', \mathcal{T}_\mathcal{U}')$ *is a* $\mathcal{T}_\mathcal{U}$*-topological hypergroupoid.*

Proof. Let $x, y \in H_2$ and $O \in \mathcal{T}'$ such that $x \circ_2 y \subset O$. Since $(H_1, \circ_1, \mathcal{T})$ and $(H_2, \circ_2, \mathcal{T}')$ are topologically isomorphic hypergroupoids, it follows that there exist an isomorphism $f : (H_1, \circ_1, \mathcal{T}) \to (H_2, \circ_2, \mathcal{T}')$. The latter implies that there exist $a, b \in H_1$ such that $x = f(a)$ and $y = f(b)$. Using Definition 12.6, we get that $a \circ_1 b \subset f^{-1}(O)$. Moreover, $f^{-1}(O)$ is open in $\mathcal{T}$ as f is continuous. Since "$\circ_1$" is p-continuous, it follows that there exist $U_1, U_2 \in \mathcal{T}$ such that $(a, b) \in U_1 \times U_2$ and for all $(u_1, u_2) \in U_1 \times U_2$, we have $u_1 \circ_1 u_2 \subset f^{-1}(O)$. Let $V_1 = f(U_1)$ and $V_2 = f(U_2)$. We have that $x \in V_1, y \in V_2$ and $V_1, V_2 \in \mathcal{T}'$ as f is open. For every $(v_1, v_2) \in V_1 \times V_2$, there exist $(u_1, u_2) \in U_1 \times U_2$ such that $u_1 \circ_1 u_2 \subset f^{-1}(O)$. The latter implies that $v_1 \circ_2 v_2 = f(u_1) \circ_2 f(u_2) = f(u_1 \circ_1 u_2) \subset O$. Thus, "$\circ_2$" is p-continuous. $\square$

Theorem 12.13. *Let* $(H_1, \circ_1, \mathcal{T})$ *and* $(H_2, \circ_2, \mathcal{T}')$ *be two topologically isomorphic hypergroupoids. If* $(H_1, \circ_1, \mathcal{T}, \mathcal{T}_\mathcal{L})$ *is a* $\mathcal{T}_\mathcal{L}$*-topological hypergroupoid then* $(H_2, \circ_2, \mathcal{T}', \mathcal{T}_\mathcal{L}')$ *is a* $\mathcal{T}_\mathcal{L}$*-topological hypergroupoid.*

Proof. Let $x, y \in H_2$ and $O \in \mathcal{T}'$ such that $x \circ_2 y \approx O$. Since $(H_1, \circ_1, \mathcal{T})$ and $(H_2, \circ_2, \mathcal{T}')$ are topologically isomorphic hypergroupoids, it follows that there exist an isomorphism $f : (H_1, \circ_1, \mathcal{T}) \to (H_2, \circ_2, \mathcal{T}')$. The latter implies that there exist $a, b \in H_1$ such that $x = f(a)$ and $y = f(b)$. Using Definition 12.6, we get that $a \circ_1 b \approx f^{-1}(O)$. Moreover, $f^{-1}(O)$ is open in $\mathcal{T}$ as f is continuous. Since "$\circ_1$" is sp-continuous, it follows that there exist $U_1, U_2 \in \mathcal{T}$ such that $(a, b) \in U_1 \times U_2$ and for all $(u_1, u_2) \in U_1 \times U_2$, we have $u_1 \circ_1 u_2 \approx f^{-1}(O)$. Let $V_1 = f(U_1)$ and $V_2 = f(U_2)$. We have that $x \in V_1, y \in V_2$ and $V_1, V_2 \in \mathcal{T}'$ as f is open. For every $(v_1, v_2) \in V_1 \times V_2$, there exist $(u_1, u_2) \in U_1 \times U_2$ such that $u_1 \circ_1 u_2 \approx f^{-1}(O)$. The latter implies that $v_1 \circ_2 v_2 = f(u_1) \circ_2 f(u_2) = f(u_1 \circ_1 u_2) \approx O$. Thus, "$\circ_2$" is sp-continuous. $\square$

Corollary 12.6. *Let $(H_1, \circ_1, \mathcal{T})$ and $(H_2, \circ_2, \mathcal{T}')$ be two topologically isomorphic hypergroupoids. If $(H_1, \circ_1, \mathcal{T}, \mathcal{T}_\aleph)$ is a $\mathcal{T}_\aleph$-topological hypergroupoid then $(H_2, \circ_2, \mathcal{T}', \mathcal{T}'_\aleph)$ is a $\mathcal{T}_\aleph$-topological hypergroupoid.*

Proof. The proof results from Theorems 12.12 and 12.13. $\square$

We present now some $\mathcal{T}_\aleph$-topological hypergroupoids.

Theorem 12.14. *Let $(H, \circ)$ be the total hypergroup (i.e., $x \circ y = H$ for all $(x, y) \in H^2$) and $\mathcal{T}$ be any topology on H. Then $(H, \circ, \mathcal{T})$ is both: pseudotopological hypergroupoid and strongly pseudotopological hypergroupoid.*

Proof. Let $(x, y) \in H^2$ and $O \in \mathcal{T}$ such that $x \circ y \subset O$. Since $x \circ y = H$, it follows that $O = H$. Take $U = V = H \in \mathcal{T}$. For every $(u, v) \in U \times V$, we have that $u \circ v = H$. Thus, $u \circ v \subset O$. Definition 12.8 implies that $(H, \circ, \mathcal{T})$ is a pseudotopological hypergroupoid.

To prove that $(H, \circ, \mathcal{T})$ is a strongly pseudotopological hypergroupoid, let $(x, y) \in H^2$ and $O \in \mathcal{T}$ such that $x \circ y \approx O$. It is clear that O is a non-empty subset of H. Take $U = V = H \in \mathcal{T}$. For every $(u, v) \in U \times V$, we have that $u \circ v = H$. Thus, $u \circ v \approx O$. Definition 12.8 implies that $(H, \circ, \mathcal{T})$ is a strongly pseudotopological hypergroupoid. $\square$

Corollary 12.7. *Let $(H, \circ)$ be the total hypergroup and $\mathcal{T}$ be any topology on H. Then the quadruple $(H, \circ, \mathcal{T}, \mathcal{T}_\aleph)$ is a $\mathcal{T}_\aleph$-topological hypergroupoid.*

Proof. The proof follows from Theorems 12.5 and 12.14. $\square$

Theorem 12.15. *Let $H = \mathbb{R}$, $(H, \circ)$ be the hypergroupoid defined by:*

$$x \circ y = \begin{cases} \{a \in \mathbb{R} : x \le a \le y\}, \text{ if } x \le y, \\ \{a \in \mathbb{R} : y \le a \le x\}, \text{ if } y \le x; \end{cases}$$

and $\mathcal{T}$ *be the topology on* H *defined by:*

$$\mathcal{T} = \{\,]-\infty, a[\mid a \in \mathbb{R} \cup \{\pm\infty\}\}.$$

Then $(H, \circ, \mathcal{T}, \mathcal{T}_\aleph)$ *is a* $\mathcal{T}_\aleph$*-topological hypergroupoid.*

Proof. By means of Definition 12.8, Theorems 12.3, 12.4 and 12.5, it suffices to show that "$\circ$" is both: p-continuous and sp-continuous.

Let $x \leq y \in \mathbb{R}$ and $O \in \mathcal{T}$ such that $[x, y] = x \circ y \subset O$. It is clear that there exist $c \in \mathbb{R}$ such that $O =\,]-\infty, c[$ and $x \leq y < c$. Take $U = V = O \in \mathcal{T}$. For every $(u, v) \in U \times V = O \times O$, we have that $u \circ v = [u, v] \subset O$. Thus, "$\circ$" is p-continuous.

Let $x \leq y \in \mathbb{R}$ and $O \in \mathcal{T}$ such that $[x, y] = x \circ y \approx O =\,]-\infty, c[$. Then $x < c$ (otherwise, we get $x \circ y \cap O = \emptyset$). Take $U = O \in \mathcal{T}$ and $V = H \in \mathcal{T}$. For every $(u, v) \in U \times V = O \times H$, we have that $u \circ v = [u, v] \approx O$ (as $u \in O \cap (u \circ v)$). Thus, "$\circ$" is sp-continuous. $\qquad\square$

Theorem 12.16. *Let* $H = \mathbb{R}$, $(H, \circ)$ *be the hypergroupoid defined by:*

$$x \circ y = \begin{cases} \{a \in \mathbb{R} : x \leq a \leq y\}, & \text{if } x \leq y, \\ \{a \in \mathbb{R} : y \leq a \leq x\}, & \text{if } y \leq x; \end{cases}$$

and $\mathcal{T}$ *be the topology on* H *defined by:*

$$\mathcal{T} = \{\,]a, \infty[\mid a \in \mathbb{R} \cup \{\pm\infty\}\}.$$

Then $(H, \circ, \mathcal{T}', \mathcal{T}'_\aleph)$ *is a* $\mathcal{T}'_\aleph$*-topological hypergroupoid.*

Proof. Let $f : (\mathbb{R}, \circ, \mathcal{T}) \to (\mathbb{R}, \circ, \mathcal{T}')$ defined by $f(x) = -x$ for all $x \in \mathbb{R}$. It is clear that f is topological isomorphism. Using Theorem 12.15, we have that $(H, \circ, \mathcal{T}, \mathcal{T}_\aleph)$ is a $\mathcal{T}_\aleph$-topological hypergroupoid. Thus, $(H, \circ, \mathcal{T}', \mathcal{T}'_\aleph)$ is a $\mathcal{T}'_\aleph$-topological hypergroupoid by using Corollay 12.7. $\qquad\square$

Theorem 12.17. *Let* $(H, \star)$ *be the hypergroupoid defined by* $x \star y = \{x, y\}$ *and* $\mathcal{T}$ *be any topology on* H. *Then* $(H, \star, \mathcal{T}, \mathcal{T}_\aleph)$ *is a* $\mathcal{T}_\aleph$*-topological hypergroupoid.*

Proof. By means of Definition 12.8, Theorems 12.3, 12.4 and 12.5, it suffices to show that "$\star$" is both: p-continuous and sp-continuous.

Let $(x, y) \in H^2$ and $O \in \mathcal{T}$ such that $\{x, y\} = x \star y \subset O$. Take $U = V = O \in \mathcal{T}$. For every $(u, v) \in U \times V = O \times O$, we have that $u \star v = \{u, v\} \subset O$. Thus, $\star$ is p-continuous.

Let $(x, y) \in H^2$ and $O \in \mathcal{T}$ such that $\{x, y\} = x \star y \approx O$. Then $x \in O$ or $y \in O$. Without loss of generality, let $x \in O$. Take $U = O \in \mathcal{T}$ and $V = H \in \mathcal{T}$. For every $(u, v) \in U \times V = O \times H$, we have that $u \star v = \{u, v\} \approx O$ (as $u \in O$). Thus, $\star$ is sp-continuous. $\qquad\square$

Example 12.5. Let $H = \{a, b\}$, $\mathcal{T} = \{\emptyset, \{a\}, H\}$ and define a hyperoperation "$\circ_4$" on H as follows:

$\circ_4$	a	b
a	a	H
b	H	b

Then, by Theorem 12.17, $(H, \circ_4, \mathcal{T}, \mathcal{T}_\aleph)$ is a $\mathcal{T}_\aleph$-topological hypergroupoid. Moreover, $\mathcal{T}_\aleph = \{\emptyset, \{\{a\}\}, \mathcal{P}^*(H)\}$.

Theorem 12.18. *Let* $(H, \circ)$ *be any hypergroupoid and* $\mathcal{T}$ *be the power set topology on* H. *Then* $(H, \circ, \mathcal{T}, \mathcal{T}_\aleph)$ *is a* $\mathcal{T}_\aleph$-*topological hypergroupoid.*

Proof. By means of Definition 12.8, Theorems 12.3, 12.4 and 12.5, it suffices to show that "$\circ$" is both: p-continuous and sp-continuous.

Let $(x, y) \in H^2$ and $O \in \mathcal{T}$ such that $x \circ y \subset O$. Take $U = \{x\} \in \mathcal{T}$ and $V = \{y\} \in \mathcal{T}$. For every $(u, v) \in U \times V = \{(x, y)\}$, we have that $u \circ v = x \circ y$. Thus, $u \circ v \subset O$. Thus, "$\circ$" is p-continuous.

Let $(x, y) \in H^2$ and $O \in \mathcal{T}$ such that $x \circ y \approx O$. Take $U = \{x\} \in \mathcal{T}$ and $V = \{y\} \in \mathcal{T}$. For every $(u, v) \in U \times V = \{(x, y)\}$, we have that $u \circ v = x \circ y$. Thus, $u \circ v \approx O$. Thus, "$\circ$" is sp-continuous. □

Theorem 12.19. *Let* $(H, \circ)$ *be any hypergroupoid and* $\mathcal{T}$ *be the trivial topology on* H. *Then* $(H, \circ, \mathcal{T}, \mathcal{T}_\aleph)$ *is a* $\mathcal{T}_\aleph$-*topological hypergroupoid.*

Proof. By means of Definition 12.8, Theorems 12.3, 12.4 and 12.5, it suffices to show that "$\circ$" is both: p-continuous and sp-continuous.

Let $(x, y) \in H^2$ and $O \in \mathcal{T}$ such that $x \circ y \subset O$. It is clear that $O = H$. Take $U = V = H \in \mathcal{T}$. For every $(u, v) \in U \times V$, we have that $u \circ v \subset O$. Thus, "$\circ$" is p-continuous.

Let $(x, y) \in H^2$ and $O \in \mathcal{T}$ such that $x \circ y \approx O$. It is clear that $O = H$. Take $U = V = H \in \mathcal{T}$. For every $(u, v) \in U \times V$, we have that $u \circ v \approx O$. Thus, "$\circ$" is sp-continuous. □

Next, we present a topological hypergroupoid that does not depend on the pseodocontinuity nor on strongly pseodocontinuity of the hyperoperation.

Theorem 12.20. *Let* $(H, \circ)$ *be any hypergroupoid,* $\mathcal{T}$ *be any topology on* H *and* $\mathcal{T}_*$ *be the trivial topology on* $\mathcal{P}^*(H)$. *Then* $(H, \circ, \mathcal{T}, \mathcal{T}_*)$ *is a topological hypergroupoid.*

Proof. We need to show that " $\circ$ " is $\mathcal{T}_*$-continuous. Define $\varphi : H \times H \to \mathcal{P}^*(H)$ the mapping that assigns the product $x \circ y$ to the pair (x, y). We have only two elements in $\mathcal{T}_*$: $\emptyset$ and $\mathcal{P}^*(H)$. Since $\varphi^{-1}(\emptyset) = \emptyset$ and $\varphi^{-1}(\mathcal{P}^*(H)) = H \times H$, it follows that φ is continuous. $\square$

Next, we present some results on T_0, T_1, T_2-topological spaces.

Theorem 12.21. *Let $(H, \circ, \mathcal{T}, \mathcal{T}_\mathcal{U})$ be a $\mathcal{T}_\mathcal{U}$-topological hypergroupoid. If $(\mathcal{P}^*(H), \mathcal{T}_\mathcal{U})$ is a T_0-topological space then $(H, \mathcal{T})$ is a T_0-topological space.*

Proof. Let $a, b \in H$ such that $a \neq b$. Since $(\mathcal{P}^*(H), \mathcal{T}_\mathcal{U})$ is a T_0-topological space and $\{a\} \neq \{b\} \in \mathcal{P}^*(H)$, it follows that there exists $M \in \mathcal{T}_\mathcal{U}$ such that $\{a\} \in M$ and $\{b\}$ is not in M (or $\{b\} \in M$ and $\{a\}$ is not in M). Without loss of generality, we suppose that there exists $M \in \mathcal{T}_\mathcal{U}$ such that $\{a\} \in M = \bigcup S_V$ and $\{b\}$ is not in M. The latter implies that there exists $V \in \mathcal{T}$ such that $\{a\} \in S_V = \mathcal{P}^*(V)$ and $\{b\}$ is not in $\mathcal{P}^*(V)$. We get now that $a \in V$ and b is not in V. $\square$

Remark 12.5. The converse of Theorem 12.21 is not always true.

We illustrate Remark 12.5 with the following example.

Example 12.6. Let $H = \{a, b, c\}$ and define $(H, \circ)$ as the total hypergroup. Define $\mathcal{T} = \{\emptyset, \{a\}, \{a, c\}, H\}$ a topology on H. Theorem 12.14 asserts that $(H, \circ, \mathcal{T}, \mathcal{T}_\mathcal{U})$ is a topological hypergroupoid. It is easy to check that $\mathcal{T}_\mathcal{U} = \{\emptyset, \{\{a\}\}, \{\{a\}, \{c\}, \{a, c\}\}, \mathcal{P}^*(H)\}$. We have that $(H, \mathcal{T})$ a T_0-topological space, whereas $(\mathcal{P}^*(H), \mathcal{T}_\mathcal{U})$ is not a T_0-topological space ($\{a, b\}$ and $\{b, c\}$ do not satisfy the condition of having T_0-topological space).

Theorem 12.22. *Let $|H| \geq 2$ and $(H, \circ, \mathcal{T}, \mathcal{T}_\mathcal{U})$ be a $\mathcal{T}_\mathcal{U}$-topological hypergroupoid. Then $(\mathcal{P}^*(H), \mathcal{T}_\mathcal{U})$ is neither a T_1-topological space nor a T_2-topological space.*

Proof. Since $|H| \geq 2$, it follows that there exist $a \in H$ such that $\{a\} \neq H$. We can find an open set $M \in \mathcal{T}_\mathcal{U}$ such that $H \in M$. Having $\{S_V\}_{V \in \mathcal{T}}$ a basis for $\mathcal{T}_\mathcal{U}$ implies that $M = \bigcup S_V$ for some $V \in \mathcal{T}$. The latter implies that $H \in S_V = \mathcal{P}^*(V)$ for some $V \in \mathcal{T}$. As a result, we get that $M = \mathcal{P}^*(H)$. The latter implies that $\{a\} \in M$. $\square$

We illustrate Theorems 12.21 and 12.22 by the following example.

Example 12.7. Let $H = \{a, b\}$ and define $(H, \circ)$ as the total hypergoup. Define $\mathcal{T} = \{\emptyset, \{a\}, \{b\}, H\}$ a topology on H. Theorem 12.14 asserts that

$(H, \circ, \mathcal{T}, \mathcal{T}_{\mathcal{U}})$ is a topological hypergroupoid. It is easy to check that $\mathcal{T}_{\mathcal{U}} = \{\emptyset, \{\{a\}\}, \{\{b\}\}, \{\{a\}, \{b\}\}, \mathcal{P}^*(H)\}$. We have that $(H, \mathcal{T})$ a T_1-topological space (and consequently a T_0-topological space), whereas $(\mathcal{P}^*(H), \mathcal{T}_{\mathcal{U}})$ is a T_0-topological space but not a T_1-topological space ($\{a\}$ and $\{a, b\}$ do not satisfy the condition of having T_1-topological space).

Theorem 12.23. *Let $|H| \geq 2$ and $(H, \circ, \mathcal{T}, \mathcal{T}_{\mathcal{L}})$ be a $\mathcal{T}_{\mathcal{L}}$-topological hypergroupoid. Then $(\mathcal{P}^*(H), \mathcal{T}_{\mathcal{L}})$ is neither a T_1-topological space nor a T_2-topological space.*

Proof. Since every T_2-topological space is a T_1-topological space, it suffices to show that $(\mathcal{P}^*(H), \mathcal{T}_{\mathcal{L}})$ is not a T_1-topological space. For all $V \in \mathcal{T}$, we have that $H \in I_V \supseteq \{H, \mathcal{P}^*(V)\}$. The latter implies that for all $M \neq \emptyset \in \mathcal{T}_{\mathcal{L}}$, we have that $H \in M$. Thus, $(\mathcal{P}^*(H), \mathcal{T}_{\mathcal{L}})$ is not a T_1-topological space. $\square$

Theorem 12.24. *Let $(H, \circ, \mathcal{T}, \mathcal{T}_{\mathcal{L}})$ be a $\mathcal{T}_{\mathcal{L}}$-topological hypergroupoid. If $(\mathcal{P}^*(H), \mathcal{T}_{\mathcal{L}})$ is a T_0-topological space then $(H, \mathcal{T})$ is a T_0-topological space.*

Proof. Let $a, b \in H$ such that $a \neq b$. Since $(\mathcal{P}^*(H), \mathcal{T}_{\mathcal{L}})$ is a T_0-topological space and $\{a\} \neq \{b\} \in \mathcal{P}^*(H)$, it follows that there exists $M \in \mathcal{T}_{\mathcal{L}}$ such that $\{a\} \in M$ and $\{b\}$ is not in M (or $\{b\} \in M$ and $\{a\}$ is not in M). Without loss of generality, we suppose that there exists $M \in \mathcal{T}_{\mathcal{L}}$ such that $\{a\} \in M$ and $\{b\}$ is not in M. Since $\{I_V\}_{V \in \mathcal{T}}$ is a subbase for $\mathcal{T}_{\mathcal{L}}$, it follows that there exists $V \in \mathcal{T}$ such that $\{a\} \in I_V$ and $\{b\}$ is not in I_V. We get now that $a \in V$ and b is not in V. $\square$

We illustrate Theorem 12.24 by the following example.

Example 12.8. Let $H = \{a, b\}$ and define $(H, \circ)$ as the total hypergoup. Define $\mathcal{T} = \{\emptyset, \{a\}, \{b\}, H\}$ a topology on H. Theorem 12.14 asserts that $(H, \circ, \mathcal{T}, \mathcal{T}_{\mathcal{L}})$ is a topological hypergroupoid. It is easy to check that $\mathcal{T}_{\mathcal{L}} = \{\emptyset, \{\{a\}, H\}, \{\{b\}, H\}, H, \mathcal{P}^*(H)\}$. We have that $(H, \mathcal{T})$ is a T_0-topological space and $(\mathcal{P}^*(H), \mathcal{T}_{\mathcal{L}})$ is a T_0-topological space.

Theorem 12.25. *Let $(H, \circ, \mathcal{T}, \mathcal{T}_{\mathcal{L}})$ be a $\mathcal{T}_{\mathcal{L}}$-topological hypergroupoid. If $(H, \mathcal{T})$ is a T_0-topological space then $(\mathcal{P}^*(H), \mathcal{T}_{\mathcal{L}})$ may not be a T_0-topological space.*

Proof. One can easily check that $\mathcal{T}$ is Theorem 12.7 is a T_0-topological space whereas $\mathcal{T}_{\mathcal{L}}$ is not a T_0-topological space ($\{a\}$ and H violate the condition of having T_0-topological space). $\square$

Corollary 12.8. *Let* $(H, \circ, \mathcal{T}, \mathcal{T}_\aleph)$ *be a* $\mathcal{T}_\aleph$-*topological hypergroupoid. If* $(\mathcal{P}^*(H), \mathcal{T}_\aleph)$ *is a* T_0-*topological space then* $(H, \mathcal{T})$ *is a* T_0-*topological space.*

Proof. The proof results from Theorems 12.21 and 12.24. □

Corollary 12.9. *Let* $(H, \circ, \mathcal{T}, \mathcal{T}_\aleph)$ *be a* $\mathcal{T}_\aleph$-*topological hypergroupoid. Then* $(\mathcal{P}^*(H), \mathcal{T}_\aleph)$ *is neither a* T_2-*topological space nor a* T_1-*topological space.*

Proof. The proof results from Theorem 12.23 and from the fact that every $\mathcal{T}_\aleph$-topological hypergroupoid both: $\mathcal{T}_\mathcal{U}$ and $\mathcal{T}_\mathcal{L}$-topological hypergroupoid. □

12.3 Topological hypergroups

Topological hypergroups in the sense of Marty are introduced by Heidari, Davvaz and Modarres [76]. In this section, we define and study the concept of topological hypergroups and we prove some properties in this direction. The results of this section are contained in [76].

Let $(H, \mathcal{T})$ be a topological space. In order to construct a topological hypergroup we need a topology on $\mathcal{P}^*(H)$. The following lemma give us a topology on $\mathcal{P}^*(H)$ induced by $\mathcal{T}$. According to Lemma 12.1, the family $\mathcal{U}$ consisting of all sets

$$S_V = \{U \in \mathcal{P}^*(H) \mid U \subset V\},$$

$V \in \mathcal{T}$, is a base for a topology on $\mathcal{P}^*(H)$. This topology is denoted by $\mathcal{T}^*$.

Let $(H, \mathcal{T})$ be a topological space. Then, we consider the product topology on $H \times H$ and the topology $\mathcal{T}^*$ on $\mathcal{P}^*(H)$.

Definition 12.9. Let $(H, \circ)$ be a hypergroup and $(H, \mathcal{T})$ be a topological space. Then, the system $(H, \circ, \mathcal{T})$ is called a *topological hypergroup* if

(1) the mapping $(x, y) \mapsto x \circ y$, from $H \times H$ to $\mathcal{P}^*(H)$ is continuous;
(2) the mapping $(x, y) \mapsto x/y$, from $H \times H$ to $\mathcal{P}^*(H)$ is continuous, where $x/y = \{z \in H \mid x \in z \circ y\}$.

Let H be a hypergroup and A and B be non-empty subsets of H. Then, $A/B = \cup\{a/b \mid a \in A, b \in B\}$.

Lemma 12.5. *Let* $(H, \circ)$ *be a hypergroup and* $\mathcal{T}$ *be a topology on* H. *Then, the following assertions hold:*

(1) *The mapping* $(x, y) \mapsto x \circ y$ *is continuous if and only if for every* $x, y \in H$ *and* $U \in \mathcal{T}$ *such that* $x \circ y \subset U$, *there exist* $V, W \in \mathcal{T}$ *such that* $x \in V$, $y \in W$ *and* $V \circ W \subset U$.

(2) *The mapping* $(x, y) \mapsto x/y$ *is continuous if and only if for every* $x, y \in H$ *and* $U \in \mathcal{T}$ *such that* $x/y \subset U$, *there exist* $V, W \in \mathcal{T}$ *such that* $x \in V$, $y \in W$ *and* $V/W \subset U$.

Proof. The proofs of (1) and (2) are similar, so we prove (1). Suppose that the hyperoperation $\circ : H \times H \to \mathcal{P}^*(H)$ is continuous and $x \circ y \subset U$ for $x, y \in H$ and $U \in \mathcal{T}$. Then, $\circ^{-1}(S_U)$ is an open subset of $H \times H$. So there exist open subsets V and W of H such that $(x, y) \in V \times W \subset \circ^{-1}(S_U)$. Thus, $\circ(V \times W) \subset \circ(\circ^{-1}(S_U)) \subset S_U$. Therefore, we have

$$V \circ W = \bigcup_{\substack{v \in V \\ w \in W}} v \circ w = \bigcup_{\substack{v \in V \\ w \in W}} \circ(v, w) \subset U.$$

Conversely, suppose that $U \in \mathcal{T}$ and $(x, y) \in \circ^{-1}(S_U)$. Then, we show that (x, y) is an interior point of $\circ^{-1}(S_U)$. Since $x \circ y \subset U$, there exist $V, W \in \mathcal{T}$ such that $V \circ W \subset U$ where $x \in V$ and $y \in W$. So, we have $(x, y) \in V \times W \subset \circ^{-1}(S_U)$. Thus, $\circ^{-1}(S_U)$ is open in H. Now, if $A \in \mathcal{T}^*$, then $A = \bigcup_{U \in \Lambda} S_U$, where Λ is a non-empty subset of $\mathcal{T}$. Thus, $\circ^{-1}(A) = \circ^{-1}(\bigcup_{U \in \Lambda} S_U) = \bigcup_{U \in \Lambda} \circ^{-1}(S_U)$. Hence, $\circ^{-1}(A)$ is open in H. Therefore, the mapping $\circ$ is continuous. $\qquad \square$

Evidently, every topological group is a topological hypergroup. In what follows, we give some other examples.

Example 12.9. Every hypergroup with discrete or indiscrete topology is a topological hypergroup.

Example 12.10. The total hypergroup H (the combination of any two elements is the set H) with every arbitrary topology is a topological hypergroup.

Definition 12.10. Suppose that X is a topological space. Let x and y be two points in X. We say that x and y can be *separated* by open subsets if there exist open subsets U and V of X containing x and y, respectively, such that U and V are disjoint. A *Hausdorff space* is a topological space in which points can be separated by open subsets.

Example 12.11. Let $(X, \mathcal{T})$ be a Hausdorff topological space. For every $x, y \in X$ we define $x \circ y = \{x, y\}$. Then, $(X, \circ, \mathcal{T})$ is a topological hypergroup.

Indeed, suppose that $U \in \mathcal{T}$. Then, $U \circ U = U$ which implies the continuity of $\circ$. Now, let $x, y \in X$ such that $x/y \subset U$. Since X is Hausdorff then there exist open disjoint subsets V and W of X containing x and y, respectively. Now, we have $(V \cap U)/W = V \cap U \subset U$. So the mapping $(x, y) \mapsto x/y$ is continuous.

Example 12.12. Consider the set of integer numbers $\mathbb{Z}$ and define the hyperoperation $\circ$ on it as follows

$$m \circ n = \begin{cases} \mathbb{E} & \text{if } m + n \in \mathbb{E} \\ \mathbb{E}^c & \text{otherwise,} \end{cases}$$

where $\mathbb{E} = 2\mathbb{Z}$. Then, $(\mathbb{Z}, \circ)$ is a hypergroup. Let $\mathcal{T} = \{\emptyset, \mathbb{E}, \mathbb{E}^c, \mathbb{N}\}$. Then, $\mathcal{T}$ is a topology on $\mathbb{Z}$ and $(\mathbb{Z}, \circ, \mathcal{T})$ is a topological hypergroup.

Some properties in topological groups do not hold in topological hypergroups. For instance:

Let G be a topological group and U be an open subset of G. Then, aU is open in G for every $a \in G$.

Now, see the following example.

Example 12.13. Consider Example 12.11. Let $X = \mathbb{R}$ with the standard topology. Then, $(\mathbb{R}, \circ, \mathcal{T})$ is a topological hypergroup, where $x \circ y = \{x, y\}$. Now, $2 \circ (0, 1) = \{2\} \cup (0, 1)$ is not open in $\mathbb{R}$.

In what follows, we prove that the fundamental group of a topological hypergroup is a topological group. First, we recall the notion of quotient space of a topological space.

Definition 12.11. Let X be a topological space and $\sim$ be an equivalence relation on X. For every $x \in X$, denote by $[x]$ its equivalence class. The *quotient space* of X modulo $\sim$ is given by the set $X/\sim = \{[x] \mid x \in X\}$. We have the projection map $p : X \to X/\sim$, $x \mapsto [x]$ and we endow $X/\sim$ by the topology: $U \subset X/\sim$ is open if and only if $p^{-1}(U)$ is an open subset of X.

Let A be a subset of the topological space X and $\sim$ be an equivalence relation on X. Then, the *saturation* of A with respect to $\sim$ is the set $\widehat{A} = \{x \in X \mid \exists a \in A, x \sim a\}$. If $\widehat{A} = A$, then A is called *saturated*.

Let $(H, \circ, \mathcal{T})$ be a topological hypergroup and β^* be the fundamental relation on H. Then, $(H/\beta^*, \overline{\mathcal{T}})$ is a topological space, where $\overline{\mathcal{T}}$ is the quotient topology induced by the natural mapping $\pi : H \to H/\beta^*$. That is $A \subset H/\beta^*$ is open in H/β^* if and only if $\pi^{-1}(A)$ is open in H.

Lemma 12.6. *Let $(H, \circ, \mathcal{T})$ be a topological hypergroup and β^* be the fundamental relation on H. Then, every saturated subset of H is a complete part.*

Proof. Suppose that A is a saturated subset of H such that $A \cap \prod_{i=1}^{n} x_i \neq \emptyset$ for a non-zero natural number n and for some $x_1, \ldots, x_n \in H$. Hence, there exists $a \in A \cap \prod_{i=1}^{n} x_i$. Then, for every $x \in \prod_{i=1}^{n} x_i$ we have $x \beta^* a$. Thus, $x \in \widehat{A} = A$. Therefore, A is a complete part. $\square$

Lemma 12.7. *Let $(H, \circ, \mathcal{T})$ be a topological hypergroup such that every open subset of H is a complete part. Then, the natural mapping $\pi : H \to H/\beta^*$ is an open mapping.*

Proof. Suppose that V is open in H. We prove that $\pi^{-1}(\pi(V))$ is an open subset in H. Let $x \in \pi^{-1}(\pi(V))$. Then, $\pi(x) \in \pi(V)$ so there exists $v \in V$ such that $\pi(x) = \pi(v)$. Since V is open so there exists an open subset U of H such that $v \in U \subset V$. On the other hand, $\pi(x) = \pi(v)$ implies that $x \beta^* y$ so there exist a non-zero natural number n and $a_1, \ldots, a_n \in H$ such that $\{x, v\} \subset \prod_{i=1}^{n} a_i$. Now, we have $v \in \prod_{i=1}^{n} a_i \cap U$ and U is complete part so $x \in \prod_{i=1}^{n} a_i \subset U \subset V$. Hence, $x \in U \subset \pi^{-1}(\pi(V))$. Thus, x is an interior point of $\pi^{-1}(\pi(V))$ so $\pi^{-1}(\pi(V))$ is open in H. Therefore, $\pi(V)$ is open in H/β^*. $\square$

Theorem 12.26. *Let $(H, \circ, \mathcal{T})$ be a topological hypergroup such that every open subset of H is a complete part. Then, $(H/\beta^*, \odot, \overline{\mathcal{T}})$ is a topological group.*

Proof. We know that $(H/\beta^*, \odot)$ is a group. We show that the mappings $(\pi(x), \pi(y)) \mapsto \pi(x) \odot \pi(y)$ and $\pi(x)) \mapsto \pi(x)^{-1}$ are continuous.

Suppose that A is open in H/β^* such that $\pi(x) \odot \pi(y) \in A$. So $x \circ y \subset \pi^{-1}(A)$. Since $\pi^{-1}(A)$ is open in H so by Lemma 12.5, there exist open subsets V and W of H such that $x \in V$, $y \in W$ and $V \circ W \subset \pi^{-1}(A)$. Thus, $\pi(V) \odot \pi(W) \subset A$. By Lemma 12.7, $\pi(V)$ and $\pi(W)$ are open in H/β^*. Hence, $\odot$ is continuous.

Now, we prove that the inverse mapping is continuous. Suppose that A is open in H/β^* such that $\pi(x) \in A^{-1}$ and let $\pi(e)$ be the identity element of H/β^*. Then, $e/x = \{z \in H \mid e \in z \circ x\} \subset \pi^{-1}(A)$. So, by Lemma 12.5, there exist open subsets V and W in H such that $e \in V$, $x \in W$ and

$V/W \subset \pi^{-1}(A)$. Thus, $\pi(x) \in \pi(W) \subset A^{-1}$. Therefore, A^{-1} is open in H/β^*. $\qquad\qquad\qquad\qquad\qquad\qquad\qquad\qquad\qquad\qquad\qquad\qquad$ $\square$

The following corollary is obtained from pervious lemma and theorem.

Corollary 12.10. *Let H be a topological hypergroup such that every saturated subset of H is open. Then, $(H/\beta^*, \odot, \overline{\mathcal{T}})$ is a topological group.*

Definition 12.12. If $(G, \times_G, \mathcal{T}_G)$ and $(H, \times_H, \mathcal{T}_H)$ are topological groups we say that $\varphi : G \to H$ is a topological isomorphism if it is a group isomorphism and a topological homeomorphism. Recall that φ is a homeomorphism if φ is one-to-one, onto and both φ and φ^{-1} are continuous.

Definition 12.13. Let G be a group and $\{A_g\}_{g \in G}$ be a collection of disjoint non-empty sets. Let $H = \bigcup\limits_{g \in G} A_g$ and for every $x, y \in H$ define $x \circ y = A_{g_x g_y}$, where $x \in A_{g_x}$ and $y \in A_{g_y}$. Then, $(H, \circ)$ is a hypergroup and it is called a (G, H)-*hypergroup*.

Lemma 12.8. *Let $(G, \cdot, \mathcal{T})$ be a topological group and $\{A_g\}_{g \in G}$ be a collection of disjoint non-empty sets and let $H = \bigcup\limits_{g \in G} A_g$. Then, $(H, \circ, \mathcal{T}_H)$ is a topological hypergroup, where H is a (G, H)-hypergroup and $\mathcal{T}_H = \{\bigcup\limits_{u \in U} A_u \mid U \in \mathcal{T}\} \cup \{\emptyset\}$.*

Proof. It is easy to see that $\mathcal{T}_H$ is a topology on H. Suppose that $A_U = \bigcup\limits_{u \in U} A_u$ is an open subset of H such that $x \circ y \subset A_U$ for $x, y \in H$. Since U is open in G and $g_x g_y \in U$, there exist open subsets V and W of G containing g_x and g_y, respectively, such that $VW \subset U$. So A_V and A_W are open in H containing x and y, respectively, and $A_V \circ A_W \subset A_U$. Thus, $\circ$ is continuous.

Suppose that $x/y = A_{g_x g_y^{-1}} \subset A_U = \bigcup\limits_{u \in U} A_u$ is an open subset of H, where $x \in A_{g_x}, y \in A_{g_y}$ and U is open in G. So there exist open open subsets V and W of G containing g_x and g_y, respectively, such that $VW^{-1} \subset U$. Hence, we have $A_V/A_W \subset A_U$. So, the mapping $(x, y) \mapsto x/y$ is continuous. $\qquad$ $\square$

Theorem 12.27. *Let $(G, \cdot, \mathcal{T})$ be a topological group and $\{A_g\}_{g \in G}$ be a collection of disjoint non-empty sets and let $H = \bigcup\limits_{g \in G} A_g$. Then, the fundamental group of (G, H)-hypergroup H and G are topological isomorphic.*

Proof. Suppose that $H = \bigcup\limits_{g \in G} A_g$. Then, for every $x \in H$ there exists $g_x \in G$ such that $x \in A_{g_x}$. Now, we define $\varphi : H/\beta^* \to G$ by $\varphi(\beta^*(x)) = g_x$. It is not difficult to see that φ is a group isomorphism.

Obviously, $\varphi(\pi(A_U)) = U$ for every open subset U of H. So φ is open and continuous. $\qquad\square$

Theorem 12.28. *Let $(G, \cdot)$ be a topological group and let H be a non-normal subgroup of G. Then, $G/H = \{xH \mid x \in G\}$ is a topological space with respect to the quotient topology induced by natural mapping $\pi : G \to G/H$. Furthermore, every open subset of G/H is the form $\{uH \mid u \in U\}$ for some open subset U of G.*

Let $(G, \cdot)$ be a group and let H be a non-normal subgroup of G. If we denote $G/H = \{xH \mid x \in G\}$, then $(G/H, \circ)$ is a hypergroup, where for all xH, yH of G/H, we have $xH \circ yH = \{zH \mid z \in xHy\}$.

Let β^* be the fundamental relation of the hypergroup $(G/H, \circ)$. Then, $(G/H)/\beta^*$ is a topological space with respect to the quotient topology induced by natural mapping $\pi : G/H \to (G/H)/\beta^*$, where $\pi(xH) = \beta^*(xH)$. By Theorem 12.28, $A \subset (G/H)/\beta^*$ is open if and only if $\pi^{-1}(A) = \{uH \mid u \in U\}$ for some open subset U of G.

Theorem 12.29. *Let $(G, \cdot)$ be a group and let H be a non-normal subgroup of it. Let β^* be the fundamental relation of the hypergroup $(G/H, \circ)$. Then, there exists a normal subgroup N of G such that the groups $(G/H)/\beta^*$ and G/N are isomorphic.*

Proof. Suppose that N is the subgroup of G generated by the set $\{g^{-1}hg \mid g \in G, h \in H\}$. Then, N is a normal subgroup of G. Now, we define $\varphi : (G/H)/\beta^* \to G/N$ by $\beta^*(xH) \mapsto xN$. Then, φ is well-defined. Indeed, if $\beta^*(xH) = \beta^*(yH)$, then there exists a non-zero natural number n and $a_1, \cdots, a_n \in G$ such that $\{xH, yH\} \subset a_1 H \circ a_2 H \circ \cdots \circ a_n H$. So $x = a_1 x_1 a_2 x_2 \cdots x_{n-1} a_n$ and $y = a_1 y_1 a_2 y_2 \cdots y_{n-1} a_n$ for some $x_1, x_1, \cdots x_{n-1}, y_1, y_2, \cdots y_{n-1} \in H$. Thus, $xN = a_1 x_1 a_2 x_2, \cdots x_{n-1} a_n N = a_1 a_2 \cdots a_n N = a_1 y_1 a_2 y_2, \cdots y_{n-1} a_n N$. So, $\varphi(\beta^*(xH)) = \varphi(\beta^*(yH))$.

Also, φ is a homomorphism. Indeed, for every $x, y \in H$ we have

$$\varphi(\beta^*(xH) \odot \beta^*(yH)) = \varphi(\beta^*(xyH)) = xyN$$
$$= xNyN = \varphi(\beta^*(xH))\varphi\beta^*(yH).$$

Obviously, φ is onto. It remains to show that φ is one-to-one. Suppose that $\beta^*(xH) \in ker(\varphi)$. Then, $x \in N$. Thus, there exist a non-zero natural number n and $g_1, \cdots, g_{n+1} \in G$ and $h_1, \cdots, h_n \in H$ such that $x = \prod_{i=1}^{n} g_i^{-1} h_i g_{i+1}$. Thus, we have $\{xH, H\} \subset g_1^{-1} H \circ g_1 H \circ g_2^{-1} H \circ g_2 H \circ \cdots \circ g_n^{-1} H \circ g_n H$. Hence, $(xH)\beta^* H$. So, $ker\varphi$ is trivial. Therefore, φ is an isomorphism. $\qquad\square$

Theorem 12.30. *Let $(G, \cdot)$ be a topological group and let H be a non-normal subgroup of G. Let β^* be the fundamental relation of the hypergroup $(G/H, \circ)$. Then, the natural mapping $\pi : G/H \to (G/H)/\beta^*$ is open.*

Proof. Suppose that A is open in G/H. So there exists an open subset V in G such that $A = \{vH \mid v \in V\}$. First, we show that $\pi^{-1}(\pi(A)) = \{nvH \mid n \in N, v \in V\}$. Let $xH \in \pi^{-1}(\pi(A))$. So there exists $v \in V$ such that $\pi(xH) = \pi(vH)$. It implies that $\{xH, vH\} \subset a_1 H \circ a_2 H \cdots a_k H$ for a non-zero natural number k and for some $a_1, \cdots, a_k \in G$. So, there exist $v_1, \cdots v_n, x_1, \cdots x_k \in H$ such that $v = a_1 v_1 a_2 v_2 \cdots a_{k-1} v_k a_k$ and $x = a_1 x_1 a_2 x_2 \cdots a_{k-1} x_k a_k$. Hence, we have

$$x = a_1 x_1 a_2 x_2 \cdots a_{k-1} x_k v_k^{-1} a_{k-1}^{-1} v_{k-1}^{-1} \cdots v_1^{-1} a_1^{-1} v \in NV.$$

Therefore, $xH \in \{nvH \mid n \in N, v \in V\}$.

Conversely, suppose that $v \in V$ and $n \in N$. Then, there exists a non-zero natural number k and $a_1, \cdots, a_k \in G$ and $h_1, \cdots, h_k \in H$ such that $n = a_1^{-1} h_1 a_1 \cdots a_k^{-1} h_k a_k$. Hence, we have $\{nvH, vH\} \subset a_1^{-1} H \circ a_2^{-1} H \cdots a_k^{-1} H \circ a_k H$. Thus, $(nvH)\beta^* vH$. So $nvH \in \pi^{-1}(\pi(\{vH \mid v \in V\}))$. Therefore, $\pi(A)$ is open in $(G/H)/\beta^*$. Therefore, $\pi^{-1}(\pi(A)) = \{nvH \mid n \in N, v \in V\}$ is open in G/H, indeed VN is open in G. $\qquad\square$

Theorem 12.31. *Let $(G, \cdot)$ be a topological group and let H be a non-normal subgroup of it. Let β^* be the fundamental relation of the hypergroup $(G/H, \circ)$. Then, $((G/H)/\beta^*, \odot)$ is a topological group.*

Proof. We know that $((G/H)/\beta^*, \odot)$ is a group. So, we prove that the mappings $\odot$ and inverse are continuous. Suppose that A is an open subset of $(G/H)/\beta^*$ such that $\pi(xH) \odot \pi(yH) = \pi(xyH) \in A$. Thus, $\pi^{-1}(A)$ is an open subset of G/H containing xyH. So, by Theorem 12.28, there exists an open subset U of G such that $\pi^{-1}(A) = \{uH \mid u \in U\}$. It follows that there exist open subsets V and W of G containing x and y, respectively, such that

$VW \subset UH$. Now, by Theorem 12.30, $\pi(\{vH \mid v \in V\})$ and $\pi(\{wH \mid w \in W\})$ are open in $(G/H)/\beta^*$ containing xH and yH, respectively, and we have

$$\pi(\{vH \mid v \in V\}) \odot \pi(\{wH \mid w \in W\}) = \{\pi(vwH) \mid v \in V, w \in W\}$$
$$\subset \{\pi(uH) \mid u \in U\} \subset \pi^{-1}(\pi(A))$$
$$\subset A.$$

Therefore, the mapping $\odot$ is continuous.

Suppose that A is open in $(G/H)/\beta^*$. Then, by Theorem 12.28, there exists an open subset U of G such that $\pi^{-1}(A) = \{uH \mid u \in U\}$. Since U^{-1} is open in G and $\pi^{-1}(A^{-1}) = \{vH \mid v \in U^{-1}\}$ we have A^{-1} is open in $(G/H)/\beta^*$. Therefore, the inverse mapping is continuous and the proof is complete. $\qquad\square$

Theorem 12.32. *Let $(G, \cdot)$ be a topological group and H be a non-normal subgroup of it. Let β^* be the fundamental relation of the hypergroup $(G/H, \circ)$. Then, there exists a normal subgroup N of G such that the topological groups $(G/H)/\beta^*$ and G/N are topological isomorphic.*

Proof. Suppose that N is the subgroup of G generated by the set $\{g^{-1}hg \mid g \in G, h \in H\}$. Then, N is a normal subgroup of G. Now, we define $\varphi : (G/H)/\beta^* \to G/N$ by $\beta^*(xH) \mapsto xN$. We have seen in Theorem 12.29 that φ is a group isomorphism. Hence, we show that φ is open and continuous. Suppose that A is open in G/N. By Theorem 12.28, $A = \{uH \mid u \in U\}$, where U is open in G. It is sufficient to show that $\pi^{-1}(\varphi^{-1}(A))$ is open in G/H. Let $xH \in \pi^{-1}(\varphi^{-1}(A))$. Then, $\pi(xH) \in \varphi^{-1}(A)$. So, there exist $u \in U$ and $n \in N$ such that $xn = u$. It follows that there exist open subsets V and W containing x and n, respectively, such that $VW \subset U$. We claim that $B := \{vH \mid v \in V\} \subset \pi^{-1}(\varphi^{-1}(A))$. If $vH \in B$, then $\varphi(\pi(vH)) = vN = (vn)N \in A$. So $vH \in \pi^{-1}(\varphi^{-1}(A))$. Hence, B is an open subset of G/H contains xH such that $B \subset \pi^{-1}(\varphi^{-1}(A))$. So xH is an interior point of $\pi^{-1}(\varphi^{-1}(A))$. Thus, $\pi^{-1}(\varphi^{-1}(A))$ is open in G/H. It concludes that $\varphi^{-1}(A)$ is open in $(G/H)/\beta^*$. Therefore, φ is continuous.

The mapping φ is open. Suppose that A is an open subset of $(G/H)/\beta^*$ and $xN \in \varphi(A)$. Then, $\beta^*(xH) \in A$. So there exists an open subset B of $(G/H)/\beta^*$ such that $\beta^*(xH) \in B \subset A$. By Theorem 12.28, there exists an open subset U of G such that $\pi^{-1}(B) = \{uH \mid u \in U\}$. We claim that $\{uN \mid u \in U\} \subset \varphi(A)$. If $u \in U$, then $uN = \varphi(\beta^*(uH)) \in \varphi(B) \subset \varphi(A)$. Hence, $\varphi(A)$ is open in G/N. Therefore, $\varphi^{-1}(A)$ is open and the proof is complete. $\qquad\square$

12.4   Topological complete hypergroups

Singha, Das and Davvaz [165] introduced the concept of topological complete hypergroups and investigated some of their properties. In this section we review their results.

In case of topological groups the translation maps are homeomorphism, but for topological hypergroups they are continuous only. It has been shown in the next lemma.

Lemma 12.9. *Let* $(H, \circ, \mathcal{T})$ *be a topological hypergroup, then the following translation maps:*

$$L_a : H \to \mathcal{P}^*(H) \text{ by } x \mapsto a \circ x \text{ and } R_a : H \to \mathcal{P}^*(H) \text{ by } x \mapsto x \circ a$$

are continuous for every $a \in H$.

Proof.   Let $U \in \mathcal{T}$ such that $a \circ x \subset U$, then by the continuity of the mapping $(x, y) \mapsto x \circ y$, $\exists V, W \in \mathcal{T}$ such that $a \in V$ and $x \in W$ and $V \circ W \subset U$. This implies that $a \circ W \subset V \circ W \subset U$. This shows that L_a is continuous on H.

Similarly, R_a is continuous. $\square$

Now, by using Lemma 12.9, we prove the following theorem:

Theorem 12.33. *Let* $(H, \circ, \mathcal{T})$ *be a compact Hausdorff topological hypergroup and* K *be a subset of* H, *then* $x \circ \overline{K} = \overline{x \circ K}$, *for all* $x \in H$.

Proof.   Using Lemma 12.9 we have $x \circ \overline{K} \subset \overline{x \circ K}$, $\forall x \in H$. To prove $\overline{x \circ K} \subset x \circ \overline{K}$, let $p \in \overline{x \circ K}$. Now, $p \in x \circ K \Rightarrow p \in x \circ \overline{K}$ but if $p \notin x \circ K$, then p is a limit point of $x \circ K$. Let $U \in \mathcal{T}$ such that $p \in U$, then $x \circ K \cap U \neq \phi$ $\Rightarrow x \circ \overline{K} \cap U \neq \phi \Rightarrow p$ is a limit point of $x \circ \overline{K} \Rightarrow p \in \overline{x \circ \overline{K}}$. Now, we show that $\overline{x \circ \overline{K}} = x \circ \overline{K}$, i.e., $x \circ \overline{K}$ is closed. Here $\overline{K}$ is compact for being a closed subset of the compact space H. Also, $x \circ \overline{K}$ is compact, since translation maps are continuous (by Lemma 12.9). So being a compact subset of a Hausdorff space $x \circ \overline{K}$ is closed. Hence, $\overline{x \circ K} \subset x \circ \overline{K}$. Thus, we conclude that $x \circ \overline{K} = \overline{x \circ K}$. $\square$

Now, we state a theorem on complete hypergroup, which will be used frequently.

Theorem 12.34. *Let* A *and* B *be non-empty subsets of a complete hypergroup* $(H, \circ)$ *such that* A *is a complete part and* $x \in H$. *Then,*

(1) $x^{-1} \circ x \circ A = x \circ x^{-1} \circ A = A$, *for some* $x^{-1} \in i(x)$;

(2) $x \circ A$ and $A \circ x$ are complete parts;

(3) $B \subset x^{-1} \circ A$ if and only if $x \circ B \subset A$, for some $x^{-1} \in i(x)$.

Proof. The proof is straightforward. $\qquad\qquad\qquad\qquad\qquad\qquad$ $\square$

We have seen in Example 12.13 that if U is an open set in a topological hypergroup, then $a \circ U$ may not be open for some a. But we think for further continuation of work we need this to be open. So we have tried this in a special class of topological hypergroups called *topological complete hypergroups*. Though the complete hypergroup has already been introduced, in this section we introduce topological complete hypergroups and investigate some of its properties.

Definition 12.14. Let $(H, \circ, \mathcal{T})$ be a topological hypergroup. Then, we say H is a topological complete hypergroup if H is a complete hypergroup.

Also, we say H is a topological regular hypergroup if H is a regular hypergroup.

It is important to mention that in this section the *completeness* and *regularity* of a topological hypergroup is different from the *topological regularity* and *topological completeness*.

Corollary 12.11. *Every topological complete hypergroup is a topological regular hypergroup.*

Now, let us develop an important tool for further work.

Lemma 12.10. *Let U be an open subset of a topological complete hypergroup $(H, \circ, \mathcal{T})$ such that U is a complete part. Then, $a \circ U$ and $U \circ a$ are open subsets of H for every $a \in H$.*

Proof. Suppose that U is an open subset of H such that U is a complete part of H and $a \in H$. Then, for some $a^{-1} \in i(a)$ we have

$$L_{a^{-1}}^{-1}(S_U) = \{x \in H : L_{a^{-1}}(x) \in S_U\}$$
$$= \{x \in H : a^{-1} \circ x \subset U\}.$$

Now, we claim that $\{x \in H : a^{-1} \circ x \subset U\} = a \circ U$. For, let $p \in \{x \in H : a^{-1} \circ x \subset U\}$, then $a^{-1} \circ p \subset U$. Now, there exists $e \in \omega_H$ such that $e \in a \circ a^{-1}$ and this implies that $p \in e \circ p \subset a \circ a^{-1} \circ p \subset a \circ U$.

For the converse, suppose that

$$t \in a \circ U \Rightarrow t \in a \circ u \text{ for some } u \in U$$
$$\Rightarrow u \in a' \circ t \text{ for some } a' \in i(a)$$
$$\Rightarrow u \in a' \circ t \subset a' \circ a \circ a^{-1} \circ t = \omega_H \circ a^{-1} \circ t$$
$$= C(a^{-1} \circ t) = a^{-1} \circ t$$
$$\Rightarrow a^{-1} \circ t \cap U \neq \phi$$
$$\Rightarrow a^{-1} \circ t \subset U, \text{ since } U \text{ is a complete part of } H.$$
$$\Rightarrow t \in \{x \in H : a^{-1} \circ x \subset U\}.$$

Hence, $L_{a^{-1}}^{-1}(S_U) = a \circ U$. Since the translation maps are continuous, it follows that $a \circ U$ is open in H.

Similarly, $U \circ a$ is open in H. □

Theorem 12.35. *Let H be a topological complete hypergroup and A, B be open subsets of H. If A or B is a complete part of H, then $A \circ B$ is open.*

Proof. Suppose that A is a complete part of H, then $A \circ b$ is open (by Lemma 12.10). Now, $A \circ B = \bigcup_{b \in B} A \circ b$, this shows that $A \circ B$ is open. □

Lemma 12.11. *Let H be a topological complete hypergroup such that every open subset of H is a complete part. Let $\mathcal{U}$ be a basis at some identity e. Then, the families $\{x \circ U\}$ and $\{U \circ x\}$, where x runs through all elements of H and U runs through all elements of $\mathcal{U}$, are basis for H.*

Proof. Let W be an open subset of H and $a \in W$. Then, there exists $a' \in H$ such that $e' \in a' \circ W$ for some $e' \in \omega_H$. Now, $e \in e \circ e' \subset e \circ a' \circ W \subset \omega_H \circ a' \circ W = a' \circ W$, since $a' \circ W$ is an open subset of H. Then, there exists $U \in \mathcal{U}$ such that $e \in U \subset a' \circ W$. This implies $a \in a \circ e \subset a \circ U \subset a \circ a' \circ W = W$ (by Theorem 12.34), i.e., $a \in a \circ U \subset W$. This shows that $\{x \circ U\}$ is a basis for H.

Similarly, $\{U \circ x\}$ is also a basis for H. □

Lemma 12.12. *Let H be a topological complete hypergroup such that every open subset of it is a complete part and $\mathcal{U}$ be a fundamental system of open neighborhoods at some identity e. Then, the following assertions hold:*

(1) *for every $W \in \mathcal{T}$ with $x \in W$, there exists $V \in \mathcal{U}$ such that $x \circ V \subset W$ and $V \circ x \subset W$;*

(2) *for every $U \in \mathcal{U}$, there exists $V \in \mathcal{U}$ such that $V \circ V \subset U$.*

Proof. (1) Suppose that $W \in \mathcal{T}$ with $x \in W$. Then, there exists $x' \in i(x)$ such that $e' \in x' \circ W$ for some $e' \in \omega_H$. Now, $e \in e \circ e' \subset e \circ x' \circ W \subset \omega_H \circ x' \circ W = x' \circ W$, since $x' \circ W$ is an open subset of H. Since $\mathcal{U}$ is a fundamental system of open neighborhoods at e, it follows that there exists $V \in \mathcal{U}$ such that $e \in V \subset x' \circ W$. This implies $x \circ V \subset x \circ x' \circ W = W$ (by Theorem 12.34).

Similarly, we can show that there exists $V \in \mathcal{U}$ such that $V \circ x \subset W$.

(2) Suppose that $U \in \mathcal{U}$, then $e \in U$. Since U is a complete part of H, it follows that $e \circ e \subset U$. So by the continuity of the map $(x, y) \mapsto x \circ y$ there exists $V \in \mathcal{T}$ such that $e \in V$, i.e., $V \in \mathcal{U}$ such that $V \circ V \subset U$. □

Theorem 12.36. *Let $(H, \circ, \mathcal{T})$ be a topological complete hypergroup such that every open subset of it is a complete part. If A and B are two non-empty subsets of H, then*

(1) $\overline{A} \circ \overline{B} \subset \overline{A \circ B}$;
(2) $IntA \circ IntB \subset Int(A \circ B)$, *where $Int(A)$ denotes the interior of the subset A.*

Proof. (1) The map $f(x, y) = x \circ y$ is continuous from $H \times H$ to $\mathcal{P}^*(H)$, then $f(\overline{A \times B}) \subset \overline{f(A \times B)} \Rightarrow \overline{A} \circ \overline{B} \subset \overline{A \circ B}$.

(2) In order to prove $IntA \circ IntB \subset Int(A \circ B)$, let $p \in IntA \circ IntB$, then $p \in a \circ b$ for some $a \in IntA$ and $b \in IntB$. Since a and b are interior points of A and B respectively then there exist $U, V \in \mathcal{T}$ such that $a \in U \subset A$ and $b \in V \subset B \Rightarrow p \in a \circ b \subset U \circ V \subset A \circ B \Rightarrow p \in Int(A \circ B)$, since $U \circ V$ is open (by Theorem 12.35). Thus, $IntA \circ IntB \subset Int(A \circ B)$. □

Theorem 12.37. *Let H be a topological complete hypergroup such that every open subset of it is a complete part. Let F be a compact subset of H and P be a closed subset of H such that $F \cap P = \phi$. Then, there exists an open neighborhood V containing some identity e such that $F \circ V \cap P = \phi$ and $V \circ F \cap P = \phi$.*

Proof. Since P is closed, it follows that for each $x \in F$ there exists an open set V_x of some identity e in H such that $x \circ V_x \cap P = \phi$. Using Lemma 12.12 we can choose an open neighborhood W_x of e such that $W_x \circ W_x \subset V_x$. Now, $\{x \circ W_x\}_{x \in F}$ is an open cover for the compact set F, so there exists $x_1, x_2, ..., x_n \in F$ such that $F \subset \bigcup_{i=1}^{n} x_i \circ W_{x_i}$. We take $V_1 = \bigcap_{i=1}^{n} W_{x_i}$. We claim that $F \circ V_1 \cap P = \phi$. It suffices to verify that $y \circ V_1 \cap P = \phi$ for each $y \in F$. Let y be an element of F. Then, $y \in x_k \circ W_{x_k}$ for some

$k \in \{1, 2, ..., n\}$. Then, $y \circ V_1 \subset (x_k \circ W_{x_k}) \circ V_1 \subset x_k \circ (W_{x_k} \circ W_{x_k}) \subset$
$x_k \circ V_{x_k} \subset H \setminus P$, by our choice of the sets V_x and W_x. This proves that
$F \circ V_1$ and P are disjoint.

Similarly, one can find an open neighborhood V_2 of e in H such that
$V_2 \circ F \cap P = \phi$. Then, the set $V = V_1 \cap V_2$ is the required open neighborhood
of e.   □

Theorem 12.38. *Let $(H_1, \circ, \mathcal{T})$ and $(H_2, *, \mathcal{T}')$ be two topological complete
hypergroups such that every open subset of them is a complete part. Let f
be a homomorphism from H_1 into H_2. Then, f is continuous if and only if
it is continuous at some identity of H_1.*

Proof.   If f is continuous, then the condition is obvious.

For the converse, let the map f is continuous at some identity e of
H_1. Let $x \in H_1$ and W be an open set containing $f(x)$ in H_2. Now,
$f(x) \in W \Rightarrow f(x) * f(e) \subset W$. Using the continuity of translation map we
have an open set V in H_2 such that $f(e) \in V$ and $f(x) * V \subset W$. Since f
is continuous at e, it follows that there exists an open set U containing e
such that $f(U) \subset V$. Now, $f(x \circ U) \subset f(x) * f(U) \subset f(x) * V \subset W$. This
shows that f is continuous on H_1.   □

Theorem 12.39. *Let $(H_1, \circ, \mathcal{T})$ and $(H_2, *, \mathcal{T}')$ be two topological complete
hypergroup such that every open subset of them is a complete part. Let f
be a good homomorphism from H_1 into H_2. Then, f is an open map if
and only if for every open set V containing some identity e_1 of H_1, $f(V)$
is open in H_2 containing some identity e_2.*

Proof.   If f is an open map, then the condition holds as $f(\omega_{H_1}) = \omega_{H_2}$.

For the converse, let the given condition holds. Let U be an open set
in H_1. To show $f(U)$ open in H_2, let $y \in f(U)$. Then, $y = f(x)$ for some
$x \in U$. Since $x \in U$, it follows that there exists a open neighborhood V of
some identity e_1 such that $x \in x \circ V \subset U$. By the given condition $f(V)$ is
open containing of some identity e_2 in H_2. Then, $y = f(x) \in f(x) * f(V) =$
$f(x \circ V) \subset f(U)$. This shows that $f(U)$ is open and hence f is an open
map on H_1.   □

Now, let us define a special kind of identity element in a regular hypergroup.

Definition 12.15. Let $(H, \circ)$ be a regular hypergroup. Let e be an identity
in H and $g \in H$. We say e is related to g if $\exists\ g' \in i(g)$ such that $e \in$
$g \circ g' \cap g' \circ g$.

We say an identity e is related to H or an related identity of H if it is related to every element of H, i.e., for every $g \in H$, $\exists \, g' \in i(g)$ such that $e \in g \circ g' \cap g' \circ g$.

Example 12.14. Consider the additive group $(\mathbb{Z}, +)$ of integers, and define the hyperoperation $\circ$ on it as $m \circ n = \, < m, n >$ the subgroup generated by m and n. Then, $(\mathbb{Z}, \circ)$ is a regular hypergroup with 0 as a related identity.

Example 12.15. Consider the set of integers $\mathbb{Z}$ with the hyperoperation $*$ on it as

$$m * n = \begin{cases} 2\mathbb{Z} & \text{if } m + n \in 2\mathbb{Z} \\ (2\mathbb{Z})^c & \text{otherwise.} \end{cases}$$

Then, $(\mathbb{Z}, *)$ is a regular hypergroup with 0 as a related identity.

Let us develop some algebraic tools which will be used later in the sequel.

Lemma 12.13. *Every subhypergroup of a complete hypergroup is a complete part.*

Proof. Let K be a subhypergroup of a complete hypergroup H. Now, $\omega_H \circ K = \omega_K \circ K = \bigcup_{x \in \omega_K} x \circ K = K$. This shows that K is a complete part of H. $\qquad\square$

Corollary 12.12. *Let K be a subhypergroup of a complete hypergroup H. Then, $\{x \circ K\}_{x \in H}$ and $\{K \circ x\}_{x \in H}$ are partitions for H.*

Proof. By the help of Lemma 12.13 we can say that K is a complete part subhypergroup of H. Since any complete part subhypergroup is invertible, it follows that $\{x \circ K\}_{x \in H}$ and $\{K \circ x\}_{x \in H}$ form partitions for H. $\qquad\square$

Theorem 12.40. *Let K be a subhypergroup of a complete hypergroup. Then, K is normal in H if and only if for every $k \in K$ and for every $x \in H$, $x \circ k \circ x^{-1} \subset K$, i.e., $x \circ K \circ x^{-1} \subset K$, where $x^{-1} \in i(x)$.*

Proof. Let K be a normal subhypergroup of H. Now, for $x \in H$ and $k \in K$, $x \circ k \subset x \circ K = K \circ x$. Then, $x \circ k \circ x^{-1} \subset K \circ x \circ x^{-1} = K \circ \omega_H = K$ (by Lemma 12.13).

For the converse, suppose that the given condition holds. Let $p \in x \circ K \Rightarrow p \in x \circ k$ for some $k \in K \Rightarrow p \in (x \circ k \circ x^{-1}) \circ x \subset K \circ x$. Therefore, we have $x \circ K \subset K \circ x$. Now, let $q \in K \circ x \Rightarrow q \in k_1 \circ x$ for some $k_1 \in K \Rightarrow q \in x \circ x^{-1} \circ k_1 \circ x \Rightarrow q \in x \circ (x^{-1} \circ k_1 \circ (x^{-1})^{-1}) \circ x^{-1} \circ x \subset x \circ K \circ \omega_H = x \circ K$ (by Lemma 12.13). Therefore, we have $K \circ x \subset x \circ K$. Hence, $x \circ K = K \circ x$ for every $x \in H$. This shows that K is normal in H. $\qquad\square$

It is observed that if K is a normal subhypergroup of a complete hypergroup H, then for every $x \in H$ with $x^{-1} \in i(x)$, $x \circ K \circ x^{-1} = K \circ x \circ x^{-1} = K \circ \omega_H = K$ (by Lemma 12.13). Hence, the above theorem can be restated as follows:

Corollary 12.13. *Let K be a subhypergroup of a complete hypergroup. Then, K is normal in H if and only if for every $x \in H$, $x \circ K \circ x^{-1} = K$, where $x^{-1} \in i(x)$.*

Theorem 12.41. *Let $(H, \circ)$ be a complete hypergroup and M, N are two normal subhypergroups of it. Then,*

(1) $(N \circ a) \circ (N \circ b) = N \circ a \circ b$, *for all $a, b \in H$;*
(2) $N \circ a = N \circ b$ *if and only if $b \in N \circ a$;*
(3) $M \cap N$ *is a normal subhypergroup of H.*

Proof.   (1) $(N \circ a) \circ (N \circ b) = N \circ (a \circ N) \circ b = N \circ N \circ a \circ b = N \circ a \circ b$.

(2) First we suppose that $N \circ a = N \circ b$. Then, $b \in \omega_N \circ b \subset N \circ b = N \circ a$.

For the converse, let $b \in N \circ a$. Then, $N \circ b \subset N \circ N \circ a = N \circ a$. Since any complete part subhypergroup is invertible, it follows that $b \in N \circ a \Rightarrow a \in N \circ b$. So, $N \circ a \subset N \circ N \circ b = N \circ b$ and hence $N \circ a = N \circ b$.

(3) Being complete part subhypergroups of H (by Lemma 12.13), M, N are invertible and hence closed. Since $\omega_M = \omega_N = \omega_H$, it follows that the intersection of M, N is non-empty. Therefore, $M \cap N$ is a closed subhypergroup of H. Since M, N are normal in H, it follows that for any $x \in H$ with $x^{-1} \in i(x)$ we have $x \circ M \circ x^{-1} \subset M$ and $x \circ N \circ x^{-1} \subset N$. So, for any $x \in H$ with $x^{-1} \in i(x)$ we have $x \circ (M \cap N) \circ x^{-1} \subset M \cap N$. This shows that $M \cap N$ is normal in H.   □

Now, we show that the component (or connected component) of an element can be obtained from the component of its related identity by using translation map in a topological regular hypergroup. Let $(H, \circ, \mathcal{T})$ be a topological hypergroup, for each $g \in H$, C_g denotes the component of g.

Lemma 12.14. *Let $(H, \circ, \mathcal{T})$ be a topological regular hypergroup. Then, for each $g \in H$, $L_g(C_e) = C_g$, where e is an identity related to g.*

Proof. $L_g(C_e)$ is a continuous image of C_e, so it is connected and $g \in L_g(C_e)$, so $L_g(C_e) \subset C_g$ as C_g is the maximal connected set containing g. Since e is an identity related to g, it follows that there exists $g' \in i(g)$ such that $e \in g' \circ g \cap g \circ g'$. This shows that $L_{g'}(C_g)$ is a connected set containing

e, so $L_{g'}(C_g) \subset C_e$. This implies $C_g \subset g \circ g' \circ C_g = L_g(L_{g'}(C_g)) \subset L_g(C_e)$. Hence, $L_g(C_e) = C_g$. □

By using Lemma 12.14, let us prove the following theorem.

Theorem 12.42. *Let $(H, \circ, \mathcal{T})$ be topological regular hypergroup and e be an identity related to H. Then, C_e is a closed (topologically) subhypergroup. Furthermore, if H is a complete hypergroup, then C_e is a normal subhypergroup of H.*

Proof. First of all C_e is a closed set, since all the components are closed. Now, we prove C_e is a subhypergroup. Let $g, h \in C_e$. Then, $g \circ C_e$ is a connected set containing g and $g \circ h$, i.e., $g \circ h \subset g \circ C_e$, so $g \circ h \subset C_e$.

Now, let $g \in C_e$, then using Lemma 12.14, $g \circ C_e = C_g = C_e$. Similarly, $C_e \circ g = C_e$, for all $g \in C_e$. Hence, C_e is a subhypergroup of H.

Now, suppose that H be complete hypergroup, for $g \in H$, $C_e \circ g'$ is connected, where $g' \in i(g)$, so $g \circ C_e \circ g'$ is connected and contains e. Hence, $g \circ C_e \circ g' \subset C_e$. This shows that if H is a complete hypergroup, then C_e is normal in H (by Theorem 12.40). □

Let us introduce topological subhypergroup.

Definition 12.16. Let H be a topological hypergroup and K be a subhypergroup of H. Let K be endowed with relative topology induced from H. Since the mappings $(x, y) \mapsto x \circ y$ and $(x, y) \mapsto x/y$ of $H \times H$ into $\mathcal{P}^*(H)$ are continuous, so are their restrictions from $K \times K$ into $\mathcal{P}^*(K)$. In other words, K endowed with relative topology is a topological hypergroup. In this case K is called a *topological subhypergroup*.

Theorem 12.43. *Let $(H, \circ, \mathcal{T})$ be a topological hypergroup such that every open subset of it is a complete part and K be a subhypergroup of H. Then, every open subset of K is a complete part.*

Proof. Let U is an open subset of K and for $n \in \mathbb{N}$, $\prod_{i=1}^{n} a_i \cap U \neq \phi$, where $a_i \in K$. Then, there exists $V \in \mathcal{T}$ such that $U = V \cap K$. Therefore, $\prod_{i=1}^{n} a_i \cap V \neq \phi$ and so, $\prod_{i=1}^{n} a_i \subset V$. Also, $\prod_{i=1}^{n} a_i \subset K$ and hence $\prod_{i=1}^{n} a_i \subset V \cap K = U$. This shows that U is a complete part of K. □

Theorem 12.44. *Let $(H, \circ, \mathcal{T})$ be a topological complete hypergroup. Then, every open subhypergroup is closed (topologically).*

Proof.   Let K be an open subhypergroup of H, then $x \circ K$ is open for every $x \in H$, since K is a complete part of H. Now, $\{x \circ K\}_{x \in H}$ is a partition for H (by Corollary 12.12). So, we can write $H = \bigcup_{x \in H} x \circ K = $ $(\bigcup_{x \in K} x \circ K) \cup (\bigcup_{x \notin K} x \circ K) = K \cup (\bigcup_{x \notin K} x \circ K)$. This implies $K = H \backslash (\bigcup_{x \notin K} x \circ K)$ and hence K is closed. $\square$

Lemma 12.15. *Every subhypergroup of a complete hypergroup is complete.*

Proof.   Let K be a subhypergroup of a complete hypergroup $(H, \circ)$. Now, for $x, y \in K$, $C(x \circ y) = (x \circ y) \circ \omega_K = (x \circ y) \circ \omega_H = x \circ y$. This shows that K is a complete subhypergroup of H. $\square$

Theorem 12.45. *Let $(H, \circ, \mathcal{T})$ be a topological complete hypergroup such that every open subset of it is a complete part and K be a subhypergroup of H. Then, K is open if and only if its interior $IntK \neq \phi$.*

Proof.   Let $IntK \neq \phi$ and $x \in IntK$. Then, there exists an open set U containing some identity e of H such that $x \in x \circ U \subset K$. Now, take any $y \in K$, then $y \circ U \subset y \circ x^{-1} \circ x \circ U \subset y \circ x^{-1} \circ K = K$, since $x, y \in K$ and K is complete (by Lemma 12.15). This shows that K is open.

For the converse, if K is open, then we have $IntK \neq \phi$. $\square$

Theorem 12.46. *Let $(H, \circ, \mathcal{T})$ be a topological complete hypergroup such that every open subset of it is a complete part and e be a related identity of H. Let $\mathcal{U}$ be the system of all neighborhoods of e, then for any subset A of H,*

$$\overline{A} = \bigcap_{U \in \mathcal{U}} A/U.$$

Proof.   Let $x \in \overline{A}$ and $U \in \mathcal{U}$, $x \circ U$ is a neighborhood of x, and hence $x \circ U \cap A \neq \phi$. This implies there exists $a \in A$ and $u \in U$ such that $a \in x \circ u \Rightarrow x \in a/u \subset A/U$. Therefore, $\overline{A} \subset A/U$ and hence $\overline{A} \subset \cap_{U \in \mathcal{U}} A/U$.

For the converse, suppose that $y \in A/U$ for every $U \in \mathcal{U}$. Then, for any open neighborhood V of y there exists $y^{-1} \in i(y)$ such that $y^{-1} \circ V$ contains e and hence $y^{-1} \circ V \in \mathcal{U}$. This implies that $y \in A/(y^{-1} \circ V) \Rightarrow y \in a/w$ for some $a \in A$ and $w \in y^{-1} \circ V \Rightarrow a \in y \circ w \subset y \circ y^{-1} \circ V = V$ (by Theorem 12.34) $\Rightarrow V \cap A \neq \phi$ and hence $y \in \overline{A}$. This completes the proof. $\square$

Remark 12.6. Let $(H, \circ)$ be a complete hypergroup and ω_H be the heart of H, for every $e \in \omega_H$ we have $e/e = \omega_H$.

For, let $t \in e/e$, then $e \in t \circ e$. Also, $e \in e \circ e \subset e \circ t \circ e \subset e \circ t \circ \omega_H = C(e \circ t) = e \circ t$. Now, we show that $t \in \omega_H$, i.e., t is a two sided identity of H. Let $x \in H$, then $x \in x \circ e \subset x \circ t \circ e \subset x \circ t \circ \omega_H = C(x \circ t) = x \circ t$. This shows that t is a right identity of H. Similarly, $x \in e \circ x \subset e \circ t \circ x \subset \omega_H \circ t \circ x = C(t \circ x) = t \circ x$. This shows that t is a right identity of H and hence $t \in \omega_H$. Also, $\omega_H \subset e/e$. Therefore, we obtain $e/e = \omega_H$.

Theorem 12.47. *Let $(H, \circ, \mathcal{T})$ be a topological complete hypergroup such that every open subset of it is a complete part and e be a related identity of H. Now, if U is an open neighborhood of e, then there exists an open neighborhood V of e such that $\overline{V} \subset U$.*

Proof. Since U is a complete part of H and $e \in U$, it follows that $\omega_H = e \circ e \subset U$. Again, $e/e = \omega_H$, this implies that $e/e \subset U$. So, by the continuity of the map $(x, y) \mapsto x/y$, there exists an open neighborhood V of e such that $V/V \subset U$. Now, by using Theorem 12.46 we have $\overline{V} \subset V/V \subset U$, i.e., $\overline{V} \subset U$. □

Corollary 12.14. *Let $(H, \circ, \mathcal{T})$ be a topological complete hypergroup such that every open subset of it is a complete part and e be a related identity of H. Then, H is locally compact if and only if there exists a compact neighborhood of e.*

Proof. Suppose that H is locally compact. Then, by the definition of locally compactness, there exists a compact neighborhood of e.

For the converse, suppose that U be a compact neighborhood of e. Then, by Theorem 12.47, there exists a open neighborhood V of e such that $\overline{V} \subset U$. Now, being a closed subset of a compact set, $\overline{V}$ is compact. So, for each $x \in H$, $x \circ \overline{V}$ is a compact neighborhood of x. This completes the proof. □

Let $(H, \circ)$ be a complete hypergroup and K be a normal subhypergroup of H. By H/K we denote the collection of all left(or right) cosets of K in H, i.e., $H/K = \{K \circ x : x \in H\}$.

Theorem 12.48. *Let $(H, \circ)$ be a complete hypergroup and K be a normal subhypergroup of H. Then, H/K forms a hypergroup with respect to the operation $\odot$ defined by $K \circ x \odot K \circ y = \{K \circ z : z \in x \circ y\}$.*

Proof.   Let us check for associativity of $\odot$ on H/K. For all $x, y, z \in H$, we have

$$(K \circ x \odot K \circ y) \odot K \circ z = \{K \circ u : u \in x \circ y\} \odot K \circ z$$
$$= \{K \circ v : u \in x \circ y, v \in u \circ z\}$$
$$= \{K \circ v : v \in (x \circ y) \circ z\},$$

$$K \circ x \odot (K \circ y \odot K \circ z) = K \circ x \odot \{K \circ u : u \in y \circ z\}$$
$$= \{K \circ v : u \in y \circ z, v \in x \circ u\}$$
$$= \{K \circ v : v \in x \circ (y \circ z)\}.$$

Since $(x \circ y) \circ z = x \circ (y \circ z)$, it follows that $(K \circ x \odot K \circ y) \odot K \circ z = K \circ x \odot (K \circ y \odot K \circ z)$.

Now, for reproduction axiom let $K \circ x \in H/K$, then we have

$$K \circ x \odot H/K = \{K \circ v : v \in x \circ y, y \in H\}$$
$$= \{K \circ v : v \in x \circ H = H\}$$
$$= H/K.$$

Similarly, we have $H/K \odot K \circ x = H/K$. Therefore, $(H/K, \odot)$ is a hypergroup.   □

Let ϕ be the natural mapping $x \mapsto K \circ x$ of H onto H/K. Then, $(H/K, \overline{\mathcal{T}})$ is a topological space, where $\overline{\mathcal{T}}$ is the quotient topology induced by ϕ, i.e., for every subset X of H, $\{K \circ x : x \in X\}$ is open in H/K if and only if $\phi^{-1}(\{K \circ x : x \in X\})$ is an open subset of H. We use the notation X/K for the set $\{K \circ x : x \in X\}$.

Lemma 12.16. *Let $(H, \circ, \mathcal{T})$ be a topological complete hypergroup and K be a normal subhypergroup of it. Let ϕ be the natural mapping $x \mapsto K \circ x$ of H onto H/K. Then,*

(1) *ϕ is a continuous;*
(2) *$\phi^{-1}(\{K \circ x : x \in X\}) = K \circ X$ for every subset X of H;*
(3) *If every open subset of H is a complete part, then ϕ is open;*
(4) *ϕ is a good homomorphism;*
(5) *If H is compact, then H/K is compact;*
(6) *If every open subset of H is a complete part, then the quotient topology is the finest topology on H/K with respect to which ϕ is continuous.*

Proof.   (1) ϕ is continuous by the definition of quotient topology.

(2) We have $K \circ X \subset \phi^{-1}(\{K \circ x : x \in X\})$ for every subset X of H. In order to prove the converse, let $y \in \phi^{-1}(\{K \circ x : x \in X\})$. Then, $\phi(y) = K \circ y \in \{K \circ x : x \in X\}$. So, $K \circ x = K \circ y$ for some $x \in X$. Then, by Theorem 12.41 $y \in K \circ x \subset K \circ X$. Thus, the equality holds.

(3) Let U be an open subset of H. We show $\phi(U)$ is open in H/K. Now, $\phi^{-1}(\phi(U)) = K \circ U$. Since U is a complete part of H, it follows that $K \circ U$ is open in H (by Lemma 12.10). Hence, $\phi(U)$ is open in H/K. This shows that ϕ is open.

(4) Let $x, y \in H$. Then, $\phi(x \circ y) = \{K \circ z : z \in x \circ y\} = K \circ x \odot K \circ y$. This shows that ϕ is a good homomorphism on H.

(5) We have $\phi(H) = H/K$. So, being the continuous image of a compact set, H/K is compact.

(6) Let $\mathcal{T}'$ be any other topology on H/K with respect to which $\phi :$ $H \mapsto H/K$ is continuous. Now, for any open subset O of H/K there exists some open subset V of H such that $O = V/K$ (by (2)). Now, $\phi^{-1}(O) = \phi^{-1}(V/K) = K \circ V$ is open in H (by Lemma 12.10). But by the definition of quotient topology, all such O's are open in the quotient topology. This shows that the quotient topology $\overline{\mathcal{T}}$ is finer than $\mathcal{T}'$. This completes the proof. □

Theorem 12.49. *Let K be a normal subhypergroup of a topological complete hypergroup $(H, \circ, \mathcal{T})$ for which open subset of H is a complete part. Then, $(H/K, \odot, \overline{\mathcal{T}})$ is a topological hypergroup, where $K \circ x \odot K \circ y = \{K \circ z : z \in x \circ y\}$ and $K \circ x / K \circ y = \{K \circ z : z \in x/y\}$.*

Proof. Let us show that the hyperoperation $\odot$ and $/$ are continuous on H/K. Suppose that $K \circ x, K \circ y \in H/K$ and $\mathcal{A}$ be an open subset of H/K such that $K \circ x \odot K \circ y \subset \mathcal{A}$. Then, $x \circ y \subset \phi^{-1}(\mathcal{A})$. Since $\phi^{-1}(\mathcal{A})$ is open in H, by the continuity of the map $(x, y) \mapsto x \circ y$, there exist the open subsets V and W containing x and y respectively, such that $V \circ W \subset \phi^{-1}(\mathcal{A})$. Now, $\phi(V)$ and $\phi(W)$ are open subsets of H/K containing $K \circ x$ and $K \circ y$ respectively, it follows that $\phi(V) \odot \phi(W) \subset \mathcal{A}$. Therefore, the hyperoperation $\odot$ is continuous on H/K.

Now, suppose that $\mathcal{B}$ is an open subset of H/K and $K \circ x / K \circ y \subset \mathcal{B}$. Then, $x/y \subset \phi^{-1}(\mathcal{B})$. Since $\phi^{-1}(\mathcal{B})$ is open in H, by the continuity of the map $(x, y) \mapsto x/y$, there exists open subsets P and Q containing x and y respectively, such that $P/Q \subset \phi^{-1}(\mathcal{B})$. Now, $\phi(P)$ and $\phi(Q)$ are open in H/K containing $K \circ x$ and $K \circ y$ respectively, it follows that $\phi(P)/\phi(Q) \subset \mathcal{B}$. Therefore, the hyperoperation $/$ is continuous on H/K and hence $(H/K, \odot, \overline{\mathcal{T}})$ is a topological hypergroup. □

Theorem 12.50. *Let $(H, \circ, \mathcal{T})$ be a topological complete hypergroup such that every open subset of H is a complete part and K be a normal sub-hypergroup of it. Let $\phi : H \to H/K$ be the natural mapping. Then, the family $\{\phi(U \circ x) : U \in \mathcal{U}\}$ is a local base of the space H/K at the point $K \circ x \in H/K$, where $\mathcal{U}$ is a base for H at some identity e.*

Proof. Let $U \in \mathcal{U}$. Then, U is a complete part of H and so, $U \circ x$ is open in H (by Lemma 12.10). Now, for every $k \in K$, $k \circ (U \circ x)$ is open in H. So, $\phi^{-1}(\phi(U \circ x)) = K \circ (U \circ x) = \bigcup_{k \in K} k \circ (U \circ x)$ is an open subset of H. Therefore, by Lemma 12.16, $\phi(U \circ x)$ is open in H/K. Now, suppose that V be an open neighborhood of $K \circ x$ in H/K. Let us take $\phi^{-1}(V) = W$, then W is an open subset of H. Since $K \circ x \subset V$, it follows that $x \in \phi^{-1}(K \circ x) \subset \phi^{-1}(V) = W$. So, there exists $U \in \mathcal{U}$ such that $U \circ x \subset W$ (by Lemma 12.12). Therefore, $K \circ x \in \phi(U \circ x) \subset \phi(W) = V$. This shows that $\{\phi(U \circ x) : U \in \mathcal{U}\}$ is a local base of the space H/K at the point $K \circ x$. $\qquad\square$

Bibliography

[1] H. Aghabozorgi, M. Jafarpour, M. Dolatabadi and I. Cristea, *An algorithm to compute the number of Rosenberg hypergroups of order less than 7*, Ital. J. Pure Appl. Math., 42 (2019), 262-270.

[2] H. Aghabozorgi, B. Davvaz and M. Jafarpour, *Solvable polygroups and derived subpolygroups*, Comm. Algebra, 41(8) (2013), 3098-3107.

[3] H. Aghabozorgi, M. Jafarpour and B. Davvaz, *Enumeration of Varlet and Comer hypergroups*, Electron. J. Combin., 18 (2011), #P131.

[4] M. Al Tahan and B. Davvaz, *On a special single-power cyclic hypergroup and its automorphisms*, Discrete Math. Algorithms Appl., 8(4) (2016), 1650059, 12 pp.

[5] M. Al Tahan and B. Davvaz, *Hyperstructures associated to Biological inheritance*, Mathematical Biosciences, 285 (2017), 112-118.

[6] M. Al Tahan and B. Davvaz, *On Corsini hypergroups and their productional hypergroups*, Korean Journal of Mathematics, 27(1) (2019), 63-80.

[7] M. Al Tahan, S. Hošková-Mayerová, and B. Davvaz, *An overview of topological hypergroupoids*, J. Intelligent and Fuzzy Systems, 34 (2018), 1907-1916.

[8] M. Alp and B. Davvaz, *Crossed modules of hypergroups associated with generalized actions*, Hacet. J. Math. Stat., 45(3) (2016), 663-673.

[9] R. Bayon and N. Lygeros, *Number of abelian H_v groups of order n*, in: N.J.A. Sloane (Ed.), The On-line Encyclopedia of Integer Sequences, 2005.

[10] R. Bayon and N. Lygeros, *Number of abelian hypergroups of order n*, in: N.J.A. Sloane (Ed.), The On-line Encyclopedia of Integer Sequences, 2007.

[11] R. Bayon and N. Lygeros, *Advanced results in enumeration of hyperstructures*, J. Algebra, 320 (2008), 821-835.

[12] R.A. Borzoei, A. Hasankhani and H. Rezaei, *Some results on canonical, cyclic hypergroups and join spaces*, Ital. J. Pure Appl. Math., 11 (2002), 77-87.

[13] G. Calugareanu and V. Leoreanu, *Hypergroups associated with lattices*, Ital. J. Pure Appl. Math., 9 (2001), 165-173.

[14] H. Campaigne, *Partition hypergroups*, Amer. J. Math., 6 (1940) 599-612.

[15] J. Chvalina, *Relational product of Join Spaces determint by quasi-orderings*, Proc. 6th International Congress on AHA and Appl., Democritus Univ. of Thrace Press, Prague, Czech Republic, (1996), 15-23.

[16] J. Chvalina, *Commutative hypergroups in the sense of Marty and ordered sets*, Proc. Summer School on General Algebra and Ordered Sets (1994), Olomouc (Czech Republic), 19-30.

[17] S. D. Comer, *Lattices of conjugacy relations*, Proceedings of the International Conference on Algebra, Part 3 (Novosibirsk, 1989), 31-48, Contemp. Math., 131, Part 3, Amer. Math. Soc., Providence, RI, 1992.

[18] S. Comer, *A remark on Cromatic Polygroups*, Congressus Numeratium, 38 (1983), 85-95.

[19] S. Comer, *Extension of polygroups by polygroups and their representations using color schemes*, Lecture Notes in Math. 1004 (1984), 91-103.

[20] S. Comer, *Polygroups derived from cogroups*, J. Algebra 89(2) (1984), 397-405.

[21] P. Corsini, *Contributo alla teoria degli ipergruppi*, Atti Soc. Pelor. Sc. Mat. Fis. Nat., XXVI, (1980), 347-362.

[22] P. Corsini, *Prolegomena of Hypergroup Theory*, Supplement to Riv. Mat. Pura Appl. Aviani Editore, Tricesimo, 1993.

[23] P. Corsini, *(i.p.s.) ipergruppi di ordine 7*, Atti Sem. Mat. Fis. Un. Modena, 34, Modena, Italy, (1986), 199-216.

[24] P. Corsini, *(i.p.s.) Hypergroups of order 8*, Aviani Editore, 1989.

[25] P. Corsini, *Feebly canonical and 1-hypergroupes*, Acta Un. Carolinae-Math. Ph., V.24, N.2, (1983), 49-56.

[26] P. Corsini, *Sugli ipergruppi canonici finiti con identità parziali scalari*, Rend. Circolo Mat. Palermo, S.II, T. 36, Palermo, Italy, (1987), 205-219.

[27] P. Corsini, *Join spaces, power sets, fuzzy sets*, Proceedings of the 5th Congress on Algebraic Hyperstructures and Applications, Iasi 1993 (M. Stefanescu ed.), Hadronic Press, Palm Harbor, FL, 1994, 45-54.

[28] P. Corsini, *Hypergraphs and hypergroups*, Algebra Universalis, 35 (1996), 548-555.

[29] P. Corsini, *Binary relations and hypergroupoids*, Ital. J. Pure Appl. Math., 7 (2000), 11-18.

[30] P. Corsini, *Binary relations, interval structured and join spaces*, J. Appl. Math. Comput., 10(1-2) (2002), 209-216.

[31] P. Corsini, *Rough sets, maps and join spaces*, Honorary Volume dedicated to Professor Emeritus Ioannis Mittas, Aristotle Univ. of Thessaloniki, Faculty of Engineering, Math div., Thessaloniki, Greece, (2000), 65-72.

[32] P. Corsini, *Hypergroupes d'associativité des quasigroupes mediaux*, Atti del Convegno su Sistemi Binari e loro Applicazioni, Taormina, 1978.

[33] P. Corsini, *On the hypergroups associated with binary relations*, Multiple Valued Logic, 5 (2000), 407-419.

[34] P. Corsini and V. Leoreanu, *Join spaces associated with fuzzy sets*, J. Combinatorics, Information & System Sci., 20(1-4) (1995), 293-303.

[35] P. Corsini and V. Leoreanu, *Hypergroups and binary relations*, Algebra Universalis, 43 (2000), 321-330.

[36] P. Corsini and V. Leoreanu, *Applications of Hyperstructures Theory*, Advanced in Mathematics, *Kluwer Academic Publishers*, 2003.

[37] P. Corsini and G. Romeo, *Hypergroupes complets et T-groupoids*, Convegno su Sistemi Binari e loro Applicazioni, Taormina, (1978), 129-146.

[38] P. Corsini and I. Tofan, *On fuzzy hypergroups*, Pure Math. Appl., 8(1) (1997), 29-35.

[39] I. Cristea and M. Stefanescu, *Binary relations and reduced hypergroups*, Discrete Math., 308 (2008) 3537-3544.

[40] I. Cristea, M. Jafarpour, S. Mousavi and A. Soleymani, *Enumeration of Rosenberg hypergroups*, Comput. Math. Appl., 60 (2010), 2753-2763.

[41] I. Cristea, *Several aspects on the hypergroups associated with n-ary relations*, An. St. Univ. Ovidius Constanta, 17(3) (2009), 99-110.

[42] I. Cristea, M. Stefanescu and C. Angheluta, *About the fundamental relations defined on the hypergroupoids associated with binary relations*, European J. Combin., 32(1) (2011), 72-81.

[43] I. Cristea and M. Stefanescu, *Hypergroups and n-ary relations*, European J. Combin., 31 (2010), 780-789.

[44] B. Davvaz, *Lower and upper approximations in H_v-groups*, Ratio Math., 13 (1999), 71-86.

[45] B. Davvaz, *TL-subpolygroups of a polygroup*, Pure Math. Appl., 12(2) (2001), 137-145.

[46] B. Davvaz, *A new view of approximations in H_v-groups*, Soft Computing, 10(11) (2006), 1043-1046.

[47] B. Davvaz, *Relations on groups, polygroups and hypergroups*, Electronic Notes in Discrete Mathematics, 45 (2014), 9-16.

[48] B. Davvaz, *Semihypergroup Theory*, Elsevier/Academic Press, London, 2016.

[49] B. Davvaz, *Polygroup Theory and Related Systems*, World Scientific Publishing Co. Pte. Ltd., Hackensack, NJ, 2013.

[50] B. Davvaz, *A new interpretation of subhypergroups of a hypergroup*, J. Korea Soc. Math. Educ. Ser. B: Pure Appl. Math., 10(3) (2003), 163-169.

[51] B. Davvaz, *Characterizations of sub-semihypergroups by various triangular norms*, Czechoslovak Mathematical Journal, 55(4) (2005), 923-932.

[52] B. Davvaz, *Rough subpolygroups in a factor polygroup*, Journal of Intelligent and Fuzzy Systems, 17(6) (2006), 613-621.

[53] B. Davvaz, *Weak algebraic hyperstructures as a model for interpretation of chemical reactions*, Iranian Journal of Mathematical Chemistry, 7(2) (2016), 267-283.

[54] B. Davvaz and F. Bardestani, *Hypergroups of type U on the right of size six*, Arab J. Sci. Eng., 36 (2011), 487-499.

[55] B. Davvaz and A. Dehghan-Nezhad, *Chemical examples in hypergroups*, Ratio Math., 14 (2003), 71-74.

[56] B. Davvaz and M. Mahdavipour, *Rough approximations in a general approximation space and their fundamental properties*, Int. J. General Systems, 37(3) (2008), 373-386.

[57] B. Davvaz, A. Dehghan Nezad and M. M. Heidari, *Inheritance examples of algebraic hyperstructures*, Information Sciences, 224 (2013), 180-187.

[58] B. Davvaz, A. Dehghan Nezad and M. Mazloum-Ardakani, *Chemical hyperalgebra: Redox reactions*, MATCH Communications in Mathematical and in Computer Chemistry, 71 (2014), 323-331.

[59] B. Davvaz and H. Karimian, On the γ_n^*-complete hypergroups. *European J. Combin.*, 28 (2007), 86-93.

[60] B. Davvaz and H. Karimian, On the γ_n-complete Hypergroups and K_H Hypergroups, *Acta Mathematica Sinica*, English Series, 24(11) (2008), 1901-1908.

[61] B. Davvaz and V. Leoreanu-Fotea, *Hyperring Theory and Applications*, International Academic Press, 115, Palm Harbor, USA, 2007.

[62] B. Davvaz and V. Leoreanu-Fotea, *Binary relations on ternary semihypergroups*, Comm. Algebra, 38(10) (2010), 3621-3636.

[63] B. Davvaz and T. Vougiouklis, *A walk through weak hyperstructures. H_v-structures*, World Scientific Publishing Co. Pte. Ltd., Hackensack, NJ, 2019.

[64] B. Davvaz and T. Vougiouklis, *Commutative rings obtained from hyperrings (S_v-rings) with α^*-relations*, Comm. Algebra, 35 (2007), 3307-3320.

[65] M. De Salvo, *Feebly canonical hypergroups*, J. Compinatorics, Inform. Syst. Scien., 15, (1990), 133-150.

[66] M. De Salvo, *Sugli ipergruppi completi finiti*, Riv. Mat. Univ. Parma, (4) 8, (1982), 269-280.

[67] M. De Salvo and G. Lo Faro, *Hypergroups and binary relations*, Multi. Val. Logic, 8 (2002), 645-657.

[68] P. Dietzman, *On the multigroups of complete conjugate sets of elements of a group*, C.R. (Doklady) Acad. Sci. URSS (N.S.), 49 (1946) 315-317.

[69] R. Engelking, *General Topology*, PWN-Polish Scientific Publishers, Warsaw, 1977.

[70] M. Farshi, B. Davvaz and S. Mirvakili, *Degree hypergroupoids associated with hypergraphs*, Filomat, 28(1) (2014), 119-129.

[71] M. Farshi, B. Davvaz and S. Mirvakili, *Hypergraphs and hypergroups based on a special relation*, Comm. Algebra, 42 (2014), 3395-3406.

[72] M. Farshi, B. Davvaz and S. Mirvakili, *On the g-hypergroupoids associated with g-hypergraphs*, Filomat, 31(15) (2017), 4819-4831.

[73] D. Freni, *Une note sur l'identite hypergroupe et sur la clôture transitive β^* de β. (French) [A note on the core of a hypergroup and the transitive closure β^* of β]*, Riv. Mat. Pura Appl., 8 (1991), 153-156.

[74] D. Freni, *A new characterization of the derived hypergroup via strongly regular equivalences*, Comm. Algebra, 30(8) (2002), 3977-3989.

[75] D. L. Hartl and E. W. Jones, *Genetics, Principles and Analysis*, Jones and Bartlett Publishers, 1998.

[76] D. Heidari, B. Davvaz and S. M. S. Modarres, *Topological hypergroups in the sense of Marty*, Comm. Algebra, 42 (2014), 4712-4721.

[77] D. Heidari, B. Davvaz and S. M. S. Modarres, *Topological polygroups*, Bull. Malays. Math. Sci. Soc., 39 (2016), 707-721.

[78] D. Heidari and B. Davvaz, *Graph product of generalized Cayley graphs over polygroups*, Journal of Algebraic Structures and Applications, 6(1) (2019), 49-56.

[79] A. Hokmabadi, F. Mohammadzadeh and E. Mohammadzade *The commutativity degree of a polygroup*, The 6th International Group Theory Conference, 12-13 March 2014, Golestan University, Gorgan, Iran, 88-91.

[80] S. Hošková-Mayerová, *Topological hypergroupoids*, Comput. Math. Appl., 64(9) (2012) 2845-2849.

[81] S. Hošková and J. Chvalina, *Discrete transformation hypergroups and transformation hypergroups with phase tolerance space*, Discrete Math., 308 (2008), 4133-4143.

[82] M. Golmohamadian and M. M. Zahedi, *Color hypergroup and join space obtained by the vertex coloring of a graph*, Filomat, 31(20) (2017), 6501-6513.

[83] A. Iranmanesh and N. Sogol, *Some special classes of divisible hypergroups*, Int. J. Appl. Math. Stat., 22 (2011), 89-98.

[84] M. Jafarpour, G. Aghabozorgi and B. Davvaz, *Solvable groups derived from hypergroups*, J. Algebra Appl., 15 (2016), no. 4, 1650067, 9 pp.

[85] M. Jafarpour, H. Aghabozorgi and B. Davvaz, *On nilpotent and solvable polygroups*, Bull. Iranian Math. Soc., 39(3) (2013), 487-499.

[86] M. Jafarpour, I. Cristea and A. Tavakoli, *A method to compute the number of regular reversible Rosenberg hypergroup*, Ars combinatorics, 128 (2016), 309-329.

[87] M. Jafarpour, V. Leoreanu-Fotea and A. Zolfaghari, *Weak complete parts in semihypergroups*, Iran. J. Math. Sci. Inform., 8(2) (2013), 101-109.

[88] M. Jafarpour and S. Sh. Mousavi, *A construction of hypersemigroups from a family of preordered semigroups*, Scientiae Mathematicae Japonicae Online, e-2009, 587-594.

[89] J. Jantosciak, *Homomorphism, equivalence and reductions in hypergroups*, Riv. Mat. Pura Appl., 9 (1991), 23-47.

[90] J. Jantosciak, *Reduced hypergroups*, Algebraic Hyperstructures and Applications (T. Vougiouklis, ed.), Proc. 4th Int. cong. Xanthi, Greece, 1990, World Scientific, Singapore, (1991), 119-122.

[91] J. Jantosciak, *Transposition hypergroups: Noncommutative join spaces*, J. Algebra, 187 (1997) 97-119.

[92] J. Jantosciak, *A brief survey of the theory of join spaces*, Algebraic hyperstructures and applications (Iasi, 1993), 1-12, Hadronic Press, Palm Harbor, FL, 1994.

[93] M. Karimian and B. Davvaz, *On the γ-cyclic hypergroups*, Comm. Algebra, 34 (2006), 4579-4589.

[94] O. Kazanci, S. Yamak and B. Davvaz, *On n-ary hypergroups and fuzzy n-ary homomorphism*, Iran J. Fuzzy Systems, 8(1) (2011), 1-17.

[95] A. Kehagias, *An example of L-fuzzy join space*, Circ. Mat. Palermo, 51 (2002), 503-526.

[96] A. Kehagias and M. Konstantinidou, *Lattice ordered join space: An Applications-Oriented Example*, Preprint.

[97] M. Konstantinidou and J. Mittas, *Contributions à la théorie des treillis avec des structures hypercompositionnelles y attachées*, Riv. Mat. Pura e Appl., 14 (1999), 83-119.

[98] M. Konstantinidou and K. Serafimidis, *Hyperstructures dérivées d'un treillis particulier*, Rend. Mat. Appl., (7), 13(2) (1993), 257-265.

[99] M. Konstantinidou and K. Serafimidis, *Sur les P-supertreillis*, in: T. Vougiouklis, (Ed.), New Frontiers in Hyperstructures and Rel. Algebras, Hadronic Press, Palm Harbor, USA, 1996, 139-148.

[100] M. Koskas, *Groupes et hypergroupes homomorphes a un demi-hypergroupe*, C. R. Acad Sc., Paris, 257 (1963), 334-337.

[101] M. Koskas, *Groupoides, demi-hypergroupes et hypergroupes*, J. Math. Pures Appl, 49 (1970), 155-192.

[102] V. Leoreanu, *Centralisateur d'un element dans un hypergroupe reversible*, Rendiconti del Circolo Matematico di Palermo, Serie II, Tomo XLIII (1994), 413-418.

[103] V. Leoreanu, *On the heart of join spaces and of regular hypergroups*, Riv. Mat Pura e Appl., 17 (1995), 133-142.

[104] V. Leoreanu-Fotea and P. Corsini, *Isomorphisms of hypergroups and of n-hypergroups with applications*, Soft Computing, 13 (2009), 985-994.

[105] V. Leoreanu-Fotea and B. Davvaz, *n-hypergroups and binary relations*, European J. Combin., 29 (2008), 1207-1218.

[106] V. Leoreanu-Fotea, P. Corsini, A. Sonea and D. Heidary, *Complete parts and subhypergroups in reversible regular hypergroups*, Accepted by Annals of Ovidius University, Math Section, 2021.

[107] V. Leoreanu-Fotea, M. Jafarpour and S. Sh. Mousavi, *The relation δ^n and multisemi-direct hyperproducts of hypergroups*, Comm. Algebra, 40(10) (2012), 3597-3608.

[108] V. Leoreanu-Fotea, B. Davvaz, F. Feng and C. Chiper *Join spaces, soft join spaces and lattices*, Analele Stiintifice ale Universitatii Ovidius, Seria Matematica, Ovidius University, 22(1) (2014), 155-167.

[109] V. Leoreanu-Fotea and I. G. Rosenberg, *Hypergroupoids determined by lattices*, European J. Combin., 31 (2010), 925-931.

[110] V. Leoreanu and Gh Radu, *Direct limits and inverse limits of hypergroups associated with lattices*, Ital. J. Pure Appl. Math., 11 (2002), 121-130.

[111] V. Leoreanu Fotea and I. G. Rosenberg, *Join spaces determined by lattices*, J. Mult.-Valued Logic Soft Comput., 16(1-2) (2010), 7-16.

[112] V. Leoreanu-Fotea and I. G. Rosenberg, *Homomorphisms of hypergroupoids associated with L-fuzzy sets*, J. Mult.-Valued Logic Soft Comput., 15(5-6) (2009), 537-545.

[113] T. Y. Lin and Y. Y. Yao, *Mining soft rules using rough sets and neighborhoods*, Proceedings of the Symposium on Modelling, Analysis and Simulation, Computational Engineering in Systems Applications (CESA'96), IMASCS Multiconference, Lille, France, July 9-12, (1996), 1095-1100.

[114] A. Madanshekaf and A. R. Ashrafi, *Generalized action of a hypergroup on a set*, Ital. J. Pure Appl. Math., 3 (1998), 127-135.

[115] F. Marty, *Sur une generalization de la notion de groupe*, 8^{iem} Congres Math. Scandinaves, Stockholm, (1934) 45-49.

[116] T. K. Maryati and B. Davvaz, *A novel connection between rough sets, hypergraphs and hypergroups*, Discrete Math. Algorithms Appl., 9(4) (2017), 1750044, 17 pp.

[117] Ch. G. Massouros and Ch. Tsitouras, *On enumeration of hypergroups of order 3*, Comput. Math. Appl., 59 (2010), 519-523.

[118] Ch. G. Massouros and Ch. Tsitouras, *Enumeration of hypercompositional structures defined by binary relations*, Ital. J. Pure Appl. Math., 28 (2011), 53-64.

[119] G. G. Massouros, *Automata and hypermoduloids*, Algebraic hyperstructures and applications (Iasi, 1993), 251-265, Hadronic Press, Palm Harbor, FL, 1994.

[120] J. McMullen and J. Price, *Reversible hypergroups*, Rend. Sem. Mat. Fis. Milano, 47, Italy, (1977), 67-85.

[121] J. McMullen and J. Price, *Duality for finite abelian hypergroups over splitting fields*, Bull. Austral. Math. Soc., 20, (1979), 57-70.

[122] R. Migliorato, *Ipergruppi di cardinalita 3 e isomorfismi di ipergruppoidi commutativi totalmente regolari*, in: Atti Convegno su Ipergruppi, altre Strutture Multivoche e loro applicazioni, Udine, 1985.

[123] R. Migliorato, *Semi-ipergruppi e Ipergruppi n-completi*, Ann. Sci. Univ. Clermont II, Sèr Math., 23 (1986), 99-123.

[124] S. Mirvakili and B. Davvaz, *Relations on Krasner (m,n)-hyperrings*, European J. Combin., 31 (2010), 790-802.

[125] S. Mirvakili, S. M. Anvariyeh and B. Davvaz, *On α-relation and transitivity conditions of α*, Comm. Algebra, 36 (2008), 1695-1703.

[126] M. Mittas, *Hypergroupes canoniques*, (Fr), Math. Balkanica, 2 (1972), 165-179.

[127] M. Mittas, *Hypergroupes canoniques values et hypervalues- Hypergroupes fortement et superieurement canoniques*, (Fr), Bull. Greek Math. Soc., 23 (1982), 55-88.

[128] M. Mittas, *Certaines remarques sur les hypergroupes canoniques hypervaluables et fortement canoniques*, Rivista Mat. Pura Appl., 9 (1991), 61-67.

[129] M. Mittas, *Sur la valuations stricte des hypergroupes polysymetriques canoniques*, (Fr), Ratio Math., 13 (1999), 5-28.

[130] J. Mittas and M. Konstantinidou, *Contributions à la théorie des treillis en liaison avec des structures hypercompositionnellles y attachées*, Riv. Mat. Pura e Appl., 14 (1994), 83-114.

[131] G. A. Moghani, A. R. Ashrafi and B. Davvaz, *On the finite join spaces associated with rough sets*, Pure Math. Appl., 14(4) (2003), 305-311.

[132] R. T. Morrison and R. N. Boyd, *Organic Chemistry*, Sixth Edition, Prentice-Hall, Inc, 1992.

[133] S. Sh. Mousavi, V. Leoreanu-Fotea, M. Jafarpour and H. Babaei, *Equivalence relations in semihypergroups and the corresponding quotient structure*, European J. Combin., 33(4) (2012), 463-473.

[134] S. Sh. Mousavi, V. Leoreanu-Fotea and M. Jafarpour, *Cyclic groups obtained as quotient hypergroups*, An. Stiint. Univ. Al. I. Cuza Iasi. Mat. (N.S.), 61(1) (2015), 109-122.

[135] S. Sh. Mousavi, V. Leoreanu-Fotea and M. Jafarpour, *R-parts in (semi)hypergroups*, Annali di Matematica Pura ed Applicata, 19(4) (2011), 667-680.

[136] T. Nakano, *Rings and partly ordered systems*, Math. Zeitschrift, 99 (1967), 355-376.

[137] J. Nieminen, *Join space graphs*, J. Geom., 33 (1988), 99-103.

[138] A. Nikkhah, B. Davvaz and S. Mirvakili, *Hypergroups constructed from hypergraphs*, Filomat, 32(10) (2018), 3487-3494.

[139] V. Novak, *On representation of ternary structures*, Math. Slovaca, 45(5) (1995), 469-480.

[140] V. Novak and M. Novotny, *Pseudodimension of relational structures*, Czechoslovak Math. J., 49(124) (1999), 547-560.

[141] M. Novotny, *Ternary structures and groupoids*, Czechoslovak Math. J., 41(116) (1991), 90-98.

[142] Z. Pawlak, Rough sets, *Int. J. Inf. Comp. Sci.*, 11 (1982), 341-356.

[143] Z. Pawlak, *Rough Sets — Theoretical Aspects of Reasoning about Data*, Kluwer Academic Publishing, Dordrecht, 1991.

[144] W. Prenowitz, *Projective geometries as multigroups*, Amer. J. Math., 65 (1943), 235-256.

[145] W. Prenowitz, *Partially ordered fields and geometries*, Amer. Math. Monthly, 53 (1946), 439-449.

[146] W. Prenowitz, *Descriptive geometries as multigroups*, Trans. Amer. Math. Soc., 59, (1946), 333-380.

[147] W. Prenowitz, *Spherical geometries and multigroups*, Canadian J. Math., 2 (1950), 100-119.

[148] W. Prenowitz, *A contemporary approach to classical geometry*, Amer. Math. Monthly, 68(1) (1961), part II.

[149] W. Prenowitz and J. Jantosciak, *Join Geometries*, Springer-Verlag, UTM, 1979.

[150] P. Račková, *Closed, refexive, invertible, and normal subhypergroups of special hypergroups*, Ratio Math., 23 (2012), 81-86.

[151] T. Richmond, *General Topology. An introduction*, De Gruyter Textbook. De Gruyter, Berlin, 2020.

[152] G. Romeo, *Limite diretto di semi-ipergruppi e iperguppi di associativita'*, Riv. Math. Univ. di Parma, 8 (1982), 281-288.

[153] J. Rotman, *An Introduction to the Theory of Groups*, Fourth Edition, Springer, 1999.

[154] I. G. Rosenberg, *An algebraic approach to hyperalgebras*, Proc. 26th Int. Symp. Multiple-Valued Logic, Santiago de Compostela, IEEE (1996).

[155] I. G. Rosenberg, *Wall monoids*, New Frontiers in Hyperstructures (Molise, 1995), 159-166, Ser. New Front. Adv. Math. Ins. Ric. Base, Hadronic Press, Palm Harbor, FL, 1996, 201-210.

[156] I. G. Rosenberg, *Hypergroups induced by paths of a directed graph*, Ital. J. Pure Appl. Math., 4 (1998), 133-142.

[157] I. G. Rosenberg, *Hypergroups and join spaces determined by relations*, Ital. J. Pure Appl. Math., 4 (1998), 93-101.

[158] I. G. Rosenberg, *Multiple-valued hyperstructures*, Proc. 28th Int. Symp. Multiple-Valued Logic, Fukuoka (1998), 201-210.

[159] R. Roth, *Character and conjugacy class hypergroups of a finite group*, Annali di Matematica Pura ed Appl., 105 (1975), 295-311.

[160] R. Roth, *On derived canonical hypergroups*, Riv. Mat. Pura e Appl., 3 (1988), 81-85.

[161] S. Sekhavatizadeh, M. M. Zahedi and A. Iranmanesh, *Cyclic hypergroups which are induced by the character of some finite groups*, Ital. J. Pure Appl. Math., 33 (2014), 123-132.

[162] K. Serafimidis, *Sur les L-hyperideaux des hypergroupes canoniques strictement réticules* (Fr), Rend. Del Circ. Matem. di Palermo, Serie II, Tomo XXXV, Palermo, Italy, (1986), 411-419.

[163] K. Serafimidis, M. Konstantinidou and J. Mittas, *Sur les hypergroupes canoniques strictement réticules*, Rivista Mat. Pura Appl., 2 (1987), 21-35.

[164] K. Serafimidis, A. Kehagias and A., Konstantinidou, *The L-Fuzzy Corsini Join Hyperoperation*, Ital. J. Pure Appl. Math., 12 (2002), 83-90.

[165] M. Singha, K. Das and B. Davvaz, *On topological complete hypergroups*, Filomat, 31(16) (2017) 5045-5056.

[166] A. Sonea, *New aspects in polygroup theory*, An. St. Univ. Ovidius, Constanta, 28(3) (2020), 241-254.

[167] A. Sonea and B. Davvaz, *The Euler's totient function in canonical hypergroups*, Indian Journal of Pure and Applied Mathematics, Accepted, June 2021.

[168] S. Spartalis, *The hyperoperation relation and the Corsini's partial or not-partial hypergroupoids (A classification)*, Ital. J. Pure Appl. Math., 24 (2008), 97-112.

[169] S. Spartalis, *Hypergroupoids obtained from groupoids with binary relations*, Ital. J. Pure Appl. Math., 16(2004), 201-210.

[170] S. Spartalis and C. Mamaloukas, *On hyperstructures associated with binary relations*, Comput. Math. Appl., 51 (2006), 41-50.

[171] S. Spartalis, M. Konstantinidou-Serafimidou and A. Taouktsoglou, *C-hypergroupoids obtained by special binary relations*, Comput. Math. Appl., 59 (2010), 2628-2635.

[172] M. Stefanescu, *Some interpretations of hypergroups*, Bull. Math. Soc. Sci. Math. Roumanie, 49(97) (2006), 99-104.

[173] Y. Sureau, *Contribution a la theorie des hypergroupes et hypergroupes operants transitivement sur un ensemble*, These d' Etat, Clermont Ferrand, (1980).

[174] R. H. Tamarin, *Principles of Genetics*, Seventh Edition, The McGraw-Hill Companies, 2001.

[175] I. Tofan and A. C. Volf, *On some connections between hyperstructures and maps*, Ital. J. Pure and Appl. Math., 7 (2000), 63-68.

[176] J. C. Varlet, *Remarks on distributive lattices*, Bull. de l'Acad. Polonnaise des Sciences, Série des Sciences Math., Astr. et Phys., vol. XXIII, 11 (1975), 1143-1147.

[177] T. Vougiouklis, *Groups in hypergroups*, Combinatorics '86 (Trento, 1986), 459-467, Ann. Discrete Math., 37, North-Holland, Amsterdam, 1988.

[178] T. Vougiouklis, *Representations of hypergroups by generalized permutations*, Algebra Universalis, 29(2) (1992), 172-183.

[179] T. Vougiouklis, *Hypergroups, hyperrings. Fundamental relations and representations*, Quaderni del seminario di Geometrie Combinatorie, (1989), 1-20.

[180] T. Vougiouklis, *Fundamental relations and hypergroup representations*, U.C.N.W. Maths Preprint, 89(14), (1989), 1-20.

[181] T. Vougiouklis, *The fundamental relation in hypergroups. The general hyperfield*, Proc. 4th Int. Congress Algebraic Hyperstructures and Applications, Xanthi, 1990, World Scientific, (1991), 209-217.

[182] T. Vougiouklis, *Fundamental relations in hyperstructures*, Bull. Greek Math. Soc., 42 (1999), 113-118.

[183] T. Vougiouklis, *Hyperstructures and Their Representation*, Hadronic Press, Inc, Palm Harbor, USA, 115, 1994.

[184] J. Zhan and B. Davvaz and K. P. Shum, *On probabilistic n-ary hypergroups*, Information Sciences, 180(7) (2010), 1159-1166.

Index

Printed in the United States
by Baker & Taylor Publisher Services

Printed in the United States
by Baker & Taylor Publisher Services